The Medieval Military Engineer

Armour and Weapons
ISSN 1746-9449

Series Editors
Kelly DeVries
Robert W. Jones
Robert C. Woosnam-Savage

Throughout history armour and weapons have been not merely the preserve of the warrior in battles and warfare, but potent symbols in their own right (the sword of chivalry, the heraldic shield) representing the hunt and hall as well as the battlefield. This series aims to provide a forum for critical studies of all aspects of arms and armour and their technologies, from the end of the Roman Empire to the dawn of the modern world; both new research and works of synthesis are encouraged.

New proposals for the series are welcomed; they should be sent to the publisher at the address below.

Boydell & Brewer Limited, PO Box 9, Woodbridge, Suffolk IP12 3DF, UK

Also in this series:

The Artillery of the Dukes of Burgundy, 1363–1477,
Robert Douglas Smith and Kelly DeVries
'The Furie of the Ordnance': Artillery in the English Civil Wars, Stephen Bull
Jousting in Medieval and Renaissance Iberia, Noel Fallows
The Art of Swordsmanship *by Hans Leckchner*, translated by Jeffrey L. Forgeng
The Book of Horsemanship *by Duarte I of Portugal*,
translated by Jeffrey L. Forgeng
Pietro Monte's Collectanea: *The Arms, Armour and Fighting Techniques of a Fifteenth-Century Soldier*, translated by Jeffrey L. Forgeng

The Medieval Military Engineer

From the Roman Empire to the Sixteenth Century

Peter Purton

THE BOYDELL PRESS

First published 2018
The Boydell Press, Woodbridge

ISBN 978 1 78327 278 5

The Boydell Press is an imprint of Boydell & Brewer Ltd
PO Box 9, Woodbridge, Suffolk IP12 3DF, UK
and of Boydell & Brewer Inc.
668 Mt Hope Avenue, Rochester, NY 14620–2731, USA
website: www.boydellandbrewer.com

A CIP catalogue record for this book is available from the British Library

This publication is printed on acid-free paper

Printed and bound in Great Britain by
TJ International Ltd, Padstow, Cornwall

FSC www.fsc.org MIX Paper from responsible sources FSC® C013056

Contents

Figures

While many of the illustrations refer to particular points in the text, others have been chosen to illuminate aspects of the book more generally.

The author and publishers are grateful to all the institutions and individuals listed for permission to reproduce the materials in which they hold copyright. Every effort has been made to trace the copyright holders; apologies are offered for any omission, and the publishers will be pleased to add any necessary acknowledgement in subsequent editions.

Acknowledgements

Many people have helped me with specific references, answers to questions or general advice. With apologies for any omissions, I wish to thank (in alphabetical order) Bas Aarts, Renaud Beffeyte, Oliver Creighton, Richard Dunn, Richard Eales, Michael Fulton, Bob Higham, Christof Krauskopf, Pamela Marshall, Derek Renn, Jiri Varhanik, Peter Vemming, Olaf Wagener and Bill Woodburn. An early outline of the contents was given as a paper at Château Gaillard 2012 and I am grateful to have been invited to present it there. Rainer Atzbach's invitation to address the "Castles at War" conference in Nyborg helped me crystallise thoughts about the history of medieval military mining, and Chas Hollwey made possible a visit to the Clearwell iron mines in the Forest of Dean. Chas, and Jeremy Cunnington, made possible a visit to see medieval building techniques reconstructed at Guédelon in Burgundy. Richard Eales, Pamela Marshall and Denys Pringle have organised study tours that greatly helped me to understand medieval castles and town defences, and many members of the Castle Studies Group have contributed directly or indirectly. Finally, the anonymous referees made numerous suggestions which have saved me from errors and greatly improved this book, for which I am deeply grateful. As always, the author is responsible for all errors that remain.

Peter Purton, 2017

Preface

WE ARE SURROUNDED BY and sometimes still use structures dating from the middle ages. Scholarship has recreated its political, economic, social and military history. Specialists debate the finer details of medieval science, education and technology. The architecture of cathedrals, monasteries, churches and castles is painstakingly examined and re-examined and their functions are explored in both academic conferences and the popular media. But there are fewer studies of the people responsible for these works. And even fewer of military engineers – who may well have been involved in both the construction and the destruction of military fortifications.

I began to ponder this even while completing my *History of the Early* and *Late Medieval Siege*. The names of engineers at war kept appearing in contemporary accounts, increasingly as more records survived. Evidently, by the time of the renaissance in Europe, a body of expertise existed that rulers called upon to fulfil their engineering needs, and some of those experts' names are well known today (if the number of exhibitions devoted to Leonardo da Vinci is any guide). It is also clear that engineering continued to be undertaken across the world even when there is no surviving record of it. But what had happened before?

I was at first anxious that it would not be possible to attempt this history because of a lack of evidence. Closer study suggested that, especially for the early centuries, it was necessary to offer rather more in the way of hypothesis than a historian ought to be happy with, and to read back from later reports into earlier times, something that also involved making unprovable assumptions. Nonetheless, there was sufficient to show that this king or that prince had successfully bridged a great river, or had built engines against a hostile fortress, or had laid out and defended a camp, or had diverted a river, and thus to identify that to have done so required having access to a degree of engineering skill that it was highly unlikely the rulers themselves possessed. That the necessary technical skills also existed can be confirmed from evidence in other fields: planning, layout and construction of buildings, hydraulic engineering (canals, dams, irrigation) or shipbuilding, all areas where archaeology has added immensely to knowledge of the past. Craftsmen built ships, houses or watermills with evident skill. Although it can rarely be shown that the people who did the one also did the other, it can be demonstrated that craftsmen (anonymous) were mobilised for military campaigns.

This book is not about technology (although it does touch on some important debates about military technology), science or architecture. It is about the people who did the military engineering. It touches on many and sometimes continuing debates on connections between book learning and the "mechanical arts" and suggests a developing overlap starting in the earliest years after the end of the western Roman empire but also continuing through times once miscalled "dark"; it also poses questions about levels of literacy among the master craftsmen from whom military engineers were drawn, and while not being able to offer conclusive answers, suggests a growing awareness (if usually very limited) of what we would now call book learning.

I can find nothing to prove or disprove that for the most part, however, knowledge was passed on by the traditional master-to-apprentice route and involved memorising the rules, except to remember that it was not always just a matter of passing on how something had always been done: innovation and change took place that can only have been the result of someone making the decision to progress from what they had inherited to something different.

Military engineering is interpreted here in a broad way because the evidence suggests it covered many disciplines. It therefore includes the design, construction and operation of medieval artillery, a central part of the duties of many of the engineers who will be encountered. A separation of artillery from other forms of engineering would not begin to take place until well after the arrival of gunpowder artillery in Europe. It involved challenges such as bridge building and laying out camps – neither of them tasks that would necessarily leave any trace, either on the ground or in the written record. Indeed, we will see surveying skills – well attested in late antiquity – become invisible in their military sense for centuries, although they must have been there. From time to time there is evidence of the construction of immense mobile siege towers that confirm the presence of skilled military engineers, although more often the record only shows that the equipment was built. All these are military engineering roles. On the other hand, it will be shown that mining was traditionally a separate operation, reliant to a large degree on the skills of those who made their daily living from this work. Although non-specialists would regularly have been used for the hard labour in a military operation, underground mining in particular (but also sometimes surface-level sapping) called for specialist skills possessed only by such folk. The inclusion of "miners and sappers" among engineering corps to the present day testifies to the relevance of this approach.

Medieval military engineers, it will be seen, were a continuing presence, and by their often anonymous contribution – often in extremely dangerous and exposed situations – made a significant contribution to the outcomes of campaigns. The noble commanders have always taken the credit. It is time to redress the balance a little.

Most of the book deals with Christian Europe, so dates are given using the Christian calendar. But I also deal with the continuing interaction with the Muslim world and discuss, with evidence from translated sources, the comparable levels of knowledge and technology there. Developments in military engineering across the known world, including Central Asia and China, are recorded, particularly at the times when they came into (sometimes violent, at other times more peaceful) contact. Where dimensions are quoted from medieval sources these are presented as they were written, reflecting the difficulty of offering precise modern equivalents: feet and inches originated in both Roman practice and other ancient civilisations and related to parts of the human body. I have offered metric equivalents of the original. Where measurements (particularly weights) matter to the interpretation, I have discussed modern equivalents. Personal names are spelt not according to any strict rule, but in order to be recognisable, and in the many cases where towns and cities and regions have changed their name I have applied the same simple approach. In many cases I have alluded to other, wider developments that provide context; readers can follow up references to relevant sources from the footnotes or bibliography.

Abbreviations

The following abbreviations for frequently cited collections or journals are used in the footnotes and bibliography.

BAR	British Archaeological Reports
CSHB	*Corpus Scriptorum Historiae Byzantinae*
EHR	*English Historical Review*
HEMS	P Purton, *History of the early medieval siege c.450–1200*, Woodbridge 2010
HLMS	P Purton, *History of the late medieval siege 1200–1500*, Woodbridge 2010
MGH	*Monumenta Germaniae Historica:*
AA	Antiquissimi Auctores
SRM	Scriptores rerum Merovingicarum
SrG	Scriptores rerum Germanicarum
SS	Scriptores
Migne *PG*	*Patrologia cursus completa: series graeca*, ed. J P Migne, Brussels
Migne *PL*	*Patrologia cursus completa: series latina*, ed. J P Migne, Brussels
ND	*Notitia Dignitatum*
PRO	Public Record Office (London) [now The National Archives]
RHC	*Recueil des historiens des croisades*
RS	Rolls Series
RIS	*Rerum Italicarum Scriptores*, ed. L Muratori (and successors)
SSCISSM	*Settimane di studi del Centro Italiano di studi sull'Alto Medioevo*

Where not given in a footnote, details of works referred to (especially multiple items on one subject) will be found in the Bibliography.

1

Military Engineers in the Middle Ages

AN UNBROKEN TRADITION OF MILITARY ENGINEERING?

THE MONUMENTS OF THE ANCIENT WORLD that have survived are eloquent testimony to the levels of engineering skills available in these societies. Each successive civilisation left evidence of its ability to build on a vast scale, to overcome geological and geographical obstacles through the application of human ingenuity and (often) the force of immense human labour, to make arid land fertile through great hydraulic projects, to cross rivers and seas. We know that these engineering skills were also harnessed for the conduct of warfare. Literate societies such as China, India and successive empires in the Middle East, and Greece then Rome around the Mediterranean, additionally left writings that confirm what they were capable of in both civilian and military contexts. Each empire influenced its neighbours and its successors. In Europe, the Middle East, central Asia and North Africa, the roots of the developments of the most recent millennia must be sought in the legacies of Rome and Persia, and their interactions with China and the kingdoms of the Indian subcontinent.

The imperial armies of Rome and its ever-present Persian rival for domination in the east were professional organisations and military engineering was integrated within them. The operation and maintenance of artillery, the erection of fortifications, the bridging of rivers and laying out of camps, the undermining of enemy fortresses were all tasks for which members of the armed forces were recruited and trained. Roman weaponry was manufactured under the supervision of salaried, skilled workers employed in arms factories (*fabricae*). The contribution of engineering to the military success of Rome was sufficiently valued for its impact to be described in classical histories, or at least clearly implied.

Not until modern times (in Europe) would a professional corps of engineers integrated into the army of the state again become common. It would be the seventeenth century before a formal body of engineers existed in the British armed forces, for instance.[1] The earliest examples would be the experts charged by a few

1 See the excellent study by W Porter, *The history of the corps of Royal Engineers*, vol. I, Chatham 1889 reprinted 1977, for the history of engineers serving the British crown up

rulers with organising a paid body of people to maintain the royal artillery in the gunpowder age from the mid fifteenth century onwards. States in western Europe and the Ottoman Turks led this development. But throughout the intervening centuries, commanders could and did routinely find the people with the necessary skills to erect fortifications, to bridge rivers, to construct and operate siege artillery and to build a sophisticated range of other siege equipment, to dig mines beneath hostile fortresses in order to create breaches in their walls, to construct ships for transport and naval warfare, and all the other numerous tasks that could not – or would not – be carried out by the knight, squire, archer or ordinary foot soldier who made up the largest parts of a medieval army. It is certain that these operations were not carried out by a professional corps similar to those that existed in the Roman empire, so who were the people who must have been recruited for the purpose? The timescale was different but the outcome was the same where a Roman empire had continued in the shape of the Byzantine empire: along with its Persian neighbour, it heavily influenced what happened in these regions, a dual inheritance taken up in the Islamic states that arose from the seventh century onwards.

This book will present the evidence to identify a continuing tradition of military engineering, and therefore necessarily of military engineers, during this thousand-year hiatus. These engineers have fallen through a gap in historiography. Who they were, their changing status, their contribution, and evidence for what skills they must have possessed, along with discussion of how such skills were first acquired and then transmitted, are the main issues discussed.[2]

Each year new studies are published covering every area of medieval history, many drawing on archaeology to advance new hypotheses or to modify or discredit established theories. Scholars of the military history of this period have progressively transformed our understanding of how war was conducted across the medieval world as it underwent many changes across the centuries, how it related to or impacted on the economy, the process of state formation, how it was seen at the time (instead of with the benefit of hindsight), the role of the soldiers, and – a specialism of particular fascination for some – military technology. However, with a few exceptions, the study of this technology has tended to be treated in isolation from social, economic and political developments and from other areas where technology was required: the shortage of evidence having been a critical obstacle in making this connection. Only in the blossoming field of castle studies has this work become established, and this necessarily for later centuries. The other largely unanswered question is who were the people responsible? With the exception of

---

to the end of the nineteenth century and the piecemeal evolution of the corps.

2 As with everything else covered in this study, the term engineer itself evolved slowly towards its modern meaning, and throughout the medieval period was used in a wide range of senses. On this see pp. 16–19 below.

a number of studies of individual engineers and of engineering in specific fields of conflict arenas, there is no history of the medieval military engineer.

A number of modern studies devoted to much wider issues help address the second question. A vast literature has appeared completely revising the history of the period now called late antiquity, and the outstanding work of people internationally, such as (in English) Peter Heather and Chris Wickham, has happily made clearer the background necessary to present this subject.[3] Their work will inform the first chapters that follow. Detailed work on the warfare conducted by the Merovingian and Carolingian realms in western and central Europe and by the Ottonian dynasty in what was to become Germany has been undertaken by B S and D S Bachrach amongst others. Their conclusions are sometimes controversial but their studies are ground breaking. Following in that tradition is Leif Petersen's massive study of siege warfare in the successor states to the Roman empire, west and east. Their readings of the limited surviving sources for these centuries (so long and so misleadingly termed the "Dark Ages") ease the way to try to recreate the role of engineering, and through that of engineers. For later periods I will make reference to other valuable studies – articles on particular events or particular individuals, and analyses of (for example) the crusades, where there is rich written evidence from all sides of the conflicts. In many of these books and articles there are references to engineers – particularly as they became recognised in contemporary accounts, and there are a few scholars who have devoted significant effort to military engineering, and military engineers, in particular regions or conflicts (such as the work of Nicholas Prouteau on the crusades and Aldo Settia on engineers from Italian cities). Their contributions, localised in time and place, have been invaluable.

In the absence of direct evidence of the continuation of any body of military engineers after the disappearance of the western Roman empire and the later transformations of the eastern, including the emergence of the Islamic caliphate, it is necessary to draw on a range of sources to produce an account. I will look at what can be learnt from economic and social contexts (particularly in relation to levels of technology and of scientific knowledge), parallel fields of activity requiring engineering skills (such as the construction of buildings and shipbuilding, in both of which archaeology makes a growing contribution), government records and contemporary histories (and particularly manuscript illustrations) of actual military campaigns in which engineering featured. Among written sources it is also necessary to consider what can be learnt from the specific field of surviving medieval military texts. All of these sources have problems of interpretation. I will address them in turn.

[3] For further details see the Bibliography.

ECONOMY, SOCIETY AND TECHNOLOGY

Engineers apply technology and it has long been established that there were significant developments in this field throughout the middle ages. Although technical progress in a pre-industrial age was subject to a very much slower rhythm than in the industrialised world, the progress when it happened was real and significant. There have been many studies of medieval technology and its development over the centuries: what was inherited from previous ages, and what emerged anew periodically to transform the way that familiar and routine tasks were done, as well as to create more exotic or esoteric devices and machines that may (or may not) have had practical uses. The economic and social background must be understood to see how it was possible for new technology to emerge or be sustained. At its most fundamental, craftsmen needed a market for their products if they were to continue making their living from their craft, and this depended on the level of surplus produced in the economy. Wealthy ruling elites would support the continuation of numerous crafts above and beyond those needed to replace or repair the peasant's tools or basic household goods. The presence of such elites – both secular and clerical – was the essential starting point. Evidence for their existence (and their wealth) informs whether there was the basis for an expansion of crafts and developments in craft technology, and archaeology can confirm the technical levels achieved. The usually invisible process of technical innovation in early medieval times was to meet some need: to increase production, for example, or to supply water in arid climates to enable the growing of crops. The presence of towns whose occupants depended on the production of agricultural surpluses which they could purchase also encouraged trade and travel, in turn one of the mechanisms for the diffusion of knowledge across the known world, the subject of study by many renowned historians of technology and culture.[4]

There have been similar studies of the development and diffusion specifically of technologies for military purposes and attempts to link these with what is known of contemporary levels of scientific awareness. The read-across is sometimes straightforward. Knowledge of surveying techniques was as relevant to laying out fortifications or camps as it was to setting out the shape of new towns or measuring land for tax purposes. Skills in building bridges were the same whether it was a permanent structure built at a river crossing vital for a town's (or a lord's) levying of tolls or a temporary one for getting an army and its impedimenta across a wide river. Construction and shipbuilding techniques reveal much about the capabilities of the people paid to carry out the work, particularly as in many cases they were

[4] Lynn White jnr is a leading western scholar of this subject; among French historians, G Beaujouan, P Braunstein and J Gimpel have been prolific writers (see Bibliography). For the Muslim world, D R Hill has provided access to an enormous literature in Arabic or Persian.

the same people who would be hired for military engineering purposes – more on this below. Sometimes the connection can only be implied and remains open to more than one interpretation. Such problems are posed when trying to explain in particular how novel types of artillery were developed, such as the trebuchet or the springald.

Despite this welcome attention, it is true to say that even now some issues remain controversial, and others – such as the date when the single most significant transformation of stone-throwing artillery (the trebuchet) first appeared, or the construction of the first gunpowder artillery – are still unknown.[5] But even the rather limited modern attention paid to military technology has not yet been accorded to the technicians who applied the technology in warfare, although as indicated the same skills and knowledge applied in this field as in the use of techniques for civilian purposes. The use of engineers for a multitude of tasks that would now be separated into distinct categories of civilian or military will be shown time and again. The development of dedicated specialists working only (or largely) on a ruler's military requirements will not be found until the demands of gunpowder weaponry eventually called into being a craft of its own, or, by the end of the fifteenth century, some engineers working only on military tasks can be identified.[6] The same will be found to be even more true of the craftsmen or artisans who carried out the engineer's instructions – carpenters, masons and smiths, to cite only those most commonly called upon. As a result of their geographical location, the wide range of people with specific skills required for shipbuilding may often have carried out only this work (though they would not distinguish between ships for naval as opposed to normal commercial purposes). Throughout, miners working under their own masters represent a group whose existing specialism was as relevant to everyday as to military purposes. But they were largely the exception.

CONSTRUCTION

In the area of architecture and construction techniques, the world of church-building has been much more closely studied than that of the construction of castles, town walls or siege equipment, despite the workers involved being often the same people. It is well known that as with the 'architects' of cathedrals (and with necessary circumspection resulting from the different sense attached to that term during this period), there were master builders of castles (again, sometimes the

5 On the origins of the counterweight trebuchet, see in particular studies by Paul Chevedden, and pp. 165–9.

6 Although occasionally it is possible to identify engineers who were employed only on a specific area such as siege engines – examples occur in the records of Henry II of England's expenditure (see pp. 140–2 and Appendix).

same people). However, with the exception of a handful whose names have been preserved, or who happen to have also served as high-ranking ecclesiastics whose saintly lives were commemorated in *vitae*, the great majority remain anonymous.[7] Nonetheless, anyone visiting the remains of a major stone castle cannot but be struck by the evident fact that many skills must have been called upon to design and oversee the construction of such a structure and all its constituent parts. Less obvious to the naked eye, because the remains surviving today give a very misleading impression of what was once there, but also significant were the skills needed successfully to erect the earthworks underneath and around what may have been a sophisticated timber castle involving carpentry skills of the highest order. Archaeology is increasingly uncovering evidence for these assertions, although investigation of medieval earthworks remains something of a poor relation, a sad reality unlikely to be reversed at the time of writing.[8] Between the patron (king, bishop or secular lord), who had decided on the building and (to varying degree) determined or influenced what the outcome they desired would be, and the builders and labourers who did the physical work stood the engineer.

The same engineer might also be the man who would have been in charge of issuing the instructions to the craftsmen who constructed the battering ram, or siege tower, or petrary or trebuchet, or who began digging the mine at a siege, or who when an army reached a river gathered the materials prior to constructing a bridge of boats or pontoons. How they gained the knowledge and skills to accomplish such tasks is one of the questions to be addressed.

THE WRITTEN WORD 1: RECORDS

While researching hundreds of medieval chronicles and the published records of governments during the middle ages that preceded the publication of my histories

7 See chapter three. Studies by (in English) J Harvey, and in German by G Bindung have led the way in dealing with architecture and building technology and those responsible for designing grand stone buildings.

8 I will discuss the rarely named "architects" of castles in the proper place but it will be important to keep in mind that the description of castles as "military architecture" that dominated writing on the subject in the nineteenth and first part of the twentieth centuries has been substantially revised and most modern students now view castles as having multiple social, political and economic functions. I believe, however, that some have gone too far in attempting to deny *any* military role, which is simply wrong, while it is also the case that those in charge of the construction had skills also used in the military sphere even if a particular building project was for other reasons. There is more on this in following chapters although the subject – and the literature – is vast and falls outside the scope of this book.

of the medieval siege,[9] I noted the names of many individual "engineers" recorded incidentally, either because of some noteworthy event during a siege that the chronicler believed made a good story for his readers – perhaps illustrating a moral or religious message – or else because they were being paid for a particular service and that payment featured in a still surviving government record. The existence of records such as the English crown's Great Roll of the Pipe from the start of the reign of King Henry II (1154–89), which records the expenditure of the royal exchequer, makes it possible to observe close up who was being paid and what they were being paid for, and is a priceless source of information on military engineers paid from English resources, even if it is sometimes frustrating since what we have is the accountant's end of the transaction (see Appendix). The absence of similar sources for other medieval realms until later years, not because they were not made, but because they did not survive, makes the challenge of proposing a thesis for what was happening much more difficult. But there is no reason to believe that the English experience was significantly different from that of any other medieval state and the evidence from the later surviving records of the French monarchy confirms this. One finding from studying these sources confirms that the terminology attached to the engineers and master craftsmen was not exact and varied according to local usage, as well as evolving over centuries to reflect changing status and new technologies. I attempt a summary of this at the end of this chapter. But the problem (for modern students) of inexact terminology in official records is only modest compared with the issues arising from using other written evidence.

THE WRITTEN WORD 2: MILITARY MANUALS

The survival of a genre of writing intended to educate and inform the elite in military matters is a great blessing to modern students but raises its own challenges of interpretation. A number of such texts will be examined more closely. They can potentially supply vital information on subjects including the level of technology then available and advice on its best use. But how far do they represent what was the practice of the time, or were they only a worthy effort by the writer to encourage contemporary rulers to follow the best practice of those that came before? Sometimes, especially in late medieval illustrated books, there are machines even the existence of which is questionable, as well as anachronistic devices long since disappeared (or had they?). The answer could, of course, be both at the same time, but posing the question is necessary and the best way to check is to analyse what actually happened on the ground – in the circumstances where

9 P Purton, *History of the early medieval siege c.450–1200*, and *History of the late medieval siege, c.1200–1500*, both Woodbridge 2010.

other evidence allows it. Here, one then encounters the common habit among other contemporaries to seek out accounts of battles or sieges found in classical writers and transpose these verbatim.

The dominant figure in this discussion was the late Roman writer Vegetius. His *Epitoma de re militari*, probably written around the year 400, remained permanently in use through to the renaissance and along with a handful of other writings made up a very modest literary canon surviving through late antiquity. Vegetius wrote unashamedly with a political agenda of seeking to encourage the Romans of his day to return to what he prized as the virtues of earlier times. His descriptions included detailed accounts of good practice in military operations and covered many aspects that required the deployment of engineering skills, surveyors and teams of miners. The questions that arise, however, are many. Had those who are known to have possessed copies of Vegetius actually read it? The only possible verification is to study what they did in practice. The bellicose Angevin count Fulk Nerra was a rare example where it is possible to compare deeds and text. If they had read the text, did they possess the means to carry out the recommended best practice? And even if they were familiar with the ideas it does not signify that their engineers were able to carry them out. In the case of Fulk, he may well have done all of these things. Whether others did so is more in doubt.

Despite all these doubts, the astonishing longevity of this text implies a continued usefulness and this can be reinforced by the frequency with which its advice was replicated later. It is striking that generations of Byzantine military manuals (beginning with the *Strategikon* in the name of the emperor Maurice, dated to 600) reproduced the same practical advice, as will be discussed. A whole series of further manuals appeared in the tenth century, coinciding with a time when the Byzantine empire experienced a long period of significant military success with a series of emperors who were genuinely skilful and experienced generals. We will look at some examples where the advice of the manuals can be compared with actual military campaigning, with the same caveat that the apparent confirmation offered by a chronicler may be no more than a circular argument. Elsewhere, and much later, are to be found a number of texts written in western or central Europe in the form of "advice to princes" on how to rule well, many of which include sections on warfare, including information reflecting what the writer believed was best practice in areas where engineers were needed. Whether this reflected current reality is one of the relevant questions to ask.

The rise of the Islamic caliphate during the seventh century and the evolution of its successor states in the Middle East, North Africa and Asia played a vital part in the development and diffusion of knowledge, skills and technology, and among its literature it also saw the production of military manuals. The same issues arise about the extent of the reality or otherwise of the advice offered. The reliance of the early Muslim caliphs on the knowledge and expertise that had existed in the lands they had conquered from east Rome and from Persia is well known, although the precise mechanisms for the acquisition and transmission of this knowledge during the age of conquest are harder to follow.

THE WRITTEN WORD 3: ILLUSTRATIONS AND CHRONICLES

Illustrations in medieval monastic manuscripts have proved to be a rich source for understanding much about buildings, building techniques and also weaponry of the period contemporary with the scribe, while diagrams preserved in Muslim handbooks are an invaluable source for appreciating what was known at that time in other areas too, such as awareness of mathematical skills required for measurement. It is not a difficult conclusion to reach that if there existed a group of people who were experts in surveying the fall of the land with sufficient precision to organise an immense irrigation system (as there were), then the same skills might have been applied in designing fortifications or preparing a siege of a town, even if we have surviving evidence only of the former.

The chronicles themselves often provide the only accounts of events, but the perils of treating them as plain truth are well known. Chroniclers did not share the concern of modern historians that what they wrote was actually what happened, only that the moral associated with it, or the heroic role of the protagonist, was clear – although the broad facts of course had to be seen to be true. The tendency to lift words from Sallust or Caesar additionally demonstrated the writer's education in the classics. At the same time, however, and with all due caution, where there is more than one account greater certainty is possible. Often there is no alternative source, unless archaeology has unearthed closely dateable confirmatory evidence.

THE WRITTEN WORD 4: SCIENCE AND TECHNOLOGY?

The vast accumulated stock of scientific knowledge from the classical world was not suddenly lost in late antiquity and early medieval times, just as aristocrats and clerics and administrators did not suddenly lose the ability to read or write as they served the new ("barbarian") kingdoms. In western and central Europe, there was unquestionably a significant falling away, with sparse continuity in the worst-affected regions. But the knowledge was not lost – think of the work of Boethius (c.480–524), Isidore of Seville (c.565–636) and Bede (c.675–735) – and gradually revived everywhere. For this study the issue is, what use was made of it? Isidore's *Etymologies*, an encyclopedia of current knowledge, in fact presented only historic information about the technology of warfare and would have been of little use even if it had been read by generals.[10] There was a period of reawakening of interest in scientific knowledge in western and central Europe during what is called the "Carolingian renaissance" in the late eighth and ninth centuries, followed by another transformation of knowledge when Arabic translations of a wider range of ancient texts (with Euclid's Geometry having the greatest possible relevance to engineering) became available in the Latin west from the eleventh century onwards. These translations were the result of the work of many scholars who flourished in environments such as the state-sponsored "house of wisdom" in Baghdad from the late eighth century onwards, but also across the Islamic world, and who not only rescued classical knowledge by copying and translating the Greek originals, but merged it with other science (such as that of India and Persia) and then built on it and took it further.

Many of these scientific works had important implications for technology. But did the theoretical knowledge that was learnt in monastic or cathedral schools, the progressive expansion of which is broadly measurable, have any impact on the "vulgar" – that is, technical – skills that craftsmen used, and on the intermediaries between them and the employer and patron, the architects and engineers? The religious authorities in charge of education in the Christian world continued to assert that knowledge of the subjects of the *quadrivium* (arithmetic, geometry, music and astronomy) was an essential part of school-learning designed to glorify God and to educate clerics, and that it was purely theoretical: *scientia* had nothing whatever to do with the skills practised by common craftsmen and artisans – the "mechanical arts". A debate has continued for decades on whether there was a definitive dividing line between this school-knowledge and its possible practical applications. Evidently, there was a link at the level of the educated clerics who went on to be the bishops recorded as "building" – signifying, in fact, designing or overseeing the construction of – cathedrals, monasteries, churches – and palaces, castles and town walls, who must have made some practical use of what they had learned. Those whose literacy made them into the clerks who populated royal chancelleries and exchequers also carried their book-learning with them, although it is harder to make a case that their scientific knowledge had practical consequences.

A long line of scientists (mostly clerics in the Christian world) produced studies and writings that took forward existing classical knowledge of the world (much reliant on the works of Aristotle) and in many cases there were major (potential) consequences for technology. Writers such as Gerbert of Aurillac (Pope Sylvester II from 999–1003) spread knowledge of the use of the abacus, for example, with significant practical applications. Later scientists such as Adelard of Bath (1080–1152) took advantage of the work of Arab scholars (the earlier role of al-Fārābī (d.950) being particularly significant) to spread awareness of more sophisticated

10 *The Etymologies of Isidore of Seville*, ed./trans. Barney et al., books 18 and 19, 364, 376–7; see also *Isidoro de Sevilla Etimologias libro XVIII*, ed. Llorca, 10–12, 106–9. Isidore's work was finished in 636 and more than one thousand copies survive.

geometry with implications for surveying and construction. Hugh of St Victor (1096–1141) argued for recognition of the importance of the mechanical arts and described their application.[11] Many other medieval scientists made contributions, some of which will be discussed as appropriate in the chapters that follow. In recent times, students of medieval science have tended to challenge rather than accept the idea that there was an unbridgeable gulf between book-learning and practical technology.[12]

Were there any links between the two realms of the literate clerical scientist and the engineer? It is apparent that the tasks of designing and building cathedrals, castles and other grand buildings could not be achieved successfully just by copying (even with a template?) the design of the last building one had worked on. It was not necessary to be familiar with Euclidian geometry (or even the incomplete and inferior version transmitted in the often copied instructions of Bede) in order to measure out parcels of land, but it still required a knowledge of measurement and some elementary geometry to be able to function as a surveyor, a planner of towns (such as the many planned towns identified in Europe and elsewhere, and fortresses of which the Anglo-Saxon *burh* is a significant and early example). Similarly, a castle builder had to be able to measure the height of the wall and the depth of foundations as well as lay out the form of any new construction and it was extremely unlikely that this would replicate the dimensions of their previous site. Another example is bridge-building: bridges could not be erected successfully without an understanding of some key questions that apply to such a structure. This knowledge may have been basic, and inherited, but it still constituted a learning that would often have to be applied and modified to cope with different situations. In bridge-building, one had to understand the geography and geology of the site and the characteristics of the waterway to be crossed, as these determined both the mode of construction and the dimensions. Otherwise the bridge would collapse or be washed away. Siege engines likewise were almost always built on site, no doubt following a clear idea of what they should look like and what they needed to do: but because each engine was new, some at least of the principles behind its operation must have been known to the builder. Similar considerations apply just as strongly to the learning acquired by educated Muslims, where clerics emphasised the religious purpose of book-learning, but where it is evident that

11 *The letters of Gerbert ...*, ed./trans. Pratt Lattin; Burnett, *Adelard of Bath, an English scientist*; Netton, *Al Fārābi and his school*; Hugh of St Victor, *Practical geometry*, ed./trans. Homann, and *The Didascalion*, ed./trans. Taylor.

12 See, for example, E Whitney, 'Paradise restored. The mechanical arts from antiquity through the thirteenth century', *Transactions of the American Philosophical Society* 80 (part 1), 1980 and for the continuing debate, works by Alessio, Bakar, Beaujean, Bennett, Bodde, Braunstein, Contreni, Crombie, Folkerts, van den Hoven, Jolivet, Long, Ovitt, Özdural, Radding and Clark, and White in the bibliography.

there were practical applications of much of the madrasa-learning, and that if the resulting engineering did not come (ultimately) from this source, then its origins remain a mystery.

One of the purposes of this study, then, is to examine what the engineers and master craftsmen needed to know to achieve what they did, and to ask where the knowledge came from. An intriguing sideline of enquiry concerns those lay rulers who were ascribed a particular awareness of military technology by contemporaries. They are few in number, and what they actually knew cannot be recovered today, but the image of a prince instructing the carpenters how to construct a siege engine, or telling the masons how thick their castle wall should be, does arise from the historical evidence. Where did they – representing as they did the tiny minority of educated lay people – obtain the necessary training to do this? Vegetius and the Byzantine military manuals did not offer this sort of detail.

It is a truism that the source of most training was from master to apprentice, who then went on to become a master, taking on an apprentice – and so on. But unless the knowledge transmitted by this route was subject to change, to innovation, how did technological progress ever come about? The diffusion of new technologies from one geographical area to another was a process that could only have taken the form of human beings conveying new ideas and methods to fellow human beings, who must have been in a position to absorb the information in order for it to spread. There are enough stories of individual engineers spurned or exiled by their original masters conveying dangerous knowledge to less cultured hostile neighbours to see how this knowledge could easily pass on, but it also means that there must have been among the recipients people able to absorb and themselves share the new ideas. One aspect in particular of the lives of medieval military engineers (as well as technicians and craftsmen) that risks appearing strange to modern readers is their mobility. This is because one of the critical flaws of many views of medieval history, particularly among non-specialists, is to read backwards from later times a whole number of concepts that are commonplace today but which, quite simply, did not exist at the time. Ideas of nationality or national identity, normal in modern history, are an anachronism for many medieval societies. When this history recounts the career path of particular engineers, whether it be the men who served both the Byzantine empire and its enemies or those who were hired by the warring city states of renaissance Italy, what would now be called treason was just not seen in such a way by contemporaries who would not understand the thinking behind it.

Similarly, there must also have been practitioners of these engineering arts who on observing the operation of, say, a trebuchet were also able to think through ways in which its efficacy might be improved. This is the only logical explanation for a century of progress in which the original and probably quite simple throwing engine developed into the most successful heavy artillery of the whole medieval period. The innovators remain nameless, but it is a reasonable hypothesis that they had some elementary understanding of the principles by which the trebuchet worked if they were to suggest ways in which these could be further improved. The number of variables that determined its performance was considerable, as modern physicists have established, and knowing their interaction was crucial.

Later, the various and continuing improvements that turned what was, at its start in the early fourteenth century, a very modest beginning for gunpowder artillery as a weapon of war into the devastating weapon that changed the face of military architecture and warfare in general by the end of the fifteenth century probably were the results of trial and error, but also must have involved practitioners in some understanding of the properties of the ingredients of the powder as well as the qualities of the metals and the techniques – and their limitations – that they used to manufacture guns. There is much detailed information available about gun makers in the later medieval period and also, from surviving records, about the people who commanded and operated gunpowder artillery, which reinforces the perception that a modern state in the fifteenth century – and especially from around 1450 onwards – saw the importance of centralising, simplifying and maintaining its resource of guns and the people who knew what to do with them. Although there are still to be found examples of itinerants promoting their abilities in the art of gunpowder weaponry to the best bidder, states such as France and Burgundy were already on the road to professionalising what was now recognised as a crucial military asset.

Were the progressive improvements in the technology of such weapons down just to trial and error? Of course, it is possible, but there are occasional hints that it was more than this.

Were engineers and craftsmen who were aware of more than the basic traditions they inherited also literate? Could they have absorbed any of the available knowledge if they had undergone education other than apprenticeship? Despite the existence and occasional expansion of the education available to lay people in towns and cathedral or monastic schools, on which there is a substantial literature, or in madrasas in the Islamic world, there is little evidence of who benefitted from this provision. It may be extremely unlikely that even a small minority of master masons or carpenters could read in their own language, let alone Latin, Greek or Arabic, even at the end of the middle ages, where there are records of literate craftsmen (such as the fifteenth-century design booklets of master masons Roriczer and Schuttermayer, edited by Shelby, or the book written by engineer and gun master Bengedans). But there is some indication that knowledge may have been passed on through this medium. Rare indeed are the surviving copies of technological writings before the renaissance, and the pearl is undoubtedly the thirteenth-century sketchbook of Villard de Honnecourt,[13] which would be a revolutionary discovery if it were unique but perhaps more realistically is the

[13] Discussed fully in chapter five.

chance survival of what may have been common at the time, and there are hints as well as commonsense deductions to suggest that conclusion. This text offers some clues to the question of how transmission occurred and it implies a lot about the ability of some engineers to make drawings, to interpret drawings and to read measurements as well as to write instructions (for whom?). Villard's notebook also includes an incomplete diagram of a trebuchet. For whose benefit was such a diagram intended? There will continue to be debate over whether even an ostensibly practical text such as this was actually written for the information and training of a master mason, a common engineer or a surveyor, or for some other purpose, and some of these questions will never be resolved. Here, I attempt to see how such knowledge might have been needed to achieve the works of military engineering that are known to have been accomplished. Gunpowder artillery offers further evidence of a consolidation of writing as the route of transmission – the story of the frequently copied *Feuerwerkbuch* is instructive.

OTHER ROUTES OF TRANSMISSION

Economic and social changes may not have much impacted the continuity of a skills base of craftsmen, since masons and carpenters (and miners) were always needed, but the mystery remains of how that higher level of skill needed by the engineer continued throughout the period. The one certainty is that it did, although the picture can become one of a travelling engineer acting like a salesman offering his services to each ruler as he travelled the roads of medieval Europe, perhaps trying to establish who next would need someone to build a siege tower. However, as more engineers are named in chronicles, the more the evidence grows that there was in the high middle ages a self-selecting group of experts hiring themselves out to the best bidder, sometimes arriving at court and offering to manufacture a machine the like of which the world had never seen before, with which the monarch was assured of success in his (and occasionally her) next siege. Italian cities seem to have been the original home to many of them, but attachment to any particular place or state was unlikely to have influenced them. One might have expected a family history to explain how they acquired their knowledge but few can be established before the fifteenth century. Even more mysterious then are those individuals where the history tells us that they really did construct something novel. Whatever the reason for the skill, having it and marketing it effectively could prove to be lucrative. Once the relationship was established, naturally enough, it could be called upon again, but there was no loyalty attaching the engineer to a particular ruler and we will find examples of those who, on capture, simply transferred their service to a new employer. Engineers – and skilled craftsmen – could find that their skill and, it appears, the difficulty in recruiting people like them offered them some chance of immunity against the horrors of war, provided they were willing to transfer their service to their captors, while other captives were either ransomed (the wealthy) or killed (the rest). The situation is less clear still in Islamic states, where we find the engineers in question materialising when so instructed by the sultan. How they were maintained is not revealed in the histories. Perhaps they found routine employment on other tasks and because they lived in cities could be mobilised rapidly for military service. The Mongol khans certainly put the artisans of captured cities to good use, sparing their lives as their neighbours were slaughtered but then pressing them into service to build and operate the siege engines needed by the conquerors at the next walled town that dared to resist. Other than these forcibly conscripted workers, there are no records of a corps of military engineers in the service of any ruler outside China, where written records of medieval dynasties confirm the existence of large units of artillery operatives. Further west, the single possible exception is Roger II of Sicily who *may* have inherited a siege corps from his Muslim predecessors. The evidence is slight, although it seems that the Norman conquerors of that kingdom were themselves among the princes well versed in the techniques of siege warfare. More likely is that rulers knew where to find the people they could call upon with the skills they required.

MINERS

Evidence for this suggestion comes chiefly from miners. Miners serving royal armies would often have come from the same area each time, but at other times they carried on their usual occupation: in England, the Forest of Dean provided groups of miners under their own masters from the first surviving records in the twelfth century up to the fifteenth; they would be paid for their travel to and from the army as well as for their service once there. There was an obvious connection between royal service and the availability of skilled practitioners, although this does not explain why the other parts of the country where there were long-established mining communities were not similarly called upon. The link to military service is less clear from the evidence for mining communities in other parts of Europe, although from the fourteenth century the miners of Liège were to find regular military service and establish a formidable reputation (and indeed to replace the Dean miners in Henry V's army in 1417[14]), and there are other places where the evidence of such connections is strong if not decisive. In the Middle East, there are similar references to miners repeatedly coming from particular locations to serve in the armies, but this may reflect no more than that these were the nearest areas where there were mines.

There are, therefore, a number of possible models for how military engineers obtained their expertise and secured their employment and for how royal

14 See p. 224.

commanders recruited them. The questions become easier to resolve as the centuries roll by and as the evidence becomes more substantial. By the end of the middle ages (not a specific time but an evolving pattern of distinct but overlapping processes varying between and in time and place), early modern states were established and operated in a world still linked to the past but substantially more developed in many areas, including their economies, state apparatuses and military capabilities. Living among them, the famous figures of the renaissance who were able to combine talents in many different spheres including the military – Leonardo da Vinci being only the best remembered among many contemporaries – were, in this area, building on the achievements of their medieval predecessors rather than creating much that was particularly novel (let alone representing throwbacks to the glories of the ancient world which, in the area of military technology, had long since been superseded). Their ability to design and construct military-related machines, buildings, fortifications, engines, bridges and so on – and there was an interesting tendency for those who specialised in siege equipment also to work in hydraulic engineering – was not based on sudden inspiration, and still less on ancient Roman models (though basic principles held true), but rather represented the evolution of practical knowledge that had been transmitted from generation to generation. They must have followed a number of routes not all of which are capable of being recovered, but which must have gone back through the recorded individuals of the fifteenth century, and the occasionally identified masters of the twelfth and thirteenth centuries, to the anonymous but real engineers who carried out the same tasks in the employ of Charlemagne and Pippin. They, in turn, did not acquire these skills without learning them from predecessors. No doubt, the further back one goes, the more often there were times and places where the transmission of knowledge failed and had to be re-imported from elsewhere. But that knowledge must have always survived somewhere. To demonstrate this unbroken progress is the main purpose of this study.

"ENGINEERS": THE EVOLUTION OF THE WORD

By the start of the early modern age, the word engineer or its form in other languages was readily understood and was commonly applied to the type of expert featured here. But a millennium before, this was far from the case. As "engineer" will be used as a shorthand descriptor for the whole millennium, it makes sense to set out the evolution and different uses of the word at the start, as well as the other terms used to describe the same people.

It is not possible absolutely to confirm the reason why the (Latin) *ingeniator* became the common word to describe people who designed and – more commonly – made engines. By the time of the renaissance, it clearly had that significance, and even in non-Romance languages it had been adopted. In German, the equivalent term was *werkmeister*, a word that is directly related to the function it describes, although that function was in fact loosely applied to building masters as well as artisans and craftsmen and those who made an engine (*machina*).[15] It has been traced back as far as Old High German with examples from the tenth century, and there were analogous words in Norse languages. *Ingenieur* was an early modern import into German from the French.[16]

There is more than one possible origin of *ingeniator*. The most likely derivation is from the Latin *ingenium*, which had the classical meaning of both a natural quality and, by further derivation, cleverness, talent, genius; these senses became further transformed to signify artifice or trickery and then developed into signifying a product of such artifice: engines.[17] A clever attempt was made a long time ago to suggest an alternative derivation based on finding the word *encignerius* in a twelfth-century Italian source, used in the sense of someone encircling a fortress (*incignere*).[18] However, if one follows the appearance of the word in Romance languages, this explanation does not seem to hold up. Numerous references have been found in Old French to *engigneurs* where the word is directly linked to the term *engin*. As in other languages, it is found describing both builders (architects) and constructors of military engines, and in the sense of the builder being "ingenious".[19] Similar usage predominates in the romances that entertained the nobility during the early part of the second millennium.[20] The same connections were found by students of both the Spanish and Italian languages, with references to Castilian sources from the eleventh century and, as we have seen, Italian from the twelfth.[21]

In Byzantine sources, a range of terms were employed, depending on the role being described. At the top end of skills (for example, when describing someone who was the equivalent of a modern architect), the word used is *mechanikos*, mechanic. Where the function is more specifically that of the creator of siege equipment, Procopius used the term *tekhnitai*, which normally means a craftsman, suggesting that this person was not the engineer in charge.[22] However, this word had a wider significance and could also mean people of both higher and lower status according to the context.

15 Diefenbach, *Glossarium Latino-Germanicarum*; Lexer, *Mittelhochdeutsches Handwörterbuch*, 3, 773.

16 Grimm, *Deutsches Wörterbuch*, 14 (1) (2), 385.

17 Latham, *Revised medieval Latin word-list from British and Irish sources*, 249.

18 Feldhaus, *Die Technik der Antike und des Mittelalters*, 277.

19 Godefroy, *Dictionnaire de l'ancienne langue française et de tout ses dialectes du IXe au XVe siècle*, III, 169; Du Cange, *Glossarium mediae et intimae Latinitatis* (new edition), 360–1.

20 See examples in Gregory, Rothwell, Trotter, *Anglo-Norman dictionary*, 908–9.

21 Corominas, Pascual, *Diccionario crítico etimologíco Castellano-Hispánico*, III; Alonso, *Diccionario medieval Español*, II; Battisti, Alessio, *Dizionario etimologico Italiano*, III.

22 Procopius, *Wars*, VIII, xi, 27 (Loeb ed. pp. 156–7).

There is an *ingeniator* recorded as a holder of lands in the English Domesday Book of 1086 (Waldin, who had ten landholdings directly from William the Conqueror in Lincolnshire),[23] but other explicit references to early engineers using that word tend to be in later accounts. Nonetheless, the link between the original and the acquired senses of the word is not unusual. In describing the siege of Bréval in 1092 by the king of France and the duke of Normandy, the early-twelfth-century chronicler Orderic Vitalis wrote that Robert of Bêlleme, who will be discussed later, "ingeniosissimum artificem adducit, cuius ingeniosa sagacitas" was to be later deployed in the Christian capture of Jerusalem. The chronicler seems to have enjoyed the playful link between Robert's ingenuity and the creation of siege engines.[24] The term engineer thereafter appears frequently in English government records, starting with payments to Gaufridus *ingeniator* in the sole surviving Pipe Roll from the reign of Henry I.[25]

For a long time, those who would fit the loose definition of a military engineer as people carrying out the many different tasks that fall within the modern scope of the term were also identified in a number of different ways, most often associated directly with their trade: master masons and master carpenters were to be found as well as engineers designing and erecting buildings or bridges (where the distinct term *pontanarius* was rare). The word *balistarius* may literally mean a crossbowman but there is some evidence suggesting that at times it included people with a wider and therefore more significant remit, alongside the possibility that it also acquired a more symbolic significance (to judge by the number who appear as tenants-in-chief in Domesday Book). In later centuries the word *a(r)(t)tiliator* was used to designate those responsible for crossbows and ballistas. Miners and sappers always had their own distinct identity recognised in a distinct vocabulary, while those responsible for measuring, *mensuratores*, also occupied a distinctive place whether or not their task at a given time had a military element to it. By the time of the gun, the various skilled craftsmen who would be involved in the manufacture of the weapons, and those responsible for operating them, would generate a whole new language of their own while also calling on existing terminology.

Other terms were also used, though more rarely. An *artifex* in classical Latin was a master in the liberal arts but also could be a master mason or carpenter. This implication of skilled worker continued alongside other more exalted meanings and the associated verb could mean designing or making.[26]

For the execution of engineering tasks, other types of worker were also needed. The Latin word *faber* was in use in sources dating from the early medieval period through to the end of the middle ages. It was often followed by a descriptive term

23 Williams, Martin, *Domesday Book*, 945.

24 *Ecclesiastical history of Orderic Vitalis*, ed./trans. Chibnall, book VIII (vol. IV, 288–9)

25 *Pipe Roll of 31 Henry I (Michaelmas 1130)*, 143.

26 Lewis and Short, *Latin dictionary*, 168.

identifying the particular specialism – carpenter, mason, smith and a multitude of specific tasks associated with shipbuilding. These people were, however, almost always of craft status, because those below them, common labourers, were described as an *operarius* or some variant based on the same root. They could be found working on any task, and again, where appropriate, they might be described as workers on a castle or a bridge as well as workers in timber or iron.[27]

It appears that engineers as so called were much less common before the turn of the millennium, which raises important questions not about the existence of the role – the various specialisms that those called engineers were later to be identified with still required people with the necessary skills to carry them out before the word engineers was applied to them – but it may give cause to doubt what if any status they had acquired, in contrast to the recognition that would be allowed them in later centuries. The absence of a specific term makes identifying them and answering such questions all the more of a challenge. The necessary starting point is the engineering provision of the later Roman empire.

27 Latham, *Revised medieval Latin word-list*, 323–4.

2

Late Antiquity and the Early "Middle Ages"

Were the "Dark Ages" Really Dark?

THE INHERITANCE

IT IS NOW GENERALLY agreed by historians that there was much continuity between the western Roman empire and its successors following the disintegration of the centralised state during the fifth century. In the east, the continuity was unbroken as the Byzantine empire survived for another thousand years. The Germanic tribal confederations, many of whom had been living in or alongside Roman society for decades, set up kingdoms of whom the Franks in northern Gaul and beyond, the Visigoths in southern Gaul then (after expelling the Vandals, who took over Roman Africa) Iberia, and the Ostrogoths who established a kingdom in Italy were the most significant. The new rulers (numerically a small minority of the population) wanted and needed to use what they found, not destroy it. Only in a few places had specific circumstances led to a collapse of the Roman infrastructure. The political and military history of the age can be traced easily through written evidence – key features were the reconquest of Africa and destruction of the Ostrogothic kingdom in Italy by east Roman (Byzantine) emperor Justinian (r. 527–65), and the expansion of the Frankish kingdom under the Merovingian dynasty following Clovis (r. 482–511) – but evidence for society and the economy is harder to find, and here archaeology makes a big contribution. Warfare and archaeology can be combined to permit a study of levels of engineering and to propose what happened to its military version. It will be proposed that although there are few references to named engineers across these centuries, they continued to operate; and that the changes to society and the economy that took place as late antiquity gave way to a recognisably different early medieval world in the seventh century appear to have had limited consequences in this field.

The seventh century also saw the sudden creation of an Islamic state as Muhammad's followers achieved extraordinary victories pushing back the frontiers of eastern Rome and destroying totally the Persian empire (by 651) before extending their power across Egypt and North Africa (whence from 711 they quickly took over Visigothic Iberia) and eastwards across central Asia to the borders of China and India. Just as the Franks on a smaller scale recognised that they needed to make use of the knowledge and skills that they found, so also the early Muslim caliphs understood that they had to make use of Christian or Zoroastrian experts living in their new empires, and did so with great success.

Written evidence for the earliest of these events does not survive in great quantities. The sixth-century accounts of bishop Gregory of Tours and of the Byzantine official Procopius are the most useful for our purpose but other surviving writings (including evidence from sources such as saints' lives – the role of the Christian church was of great importance, representing as it did the only institutional continuity with the western Roman state) can contribute too.

THE WRITTEN WORD

Gothic kings still demonstrated their interest in the inherited learning of antiquity – Cassiodorus' official letters confirm interests in machinery, in arithmetic and geometry, in architectural skills and in surveying, and referenced great inventive minds such as Archimedes.[1] In east Rome, there is no doubt that the ability to read and write remained widespread among people of the ruling classes and those that served them, including in the military and administrative structures.[2] Some writings never disappeared, of which Vegetius is of the greatest significance here.[3] Vegetius' *Epitoma de re militaris*, from around 400, was itself based on earlier writers (historians such as Sallust and administrators such as Frontinus), with its main focus being to emphasise how vital to success were the professionalism, training and discipline of an army and the range of skills required of its commander. But the book also devoted considerable attention to areas in which engineering skills were paramount, such as siege warfare, and, as part of the discipline of any army, to the question of setting up camps. In his book four, Vegetius described the weaponry available to the Roman army in conducting a siege, either as attacker or defender, and discussed weapons such as the ballista and the *onager* (one-armed stone-throwing torsion catapult that had been standard equipment since the third century) (see pp. 43–5). His comprehensive recommendations included advice on

1 P Riché, *Education and culture in the Barbarian west sixth through eighth centuries*, trans. J Contreni (1976), 6–7, 68–9. Barnish (ed./trans), *Selected Variae of Magnus Aurelius Cassiodorus Senator* (being *Variae* vol. I, number 45), 21–3, explains how Gundobad, the Burgundian king, had requested a water-driven mechanical clock from Theodoric.

2 For a summary of the evidence, M Mullett, 'Writing in early medieval Byzantium', in R McKitterick (ed.), *The uses of literacy in early medieval Europe* (1990), 156–85.

3 C Allmand, *The* de re militari *of Vegetius* (2011), analyses the history with particular emphasis on the later middle ages but showing its continuous use throughout.

sapping and mining. His advice on fortifications was highly practical.[4] Vegetius' work continued to be copied – the oldest known manuscript today is in a collection of texts thought to date from the sixth century.

Vegetius was copied in Constantinople too,[5] but the most striking confirmation that what he proposed remained the standard – although it can be argued that this was likely to be based on experience and a continuing military culture rather than on having read this work – is to be found in the *Strategikon* ascribed to the emperor Maurice and dated to the year 600. This work covers similar ground with similar advice and was rooted in the theoretical best practice of the age. The information it provides for Byzantine military organisation and tactics but also for the continued integral role of engineers within the armies of eastern Rome at the turn of the seventh century is invaluable, as evidence from actual campaigns suggests its continued use by Byzantine commanders.

The transmission of classical knowledge is a topic the relevance of which to our subject depends on whether one believes that scientific awareness had any practical impact. In this age, it can be shown that the level of understanding revealed in contemporary writings was reflected in what is known from archaeology or written sources to have been the practice of the sixth century. This applies particularly to those parts of mathematics essential to building, where the role of the master builder, the architect described by Isidore as emulating the description of his role in the teachings of St Paul,[6] and therefore very well known to educated people lay and ecclesiastical, must have involved at least a basic understanding of geometry and measurement. The trickier question will be the extent to which this continued into the following centuries.

At the beginning, however, evidence suggests that the social and economic conditions in which such skills were called for were present, or the skills could be obtained from more cultured neighbours. Neither seas nor waterways, mountains nor forests, state boundaries nor (as will be clear later) religious differences stood in the way of trade, travel and transport, including the transmission of technologies and the technicians with the requisite skills.

4 In book IV, cc.1–30. Translated by N Milner, *Vegetius: Epitome of military science* (1993), 113–31.

5 Vegetius was often copied with the works of Frontinus, a first-century writer, but while the latter's *Aqueducts* contained much relevant to hydraulic engineering, his *Strategemata* did not, unless implied in his advice on diverting streams: Frontinus, *The stratagems and the aqueducts of Rome*, trans. E Bennett, ed. M McElwan (1925), esp. 226–9.

6 Isidore of Seville, *Etymologies*, ed. Barney, 376–7.

ECONOMY AND SOCIETY IN THE "DARK AGES"

Current orthodoxy proposes, with Chris Wickham, a "very substantial simplification in post-Roman material culture in the fifth to seventh centuries", particularly drastic in Britain and the Balkans, with only Syria, Palestine and Egypt (all under Byzantine rule) exempt,[7] with resources reduced and structures "simplified". But there was sufficient continuity to ensure Roman-style government systems in the major kingdoms (the Franks and Visigoths) for several centuries.[8]

In Italy from the fifth century on, a combination of factors, including military expenditure, relentless war in the sixth century and massive tax evasion, contributed to a collapse of investment in infrastructures and urban decline in favour of the countryside, but with some important exceptions. An overall decline in building technique has been posited, based on the argument of declining skills.[9] In the Visigothic kingdom (Iberia and southern France), the state continued to function and continued to use imperial Roman titles. However, partly because it was unable to maintain the collection of taxes, it could no longer pay for a professional army but was instead dependent for its personnel on a small entourage around the king, retinues supported by the great landowners, and the continuation of a system of obligatory public duties to the state. There was no decline in available craft skills – indeed, it is argued that Visigothic metalworkers were superior in skill to those they now lived alongside[10] – but their relationship with the military structures of the kingdom must have altered as a result of the changed character of the state's military functioning. What consequences did this have for there being a professional engineering corps?

Merovingian Gaul and Burgundy offer evidence consistent with the continuity of craft-based industry, sometimes in a few cities that continued to have important trading or administrative functions, while others shrank: but this was accompanied by the transfer of the crafts to rural settings, and monasteries acquired important

7 C Wickham, *The inheritance of Rome* (2010), 9. Also useful is the compendium of current evidence presented in J Graham-Taylor, M Valor (eds), *The archaeology of medieval Europe, vol. I: eighth to twelfth centuries AD*, published in 2007, and covering non-Byzantine Christendom.

8 Wickham, 74.

9 E.g. F Marozzi, 'The destinies of the late antique Italies', in R Hodges, W Bowden (eds), *The sixth century: production, distribution and demand* (119–59).

10 F Retamero, 'As coins go home: towns merchants, bishops and kings in Visigothic Hispania', 284; P Diaz, 'Visigothic political institutions', 339; and G Ripoll López, 'Symbolic life and signs of identity in Visigothic times', 410–11; all in P Heather (ed.), *The Visigoths from the migration period to the seventh century: an ethnographic perspective* (1999), 271–320, 321–72 and 403–46 respectively.

economic roles too.[11] They became increasingly prominent as their number expanded (up to 550 across the Frankish realm) during the seventh century, many of them part of the royal fisc. With good communications, many became centres of production, while local villages needed to supply their needs, including for craftwork.[12] Archaeology suggests the continuation of skills in carpentry – including the continued existence of specialised craftsmen, who could only survive in the trade if there was a need for them.[13] Law codes and legal records established the status and the value attached to craftspeople, and in the process confirmed their continued role in society, with evidence that even when some of them were legally unfree, their economic importance was recognised, as it appeared to have been in the laws established in the successor states in the west.[14]

The level of continued craft skill is also attested by archaeological finds in different areas. Windmills and watermills were important to local economies and involved a considerable level of "elaborate engineering" (Hodges' phrase) to construct and maintain. From this period there are seventh- and eighth-century examples in Anglo-Saxon England and Merovingian Gaul, extending to its far eastern frontiers, with evidence of sophisticated hydraulic engineering that was presumably based on the preservation of classical knowledge and skills, and references to a floating mill on the Tiber in Rome.[15]

11 R Hodges, *Dark age economics: a new audit* (2012), especially 90–4. The evidence for Burgundy is presented in D Boyson, 'Romano-Burgundian society in the age of Gundobad: some legal, archaeological and historical evidence', *Nottingham Medieval Studies*, XXXII, 1988 (91–118).

12 S Lebecq, 'The role of monasteries in the systems of production and exchange of the Frankish world between the 7th and the beginning of the 9th centuries', in I Hansen, C Wickham (eds), *The long eighth century* (2000), 121–48.

13 F Damminger, 'Dwellings, settlements and settlement power in Merovingian southwest Germany' (33–106); G Halsall, 'Social identities and social relationships in early Merovingian Gaul' (141–75); and S Loseby, 'Gregory's cities: urban functions in sixth century Gaul' (239–84), all in I Wood (ed.), *Franks and Alemanni in the Merovingian period: an ethnographic perspective* (1998).

14 D Claude, 'Die Handwerker der Merovingerzeit' (204–66) and H Nehlsen, 'Die rechtliche und soziale Stellung der Handwerker in den germanischen Leges' (267–83), both in H Jahnkuhn et al. (eds), *Das Handwerk in vor- und frühgeschichtlicher Zeit*, I (*Abhandlungen der Wissenschaften in G. Philologisch-Historische Klasse, dritte Folge 122, 1981*).

15 Hodges, *Dark Age economics*, 59–63 and the sources listed there; D Lohrmann, 'Le moulin à l'eau dans le cadre de l'économie rurale de la Neustrie', in H Atsma (ed.), *La Neustrie. Les pays au nord de la Loire de 650 à 850* (1989), I, 367–404, here 396–98; S Stelzle-Hügin, '"Renovatio imperii" on the Münsterhugel of Basel? A reappraisal of mechanical mortar-mixers', at http://medieval-europe-paris-2007.univ-paris1.fr/stelzle-huegin.pdf (accessed 2/1/2014), discusses evidence for the use

The eastern Roman empire also underwent significant changes during this period that would have an impact on the continuity of military structures. Nonetheless, the great cities remained as densely populated centres – 400,000 people in Constantinople (before the terrible plague of 541), 100,000 each in Antioch, Alexandria and Thessaloniki, with evidence of continued investment in building during the sixth century, work and materials to sustain many craftsmen, watermills, large timber oil presses in Asia Minor (operated by counterweights), and significant state expenditure on arms. The plague years of the sixth century had serious impact, and then in the years following 632 the empire faced enormous losses of territory to Islam and continuous pressure on all of its reduced frontiers. There must have been consequences for engineering as well. It is suggested that the scale and scope in particular of public building reduced with the drop in central resources, because such projects were capital-intensive. That the consequences might have included a reduction in the numbers of skilled engineers and possibly a reduction in their level of skill is hinted at in the story of the great aqueduct built for Constantinople in the fourth century by Emperor Valens, which was destroyed during the Avar siege in 626. It was only restored because of a drought in 766/7, when, according to the chronicler Theophanes, the emperor had to summon skilled workers from the Asian provinces to accomplish the task.[16] It is important not to overdramatise this, as military necessity created a continuing need for military engineering, as will be discussed below, and there is evidence that fine crafts continued to be produced.[17]

Despite the shrinkage of imperial power and wealth, imperial armies continued to be paid centrally even if their structures had to be modified to take account of new realities.[18] There has been no study conducted of the impact of changes over these critical years in the provision of military engineering. The work was still

---

of mechanical mixers for mortar for building work. A "solidly constructed" watermill dendrochronologically dated to 696/7 (and repair over the following century) was found in (modern) Bavaria: W Czysz, 'Eine bajuwarische Wassermühle in Paartal bei Dasing', *Antike Welt*, 25/2, 1994 (152–55).

16 Theophanes, ed. Mango and Scott, 608.

17 M Lewis, 'Antique engineering in the Byzantine world', in L Lavan et al. (eds), *Technology in transition AD 300–650*, Late Antique Archaeology vol. 4 (2007), 367–78; J Haldon, 'Information and war', in A Sarantis, N Christie (eds), *War and warfare in late antiquity* (2013), 373–93, esp. 376–7.

18 A Laiou, C Morrison, *The Byzantine economy* (2007), 26–42; J Haldon, 'Administrative continuities and structural transformation in East Roman military organisation ca. 580–640', in F Vallet, M Kazanski, *L'armée romaine et les barbares du IIIe au VIIe siècles* (1993), 45–54. P Sarris, *Economy and society in the age of Justinian* (2006) argued that persistent tax evasion weakened imperial revenues (with serious consequences for the loyalty of the army) and that the emperor failed to reverse the collapse of imperial control over provincial government (summarised on pp. 228–34).

done, but by whom? Procopius, describing one of the repeated Roman–Persian wars, in discussing the design of a new, easily portable type of battering ram at the Byzantine siege of Petra in the Caucasus in 550, made the helpful comment in passing that this had never previously been devised despite the fact "that there have always been and now are great numbers of engineers in both countries" (that is, east Rome and Persia).[19] Even if the technical skills base was sustained across Europe, the case for continuity of professional military engineers in the eastern empire is far easier to demonstrate than it is further west.

This reference draws attention to the comparable engineering skills and technology available to the great power represented by the Sasanian monarchy of Persia, although contemporary records are in short supply and it is often necessary to rely on much later writing from the Islamic era. The largest siege operations of the Middle East have been uncovered using satellite imaging of the Sasanian siege of Hatra in 240–41, revealing a double-walled circum- and contravallation 24 kilometres long in total provided with 45 ballista platforms.[20] The Sasanian king Chosroes I (531–79) is credited with a modernisation of his state including establishing professional armies and extensive fortifications on the frontiers, implying the existence of the infrastructure and building and design skills required. Massive irrigation systems were built during this era: rivers diverted and canals dug – something only possible with substantial engineering skills.[21] When the old regimes collapsed to the invading armies led by Muhammad's successors, it was hardly a surprise that the new rulers would rely directly on the knowledge and skills already present in the lands that they now ruled. It may be over-simplistic to assume total ignorance among the Arab conquerors: after all, Arab tribes had for long related to both of their overwhelmingly powerful imperial neighbours and sometime rulers, with considerable evidence of interaction including in the adoption of building skills. Early Muslim fortresses and palaces in Syria were clearly based on the design of the former Byzantine rulers[22] and there is also much to suggest that the well-known "desert castles" of the Ummayad caliphs were built by Christian craftsmen.[23]

19 Procopius, *Wars*, book VIII, xi, 25–30; Loeb edition at pp. 156–7. See above, chapter one, for the word translated as "engineers".

20 S Hauser, D Tucker, 'The final onslaught. The Sasanian siege of Hatra', *Zeitschrift für Orient-Archäologie*, 2, 2009, 106–39.

21 Z Rubin, 'The reforms of Khusro Anūshirwān', in A Cameron (ed.), *The Byzantine and early Islamic near east: III, States, peoples and armies* (1995), 227–98; J Howard-Johnston, 'The two great powers', in same vol., 157–226; P Mohebbi, *Techniques et ressources en Iran du 7e au 9e siècle* (1996), 32, 48.

22 G Fisher, *Between empires. Arabs, Romans and Sasanians in late antiquity* (2011), summarises the argument, with references, 201–5.

23 The iconography and design features of these buildings are widely agreed by experts to reflect local, Christian ideas and building methods. I was fortunate to visit the 'desert castles' in 2016 with the Castle Studies Group and find this argument is convincing.

The subsequent flowering of science and culture under the rule of the caliphs represented a fusion of existing knowledge and the re-employment of Christian, Zoroastrian, Jewish and other scholars and administrators. An additional influence was the learning of India, as the early Arab conquests had extended to what are now Afghanistan and the valley of the Indus. The conquerors encountered a culture where iron and steel manufacture were well established, cities walled in stone and brick were commonplace, and mathematical knowledge – and in particular geometry – was a long tradition.[24] Indian mathematics had an influence on the development of Arab maths, and geometrical skills had significance for hydraulics and surveying. In addition, Persia and India had seen the invention and use for many centuries of a simple device used to lift water, the noria.[25] It has been suggested by a number of students that this lever-like machine looks remarkably like a stone thrower of the misleadingly called "traction trebuchet" type (see below) and may possibly have inspired its invention (if, indeed, it was invented separately in the Middle East rather than having been imported from China).

The early Muslim rulers succeeded, despite being a small minority, in absorbing the knowledge and skills they did not possess without losing their specific identity. The result was an amalgamation of existing science and philosophy to which the requirements of the new regime for skills, in particular in measurement and counting, added new advances: the magnificent architectural features of the new mosques erected in cities such as Damascus as early as the seventh century are testimony to this fusion. Within a hundred years of the conquests, but certainly starting under the Ummayad caliphs, there was widespread basic literacy in the cities, and schools (and tutors for the wealthier elites) provided the more advanced knowledge required by government.[26] The caliphs' military commanders also benefitted from this provision.

24 N Srinivasan, 'Technology of iron-making in ancient and medieval India', in K Venkatachari (ed.), *Ananthacharya Indological Research Institute series XV, Technology* (1984), 132–50; B Sengupta, 'Engineering and architecture in ancient and medieval India', in P Ray, S Sen (eds), *The cultural heritage of India: VI, Science and technology* (1986), 205–19. The mathematical treatise of Ārqabhaṭa I (born 475) was acknowledged by many later scholars. A school established in 628 by the renowned and ground-breaking mathematician Brahmagupta specialised in geometry and astronomy, which were then closely associated disciplines everywhere – see T Sarasvati Amma, *Geometry in ancient and medieval India* (1979), 1–11.

25 I Habib, *Medieval technology exchanges between India and the Islamic world* (1985), 4–5; Sengupta, 'Engineering', 210.

26 An excellent summary in A K Bennison, *The great caliphs. The golden age of the 'Abbasid empire* (2009), esp. 163–7.

CONTINUITY IN ARMS MANUFACTURE, ARMIES, BUILDING AND SURVEYING?

The evidence therefore supports the case for continued craft skills, signifying their continued transmission from one generation to the next; and that significant levels of engineering capability continued to be available in many places.

The armies of imperial Rome had been supported by arms factories, where labourers operated under the direction of skilled craftsmen and administrators. The workshops concerned, known as *fabricae*, were listed in the late Roman document recording the units of the army in the western and eastern sections of the empire: and the *Notitia Dignitatum* (*ND*) has remained a source of debate, with some scholars doubting that it reflected reality.

Assigned to around 400, the *ND* identified 15 arms factories in the eastern and 20 in the western empire. If the list is complete, some whole provinces were unprovided for, where it has to be assumed either that craftsmen attached to the unit had been responsible for the work – Vegetius confirmed that this had been the practice previously[27] – or that private suppliers were used. By the fourth century, it is likely that, in contrast to the time of the legions, most units of the Roman army did not have artillery, and the *ND* suggests that there were specific units of *ballistarii*.[28] This signified men who operated static bolt-shooting artillery, rather than crossbows which would be the medieval sense of the word. The answer to the question of what happened to the *fabricae* has both a direct and an indirect bearing on the question of the continuity of Roman military engineering skills and the type of weaponry used by Roman armies that required specific skills, in particular artillery.[29]

27 S James, *Excavations at Dura Europos 1928–37. Final report VII. The arms and armour and other military equipment* (2007), 209–10, 214–15. Finds included many ballista bolts in the process of manufacture. Dura Europos had been abandoned after capture by the Persians in 256; Vegetius book II, c. 11 (p. 43).

28 S Johnson, *Late Roman fortifications* (1983), 79; P Southern, H Dixon, *The late Roman army* (1996), 159–60.

29 In a large literature, see: D Hoffmann, *Das spätrömische Bewegungsheer und die Notitia Dignitatum* (1970), and C Neira Faleiro, *La Notitia Dignitatum, nueva edición crítica y comentario histórico* (2008) for the text; S James, 'The "fabricae". State arms factories of the later Roman empire', in J Coulston (ed.), *Military equipment and the identity of Roman soldiers. Proceedings of the 4th Roman equipment conference*, BAR International Series 39 (1988), 257–86, for a detailed analysis, with the additional references listed there. M Clauss, *Der magister officiorum in der Spätantike* (1980) discusses the roles of the office, esp. 51–54, 124–8; A Bock, J Dunlop, *Two studies in later Roman and Byzantine administration, Part I: The master of the offices in the later Roman and Byzantine empires* (1924), 87–9, is still useful.

Only two factories specialising in the production of any kind of artillery (*ballisteria*) were listed, and both were in the western empire (at Autun and Trier).[30] Factory locations may have been based on factors including access to raw materials and the necessary skilled labour as well as connectivity to the armies they were designed to support. Regulations covered the workers, who were treated in the same way as soldiers, with pay and conditions laid down. They enjoyed a higher status than civilian artisans but were tied to the service. Excavated remains of what appear to be arms factories from earlier periods seem to confirm their role as official, organised and specialised workplaces set up to support the front-line army.[31] They were under the control of a very high-ranking official with numerous responsibilities, the *magister officiorum*. How much of this infrastructure continued in late antiquity, and how much made it through the sixth century? There is evidence for continuity of arms manufacture in the first years of the successor states. Cassiodorus' collection of official letters appears to demonstrate that it continued to be a strongly regulated operation, but there is no reference to factories in the western empire by this time (the early sixth century).[32] In the eastern empire, the last reference to an arms factory (in Nicomedia) dates from 612, while laws enacted by Justinian incidentally confirm the existence of private arms factories, although it is thought possible that an official factory remained in Constantinople for many centuries after.[33] Scholars have identified no explicit references to arms manufacture in the empire between then and the

30 It is also curious that the only references to artillery manufacture in the *ND* are to the ballista, leaving unclear who was responsible for producing the stone-throwing *onager* (see below). Perhaps this equipment was manufactured by soldiers rather than craftsmen.

31 H von Petrikovits, 'Militärische Fabricae der Römer', *Actes du IXe Congrès International d'études sur les frontières romaines*, 1972, ed. D Pippidi et al., 1974 (399–407).

32 Cassiodorus, 'Variae' book VII, numbers XVIII and XIX, in *MGH* AA XII, ed. Mommsen, 213–14. These variae have not been translated.

33 On this, James, 'Fabricae', 286. The reference to Nicomedia actually describes three people as *fabricenses*, the Latin term simply transposed into Greek, and translated here as "armourers" though the implication (no more) is that they worked in a *fabrica*: A-J Festugière, 'Vie de Théodore de Sykéôn', *Subsidia Hagiographica* 48, 1970, s.159.46 (p. 140 of the translation, p. 134 of the Greek text). Nicomedia was one of the *fabricae* listed in the *ND*. The existence of a *Collegium* of *fabricenses* alongside similar structures for other crafts in Constantinople is claimed by Petrikovits, 'Die Spezialisierung des römischen Handwerks. II, Spätantike', *Zeitschrift für Papyrologie und Epigraphik*, 43, 1981, 285–306, at 291, though no reference is provided for the claim. For Justinian's law see R Schoell, G Kroll (eds), *Corpus iuris civilis*, 3: *Novellae*, Novel 85 (pp. 414–15), confirming incidentally that *ballistarii* were located in city garrisons.

eighth century, although it is obvious that some facility for manufacture must have continued.[34]

If the *fabricae* disappeared, this would be consistent with other evidence for the changes that took place in military organisation in the west. Information for the fifth century is indirect. Between 400 and the last quarter of the century, a period that saw numerous military conflicts in Gaul and Italy, the Roman army in the west as listed in the *ND* disappeared. A force totalling half a million soldiers, paid out of taxation and consuming half of the imperial budget, ceased to be in one half of the empire.[35] This is consistent with a collapse of the enormous bureaucratic system once the centralised administration and the taxes that funded it ceased to operate (over time). The impact of the collapse of centralised tax collecting was not as severe as might be thought, because of the long-standing system of compulsory levying of public duties, done locally, on all citizens (the *munera publica*), alongside taxes. These included obliging city administrations and landowners to send artisans for service, for example, working on everything including bridges and fortifications. The successor kings relied on their retinues and those of their leading men, but also the continuing obligations of the *munera*.[36] In Gaul, Clovis inherited and maintained parts of the late Roman military infrastructure, including some taxation, but this could not be sustained, and the evidence for the maintenance of arms factories is not there.[37] Individual craftsmen working on the estates of the aristocracy are most likely to have filled the gap. But arming the soldiers of a great man's retinue was not the same as replacing the skills and experience levels required for the manufacture, and then the maintenance, of a piece of Roman artillery, not

34 J Haldon, *Byzantine Praetorians. An administrative, institutional and social survey of the Opsikion and Tagmata c. 580–900* (1984), 318–22 discusses how the army was armed during these centuries. The basic picture had been established long ago by J Bury, *The imperial administrative system in the ninth century* (1911), who identified the change in responsibility for the factories but implicitly confirmed that they continued to exist (98–100).

35 Southern and Dixon, *The late Roman army*, 179–80; Wickham, *The inheritance of Rome*, 32–3; M Whitby, in *Cambridge Ancient History*, XIV, 298–300; W Pohl, 'The barbarian successor states', in L Webster, M Brown (eds), *The transformation of the Roman world* (1997), 44.

36 There is very little information on *munera*: the best discussion is L Petersen, *Siege warfare and military organisation in the successor states (400–800AD). Byzantium, the West and Islam* (2013), 67–74.

37 The leading proponent of continuity is B S Bachrach. See here his *Merovingian military organisation 451–751* (1972), 4, where he states that Clovis took over a Roman arms factory at Soissons in 486. His source is Wallace-Hadrill, *The long-haired kings*, which states that this was where, in 454, "the Romans had not long since manufactured arms" (p. 159). Does this mean that Clovis took over a functioning factory more than thirty years later?

to mention its operation, although even here it is possible to identify one case of continuity of manufacture.[38]

The ultimate proof of continuity lies in the evidence for the continued use of Roman artillery under the Merovingian kings, to confirm whether artillerymen were among the engineers who continued to function in the seventh and eighth centuries (as will be seen below). There is no doubt of continuity of necessary engineering skills in other areas of life, and these may also have applied in the military sphere.

BUILDING, GEOMETRY, THE SCIENCE OF MEASUREMENT AND MILITARY CAMPS

Actual buildings dating from late antiquity are material evidence for the continuation of classical knowledge and skills in this area. In addition, the preservation of the writings of Procopius on the buildings erected at the command of emperor Justinian offers a contemporary account by someone close to the centre. Procopius' description of the construction of the still-standing church of Hagia Sophia in Constantinople under the direction of two named master builders, Anthemius of Tralles, the *architekton*, and Isodorus the Milesian, and of its reconstruction following the collapse of the original dome in 557, stands out for its detail (even if it betrays that he was not an expert in the subject) and as evidence that the science of Roman monumental architecture had survived into the sixth century.[39] The proportions of the dome of Hagia Sophia used the ratio of $1:\sqrt{2}$ just as the dome of the Pantheon in Rome, and the use of architectural drawings at this time is evidenced by references to their existence: similar references confirm that such drawings continued to be used into early medieval times, and even as far away as Scotland.[40] Isidorus was the equivalent of a professor of geometry or "mechanics"

38 Biship Nicetius of Trier maintained a *castellum*, his city's defences and a ballista according to a verse encomium: Venanti Fortunati Carminum, lib. III, *MGH* AA IV, pt 1, pp. 64–5. Nicetius (c.525–60) may have exploited the skills once provided by the workforce of the previous *fabrica* in Trier?

39 Procopius, *Buildings*, book I, i, pp. 12–33. The roof of the spectacular mausoleum of Theodoric that can still be seen at Ravenna was a single piece of stone nearly 12 metres across and was "a tour de force in terms of quarrying, carriage and construction": R Stalley, *Early medieval architecture* (1969), 64.

40 See: P Kidson, 'Architectural proportion', in J Turner et al. (eds), *The Dictionary of Medieval Art*, II (1996), 343–52 and W Kleinbauer, 'Pre-Carolingian concepts of architectural planning', in M Chiat, K Reyerson (eds), *The medieval Mediterranean: cross cultural contexts* (1988), 67–79 and the examples of references to drawings and plans gathered there. Adamnan, abbot of Iona (679–704) refers to the *figuras* of the church of the Holy Sepulchre and three other churches in Jerusalem sketched by the

(in the architectural sense), the author of a commentary on Archimedes and the writer of supplements to Euclid's geometry. Anthemius' writings have survived less well, but it is known that he was the author of texts including one on "remarkable mechanical devices". These experts were applying book-learning to practical outcomes.

A century and a half later, the Northumbrian monk Bede, familiar with much classical knowledge, called upon Isidore, Pliny the elder, Vegetius and others for his scientific writing. This included reproducing the methods used by the ancients to measure heights using triangles and shadows and even applying this to the measurement of latitude. Geometrical awareness was important to building, and was clearly not kept hidden within monastic walls, as Bede's own description of the church-building works of Bishop Benedict in 674 confirms, and if the latter had to import the expertise of masons and glass-makers based across the Channel to erect his church at Monkwearmouth, it confirmed that this knowledge did exist in Merovingian Gaul (evidence such as that for Bishop Desiderius building churches, houses, gates, towers and walls in his see of Cahors is among that confirming continued construction work in Gaul[41]), even if it was hitherto unknown to practitioners in Northumbria, the then dominant Anglo-Saxon kingdom.[42] As will be seen, Anglo-Saxon knowledge extended to surveying as well.

These individuals were highly educated and cultured theorists as well as practitioners, who could stand alongside their classical predecessors. The names of those who oversaw the construction of the other churches and the secular architecture erected in the Byzantine capital during Justinian's reign are not known but the quantity of building done shows there must have been others, too, sufficiently skilled to lead this work. Procopius gave the names of only five architects in all.[43] He also described numerous works of fortification carried out in all provinces of the eastern Roman empire, and if archaeology has undermined the contention that all of this mighty effort was actually constructed at this time, sufficient remains attributable to the sixth century to confirm the view that a high level of skill, largely if not exclusively based on inherited knowledge, was shown in military architecture as well.

---

traveller Arculf on wax tablets: D Meehan (ed.), 'Adamnan's De Locis Sanctis', *Scriptores Latini Hibernicae*, III (1958), 46–7. The article reproduces the images from a MS of the text.

41 *Vita Desiderii Cadurcae Urbis episcopi, MGH* SRM IV, ed. Krusch, esp. 574–5 (he was bishop between c.626 and 650).

42 W Stevens, *Bede's scientific achievement* (1985), 6, 9; Bede, *Historia Abbatum*, ed. Plummer, 368.

43 G Downey, 'Byzantine architects. Their training and methods', *Byzantion* XVIII, 1948, 99–118, remains the classic summary of the learning possessed by these talented individuals and the evidence for their knowledge and work.

Geometric principles also underlie the only area of engineering in which there is an unbroken continuity of written material running from the days of the ancient empires through late antiquity and the medieval centuries: surveying. The skill of measuring land was important to all rulers so people able to do it continued to be found employment.[44] With landownership and land produce the basis for both ancient and medieval economies, measuring it fairly and with a degree of accuracy was essential, not only from the viewpoint of the landowners, but also directly in the interests of those states that drew income from taxing land – and therefore also in the interests of those whose lands might be exempt from taxation The skills of measurement were equally important for those establishing towns, something of importance in western Europe only in later times, but significant throughout in the Muslim world. Measurement was of course crucial to designing, planning and organising the construction of a new building. The link to military purposes was through a number of activities requiring the same knowledge and the same practical skills. A major role was reserved for engineers who knew how to measure out and plan a camp for an army in the field, for which the practice of ancient Rome was the basis. But also useful to a commander was having someone who could measure (from a distance) the height of a wall so that siege ladders could be constructed. What were the sources of such skills, how much did they require surveyors or engineers to actually understand, and what practical examples survive to show that such skills definitely existed?

A group of Latin texts dating from the days of the early Roman empire but frequently added to in later years have survived that spell out the science of measurement for land surveyors. Known as the *corpus agrimensorum*,[45] the body of surveyors' texts and analysis of their origins and contents give a good picture of what Roman surveyors needed to know. Some groups of manuscripts dated originally from fifth-century Italy but were based on earlier writings. The empire was run in practice from local towns and it is a reasonable proposition that there was regular employment for a large body of surveyors. This in turn called into being texts to instruct them. Copying continued into late antiquity and the early middle ages. Analysis of the instructions shows, among other things, common approaches to systems of measurement and the use of various pieces of equipment to enable the task, including measuring where the land was hilly or crossed by waterways. This required knowledge of geometry. It is not clear exactly how much, and how far a surveyor could have managed with the most basic understanding of shapes and

44 On the high value attached to Roman surveyors, R Browning in *Cambridge Ancient History*, XIV, 880.

45 B Campbell, 'Writings of the Roman land surveyors', *Journal of Roman Studies Monograph* 9, 2000; L Toneatto, *Codices artis mensoriae, I: I manoscritti degli antichi opuscoli Latini d'agrimensura (V–XIX sec.)*, I, 2–13. J Peyras (ed.), *Arpentage et administration publique à la fin de l'antiquité* (2008), introduction, vi–x, xiv–xviii.

angles, but it has been argued that the texts were rooted in the geometric principles enunciated by Euclid and reproduced by Boethius.[46] Attention has been drawn, for example, to the remarkable uniformity of the size and shape of settlements identified from Merovingian Gaul, with some evidence of the use of a standard foot as a measure.[47]

Similarly, in Anglo-Saxon England, it has been proposed that settlements in Kent, Northumbria, Mercia and East Anglia were laid out "with the aid of pre-surveyed, geometrically precise grids, employing a short perch (about 4.6 metres) in which a frequent base module was the four-perch square". It has been concluded from archaeological investigation of the palace at Yeavering (around 600) and the church of Sts Peter and Paul at Canterbury (597–618) that sites were laid out in grids using the rods and devices for measuring right angles that were the basis of the methods of the *agrimensores*, and that if the first examples were executed by borrowed Frankish engineers, the knowledge seems to have spread rapidly. Whether or not this particular hypothesis of the basis of the grid-plan holds up, it is clear that there were surveyors in Anglo-Saxon England by the seventh century.[48]

Alongside the *corpus agrimensorum* are found instructions for those responsible for military camps. The otherwise unknown author is identified as pseudo-Hygin, who wrote a text entitled *De munitionibus castrorum* ("on the fortification of camps"). The original text from the second century was copied in the sixth century collection. The author was the only one of the surveyors to be called a *gromaticus* – from the cross-shaped instrument called the *groma*, used for marking out angles and straight lines[49] – rather than a surveyor. The text ascribes this term to the fact that the *groma* was the instrument used to start the laying out of a Roman army camp. Scholars have identified that the writer was a genuine technician completely familiar with the subject from his use of language, and the fact that

46 M-T Zenner, 'Imaging a building', in J Contreni, S Casciani (eds), *Word, image, number*, Micrologus Library 8, 2002 (219–46), 219–25.

47 Damminger, 'Dwellings', 69, points to regular plots of 600–800 hectares.

48 J Blair, 'Grid-planning in Anglo-Saxon settlements: the short perch and the four-perch module', *Anglo-Saxon Studies in Archaeology and History* 18, 2013 (18–61) and the sources cited there. Part of a device that might have been a damaged *groma* was found in a grave at Yeavering according to the report published by Hope-Taylor in 1977.

49 See the description of the *groma* and other Roman surveying instruments in Glick et al. (eds), *Medieval science, technology and medicine: an encyclopedia* (2005), article *agrimensores* by A Pikulski, 9–10. On the Roman inheritance and its relationship with cartography (along with evidence that map-making itself did disappear after around the year 500): O Dilke, 'Roman large scale mapping', in J Harley, D Woodward (eds), *The history of cartography* (1987), I, 212–33; O Dilke, 'Cartography in the Byzantine empire', in the same, 258–75.

there is continuity in the copying of this manual suggests continued interest in its contents.[50]

It is well known that part of the disciplined functioning of a Roman army on the march included establishing a properly laid out, protected encampment each night, especially in potentially hostile territory. The task of laying it out fell to an engineer (it seems, a *gromaticus*) who could mark out the dimensions in such a way as to encompass all who had to be inside, according to a long-established basic plan organised around the commander's quarters, including provision for the sick, for the animals, and space for workshops where arms or other equipment could be serviced or repaired.[51] While the principles involved were unchanging and while soldiers did the actual work of erecting the camp, knowledge of the principles of the design alone would not have been enough to ensure that the result was what was required, particularly where the terrain was not level.

Vegetius set out exactly these requirements in book three of *De re militari*. He wrote that the surveyors (*agrimensores*) needed to "calculate the square footage defined by the size". The shape depended on the terrain.[52] The *Strategikon* stated that an army required surveyors "who measure and set up camps". If he was marching through an unknown region, the commander must send out surveyors "to survey the entire location where the camp is to be made and to apportion certain sections" to each regiment. Even at the level of the regiment (*tagma*) it was the duty of the commander to "select two reliable men as surveyors" to pick the campsite.[53] This wording leaves it unclear whether these were professional surveyors, but the rest, like that of Vegetius, is explicit that these are engineers serving in the army with a highly specialised responsibility. This tradition would continue in later Byzantine military manuals in unaltered form.

It is evident that armies set up on the imperial Roman model made use of engineers whose responsibility it was both to identify appropriate locations for the camp and to lay it out. Did armies of east Rome always follow the advice? Did they have surveyors on their establishment? According to Menander, it required Maurice (while still just a general) to reinstate the practice which his army had forgotten.[54] Other chronicle accounts of warfare refer to camps but provide insufficient detail to make it clear who had responsibility for laying them out. But the long continuity of the recommendations showed that this was what military commanders were expected to do. What of the successor states in Italy and the west? It is likely that

50 M Lenoir (ed.), *Pseudo-Hygin. Des fortifications du camp*, introduction x–xv. G Chouquer, F Faury, *L'arpentage romain* (2001), 24–9.

51 Lenoir, *Ps-Hygin*, 3–4.

52 Vegetius, book III, c. 8 (76–7).

53 Maurice, *Strategikon*, ed/trans. G T Dennis, I, c. 3 (15–16); c. I, 9 (21); VII, c. 17 (77).

54 *The history of Menander the Guardsman*, ed. Blockley, frag. 23, 2–3 (discussing campaign of 578). Other references to entrenchment at frag. 26, 1 (pp. 232–3).

people with surveying skills were available in the fifth and sixth centuries in Italy, southern Gaul and Iberia, and in North Africa, or else why continue to copy the texts making up the *corpus agrimensorum*? It is possible – this is a contentious point – that new generations of units of the imperial army, settled as "colonists" in various locations, maintained earlier traditions of organisation. There is evidence that Frankish and Visigothic leaders did, of course, set up camps around besieged cities during their many wars, and evidence from a Frankish expedition to Italy indicates that on this occasion at least they followed the Roman practice.[55]

OTHER EVIDENCE OF MILITARY OPERATIONS

The little (and none of it contemporary) information about the nature of Muhammad's army does imply that any engineering skill used by the early Muslim forces was imported. When in 627 Muhammad held Medina against his enemies, his defence was reinforced with the knowledge offered by a Persian. His capture by siege of Ta'if in 630 followed his sending of followers to Yemen to "learn the use of the testudo, the catapult and other instruments" but since they had yet to return, he turned to a Persian named Salman (perhaps the same man as at Medina) for advice on the building of a stone-throwing catapult (a *manjanīq*) and a cowhide shelter to cover an approach to the gate. His men may have been able to construct the instruments according to instructions but did not know how to use them effectively, so the attack failed. This information is helpful in confirming the diffusion of technical knowledge, and of engineers, around the region, as well as the role of Yemen, which had become a Persian domain and had developed an urban culture.[56] But there is no sign of any engineering technology in use by the Arabs in their successful seven-month siege of Babylon (Egypt) in 640–41.[57]

Reconstructing early medieval military operations from the available sources is a challenge. The few contemporary written accounts were rarely from eye-witnesses. Accounts written by people without military experience – such as the clerics who made up the majority of the literate population – can be just as misleading the more

55 Agathias, *The histories*, trans. Fremdo, II, 6 (p. 35); see also Gregorii episcopi Turonensis, *Liber vitae partum*, *MGH* SRM I, ed. Krusch, Arndt, 678 describing a campaign in the Auvergne: Theuderic "cumque in Arthonensis vici prata castra metasset"; compare Gregory, *History of the Franks*, trans. Thorpe, book II, 32 (Clovis encamped around Vienne) (p. 146), and "suburbana castra fixit" at the siege of Clermont in 524.

56 Ibn Isḥāq, *The life of Mohammad*, 587–91; on Yemen: Chelhod et al., *L'Arabie du Sud*, I, 41–8, 195–215. Gabrieli, *Muhammad and the conquests of Islam*, 114. On Muslim construction, K Cresswell, *Early Muslim architecture*.

57 A J Butler, *The Arab conquest of Egypt and the last thirty years of the Roman dominion* (1978), 238–74.

they appear to be detailed, in the same way that modern journalists commenting on wars are prone to blunders rooted in their ignorance of the realities of their subject. When it comes to reconstructing the role of engineers – whether this takes the form of those erecting camps or field fortifications, or building siege equipment, or preparing and operating artillery, or perhaps charged with getting an army across a wide river – there is usually silence. All that can be done is to identify from the sources what happened, and then to conclude what skills were needed to achieve the tasks called for by the commander, who was often the only individual identified by name.

(a) Bridge-building

Some of the stone bridges built in Roman times still exist, demonstrating the engineering skill and awareness involved in their construction. Knowledge at least of the effect of running water and how to create foundations in different circumstances would have been needed to erect timber bridges, probably the commonest form, but, of course, leaving no trace.[58] There is the visual evidence of Roman imperial practice preserved on "Trajan's column", the carved record of the emperor's Dacian campaign in the third century that shows his soldiers and engineers bridging streams, constructing canoes to cross rivers and erecting a substantial bridge across the Danube[59] [Fig.1; compare Fig. 2]. With the reduction in state revenues, infrastructure that required continuous maintenance which continued to be the responsibility of the state must have relied increasingly on the relevant public duty, and its enforcement, which was everywhere and always unpopular.

According to Ammianus Marcellinus, the emperor Valentinian (in 369) instructed men "skilled in hydraulic work" to undertake the massive job of diverting the river Neckar. The labour was provided by a great number of soldiers but under the direction of these engineers. Three years later, the same ruler had a bridge built across the river Rhine, constructed by means of a bridge of boats with its roadway formed of timber beams.[60] The need to cross rivers

58 Although it is interesting that Vitruvius, whose text *De architectura* devoted considerable attention to the qualities of water and to how to construct aqueducts, did not discuss bridge building. However, his advice on constructing harbours by sinking coffer dams (see glossary) also applied as one of the possible methods for building bridges (book V, chap. xii).

59 I Richmond, 'Trajan's army on Trajan's column', *British School at Rome*, 1982, particularly 21, 26–7, 30, 35.

60 Ammianus Marcellinus, book XXVIII, 2, 1–4; book XXIX, 4, 2 (Loeb edition vol. III, 122–3, 240–1). There are several further references to the emperors' armies crossing major rivers by constructing bridges of boats.

FIG. 1 Roman soldiers cross a bridge of boats, shown on Trajan's column, photo by permission of Roger B. Ulrich

remained a constant for military operations, of course, and the traditional Roman methods were spelled out by Vegetius. If there was no ford, he proposed either driving in piles across which boards should be placed; or for temporary structures, a bridge of barrels tied together and covered, again, with boards. He recommended commanders to convey *monoxyli*, "which are shallow ... hollowed-out canoes", with the army, along with planks and nails. "The bridge thus speedily constructed", tied together by ropes kept for the purpose, "provides the solidity of masonry architecture in quick time." The continuity with the activities described by Ammianus Marcellinus is evident, as is the judgement that this superficially straightforward operation would actually have to be done under the direction of people qualified and experienced in organising getting an army across a large stretch of flowing water.[61]

The standard Roman practice continued in the eastern empire, and the advice offered by the mid-sixth-century "anonymous" was essentially the same as the advice provided by the emperor Maurice in 600. The latter text stated: "A bridge can be constructed either with wooden beams ... or on pontoons, with towers erected at both ends", the towers relating to the dangers of doing the work in the vicinity of an enemy. Timbers, struts and pontoons needed to be gathered before the bridging operation began, he stated in another section, and ballistas placed on boats in the river as the bridge advanced towards an enemy-held bank.

61 Vegetius, book III, 7 (ed./trans. Milner, 75–6).

Maurice recommended what he calls "floating bridges" for getting across the many unfordable rivers in the Balkans.[62] It is interesting that the "barbarian" Avar ruler managed to build a bridge across the river Save near Sirmium, using his whole army as labourers, in 579.[63] Procopius confirmed that the Persians were similarly skilled in building bridges consisting of timber roadways placed on a bridge made of boats tied together in his accounts of the wars of King Chosroes. He commented that the Persians could cross all rivers without difficulty because they have "in readiness hook-shaped irons to fasten together long timbers to improvise bridges".[64] Both the Persian army and Emperor Heraclius were recorded making use of bridges of boats across the rivers Tigris, Euphrates and the Lesser and Greater Zab in the successful Byzantine campaign in 625–27, where there is also a reference to watch towers having been built to protect a bridge, and to a bridge being disconnected from the opposite bank and brought back to the near side.[65]

There is, in contrast, an absence of reference to bridge-building of any kind in western Europe over several centuries. There is no reference in Bede's writings to bridges. It remains a matter of dispute whether the issue of maintaining bridges was dealt with by it being made a communal responsibility in the mid eighth century, as elsewhere. Certainly, fords appear to have been a routine way to cross rivers. In such conditions, there would be no call for people with bridge-building skills. An eighth-century manuscript of the technical treatise known as the *Mappae Clavicula* contains instructions about building bridge foundations in water (based on the Roman method) but there is no evidence that it reflected contemporary practice.[66] Did Merovingian kings continue to maintain Roman bridges as a communal responsibility? Did they get their armies across rivers where bridges did not exist by following traditional Roman practice, in which case there must have been some engineers familiar with how to do it? The evidence is mixed and

62 Maurice, *Strategikon*, ed. Dennis, book IX, 1, 4; book XII, 21. Pages 94, 122, 124, 157–8; 'The anonymous Byzantine treatise on strategy', in *Three Byzantine military treatises*, ed. Dennis, 63.

63 Menander (ed. Blockley), frag. 25,1 (pp. 220–1). Menander described a technique using cutwaters at another river crossing in the Balkans made by the Avars (frag. 38, p. 247).

64 Procopius, *Wars*, book II, xii, 3–5 and xxi, 21–4; pp. 362–5, 446–7.

65 Theophanes, *Chronicle*, ed. Mango and Scott, 449–50.

66 N Brooks, 'Medieval European bridges', *Haskins Society Journal* 7, 1997 (11–29), reprinted in Brooks, *Communities and warfare 700–1400* (2000), 6–10; D Harrison, *The bridges of medieval England* (2004), 34–6; M Boyer, *Medieval French bridges* (1976), 143–4; J Mesqui, *Le pont en France avant le temps des ingénieurs* (1976), 144. The *mappae claviculae* was translated by C Smith and J Hawthorne in *Transactions American Philosophical Society* 64, 1974. The eight-century text was reproduced and translated by V Mortet, 'Une formulaire du VIIIe siècle pour les fondations d'édifices et de ponts', *Bulletin Monumental*, 71, 1971 (442–65).

FIG. 2 A 'flying bridge' to cross the river Jordan in late antiquity. The Madaba Mosaic map, Jordan. (Photo: author)

there is only one unequivocal statement from events reported by Bishop Gregory. In 580, his enemies began to cross the Loire upstream from the bridge that "used to be supported on two pontoon boats" using a ferry. King Chilperic posted guards on a bridge across the Orge near Paris, and King Gundobald managed to cross the Garonne on the way to Comminges but his men had to swim. In 591, King Guntram's army sent against the Bretons managed to cross both the Vilaine and the Oust, where they "constructed bridges", but on their return some of their troops waded across the Vilaine.[67] When across western Europe there was renewed bridge work from the eighth century the evidence does not confirm or deny continuity with traditional Roman methods [Fig. 2].

67 Gregory of Tours, *The history of the Franks*, book V, c. 49 (page 317), book VI, c. 18 (p. 349), book VII, cc. 34–5, pp. 417–18, book X, cc. 9–10, pp. 556–8.

(b) Artillery and Other Siege Equipment

Procopius confirmed that the siege equipment used by the army of Belisarius was that which had been in continuous use from the days of the later Roman empire. His detailed accounts of Justinian's wars against the Persian king Chosroes support the case that there was a corps of engineers as a regular part of both Roman and Persian armies. Not only on the eastern frontier either: in one particularly useful passage in which he recounted Belisarius' preparation to defend Rome against the counter-attack launched by the Gothic king Vittigis in 537–38, a range of siege equipment and artillery was described – mobile wooden siege towers, erected by the besieging Goths and propelled by oxen, and the ballistas and *onagri* mounted by the Romans on the walls of the city to defend it. The method used by the Goths to calculate the correct height of their towers and of their scaling ladders was not exactly mathematical: apparently they did it by counting the number of courses in the masonry (although how Procopius could know that remains untold). The Roman ballista as described was operated by tension, a rope placed to propel a moving "shaft" on which was placed the missile, half the length but four times the width of a normal arrow. The description matches those found at Dura Europos three centuries earlier. The *onager* was briefly described as resembling a sling. In preparing the city to withstand siege, Belisarius had also rigged up boats to mount replacement watermills relying on the force of the Tiber flowing through the Aurelian bridge. When the Goths attempted to thwart this by rolling trees and dead Romans downriver to break the water wheels, the talented Roman commander had an iron spiked chain placed above the bridge to intercept the gruesome traffic, and restore sufficient water and the ability to mill flour to the citizens. While it is possible that Belisarius himself possessed the knowledge to organise all these tasks, Procopius' own words confirm that his army (and perhaps the city itself) included engineers able to carry out his orders with so much effect.[68] Certainly, there can be little doubt that the engineering (including artillery) skills possessed by Justinian's armies continued in use in east Rome into the seventh century.

The late sixth century witnessed the introduction of the manually operated lever artillery that would remain in constant use for the following eight centuries, most conveniently named the mangonel. The origin of the term is the Greek *mangana*, a generic term for construction machinery,[69] and the stone thrower acquired many

68 Procopius, *Wars*, book V, XXI (engines and tower), book V, XIX (water works), Loeb edition, III, 200–7, 184–3.

69 See Petersen, *Siege warfare and military organisation in the successor states*, 119 and Beekes, *Etymological dictionary of Greek* (2010), vol. 2, 887–8. It is suggested that the same word in Latin, *manganum*, meant 'machine' and its application to a war engine came from the Greek: A Ernout, A Meillet, *Dictionnaire étymologique de la langue Latin* (1967), 383.

and varied names to the confusion of later historians, some of whom continue to argue that the mangonel was actually the descendant of the torsion-driven *onager*. Contemporary accounts are often unclear whether they are using a generic term for a stone thrower (such as mangonel became) or intend a precise meaning. For modern students, it has not helped that in recent decades, the original manually operated lever catapult that I will call generically a mangonel has now been dubbed the "traction trebuchet". The problem is that trebuchet is a term applied properly only to the counterweight weapon introduced centuries later.[70] "Traction trebuchet" would have meant absolutely nothing to contemporaries. Because siege artillery and the engineering skills required to build and operate it will play an important part in this study, it is worth establishing what kind of weaponry is being talked about, and then considering what weapons were in use in late antiquity and the early medieval period.

It is firm fact that Roman armies originally used a torsion-powered stone-throwing catapult called a ballista, with devastating effectiveness in siege warfare (but not powerful enough to overcome properly built stone walls), in the days of the early empire. But this was an immensely complicated weapon to construct, and at some point during the third or fourth century the ballista became the smaller, but still static, tension-driven and crossbow-like bolt thrower described by Vegetius and Procopius. Its place as a stone thrower was taken over by the *onager*. This was made up of a strong timber arm wrapped in a set of twisted skeins of sinews held in a very strong wooden frame. The torsion of the twisted sinews propelled the arm upwards and released a stone from the sling at its end as it reached the appropriate height. The arm was brought to rest by what must again have been a strong timber horizontal beam. It then had to be wound down again by its crew, against the resistance of the sinews, so that it could be held fast while being reloaded. A detailed although technically illiterate description was provided by the Roman historian Ammianus Marcellinus in his accounts of warfare against the Persians in the fourth century and the main elements of the engine can be confirmed by the description later given by Vegetius.[71]

Contained within the words chosen by such authors are clues to the characteristics of the artillery. The *onager*, as a stone thrower, depended for its power on the significant kinetic energy made by the torsion of the skeins of sinew and its capacity would increase with its dimensions, as the size of the missile necessarily determined its effectiveness. The skill involved in operating both types was evident in Vegetius' words: "If it is tuned in accordance with mechanical art and aimed by

70 Although perhaps inevitably this is also disputed: see below.

71 Ammianus Marcellinus, book 23, 4.4–7 (Loeb, 326–9); Vegetius, book 4, 22 (Milner, 126). The classic study of the whole subject of ancient artillery is E Marsden, *Greek and Roman artillery*, 2 vols (1969); now updated by T Rihll, *The catapult: a history* (2007), especially pp. 249–61.

trained men who have worked out its range in advance", it will penetrate whatever it hits (ballista, in its earlier form as a torsion weapon), and "the larger the machine, the bigger the stone it hurls like a thunderbolt. ... Heavy stones thrown by the *onager* destroy not only horses and men but also the machines of the enemy." Tracey Rihll hypothesised that Vegetius' words signified that the precise mathematical formulae originally used to determine the proportions (and therefore the replicable qualities) of Roman artillery and reflected in analysis of archaeological discoveries of catapult ironwork (springs, washers etc.) may have been forgotten by the late fourth century, while the qualifying term "if it is tuned" implied that this was not always the case.[72] The finding of what may have been catapult shot at Dura Europos suggested that the normal size was not more than 7 kilograms, a significant missile but not one that would seriously damage stone walls.[73]

It seems therefore that by around 400 some of the earlier, more scientific, knowledge had been lost, and the result was artillery which was less likely to be built according to standard and uniform traditional formulae, and more likely to be made ad hoc. The people capable of constructing and operating artillery may have become fewer, and the infrastructure to transmit the knowledge less. Procopius' evidence suggests that the *onager* continued in use during the sixth century while the *Strategikon* and Isidore's *Etymologies* show that both in the east Roman army and in a Visigothic court, at the opposite end of the Mediterranean, late Roman knowledge continued to be copied.

References in the historical accounts leave no certainty as to how long the artillery engineering skills of late Rome continued under Clovis and the Merovingians, or under the kings of Burgundy before their absorption by their more powerful neighbour.[74] The terminology used by Gregory of Tours was inexact. It has been suggested that the Franks absorbed Roman military skills in part through recruiting new generations of Roman soldiers: some very limited references suggest that colonies of soldiers and their families would have continued to offer their services to the ruling power. Procopius referred to such units continuing into the time of Clovis. It is likely that arrangements whereby armies continued to be paid persisted under the first Merovingians, with landowners and bishops taking over the former central government responsibility for making sure it happened.[75]

72 Rihll, *The catapult*, 248.

73 James, *Excavations at Dura Europos. Final report VII* (2007), 214–15.

74 Petersen, *Siege warfare and military organisation in the successor states*, follows the Bachrach thesis. The claim that "in terms of labor, materials, finance and organisation, then, East Roman and Ostrogothic warfare were fundamentally similar" (p. 161) is typical of the approach.

75 Bachrach, *Merovingian military organisation 451–751*, 4–21; the same, 'A note on Alites', *Byzantinische Zeitschrift*, 61 (1968), 35, with the reference to Procopius, *Wars*, book I, 12). More broadly, J Durliat, 'Armée et société vers 600. Le problème des soldes', in F

In the long period from the middle of the fifth century until the wars of Charles Martel in the early eighth century there is little firm evidence from which to draw conclusions about the presence in the many sieges (in a region well filled with cities with Roman walls) of siege artillery or therefore of siege engineers. The town of Vienne (Isère) fell to siege by the Burgundian king Gundobad in 512 when an engineer (*artifex*, according to Gregory) who had been expelled by the defenders showed the besiegers how to get in by breaking open an iron gate on the aqueduct. This was the man who looked after the town's water supply, and he was apparently not rated as more important than the other "useless mouths" consuming limited provisions. Later, and contemporary with his own time, Gregory described how the Merovingian king Guntram (ruled 561–92/3) sent Count Leudigisel to attack the walled town of St Bertrand de Comminges (Haute-Garonne) in 585 and prepared wagons fitted with battering rams to knock down the walls: but they were driven off by flaming barrels of pitch and fat.[76] This is the only attack on a fortified site described in sufficient detail to enable an understanding of the capabilities of the forces involved. There are no references to artillery. Did the *onager* survive in Frankish Gaul? The silence of the evidence cannot be entirely ignored on the basis of the circular argument that the Franks maintained continuity with Rome so must have maintained the military technology.

Similarly, further east, references to events in Byzantine history of the seventh century in particular are frustratingly vague. The manually operated stone-throwing catapult, generically called a mangonel, or petrary, or petraria or some other variant, and in Greek sometimes using the word for stone thrower, *petrobolos* – was definitely in use from either shortly before or shortly after the year 600, having possibly been imported from China, perhaps courtesy of the Avars whose arrival in the Balkans added further to the woes facing Constantinople at this time. Alternatively, of course, it might have been invented independently in Europe, North Africa or the Middle East, as the lever principle on which it operated was familiar enough from building sites or hydraulic engineering.

The basic principles of design of the mangonel were simpler to execute than those of the *onager*, and its potential for being scaled-up as a stone thrower was greater. It required the services of a master carpenter to erect the timber frame – the numerous medieval illustrations show it as having either a single upright or two, for larger versions – and carve a suitable throwing arm that used the upright(s) to mount a pivot; a sling; and ropes to enable a number of people to pull down the arm from the front end in order to swing the throwing arm upwards and release the

Vallet, M Kazanski, *L'armée romaine et les barbares du IIIe au VIIe siècle* (1993), 31–2, 34–6. Durliat argues a similar position with the Visigoths and Lombards.

76 Gregory of Tours, *The history of the Franks*, book II, 32–3 (Vienne); book VII, 36 (Comminges), pp. 419–24. For a discussion of the evidence for this period in western Europe, Purton, *HEMS*, 10–16.

missile. It required knowledge of the correct proportions of the different elements, of the effect on range of the size of the missile in order to decide where to erect the engine, and awareness of the strength of the timbers used [see Figs 3 and 4]. These could be learned from seeing an engine working, but some knowledge would be needed to build one from scratch that would function efficiently. The *onager* additionally required animal sinews to provide its propulsion, and they required plaiting and twisting together to provide the torsion, and, as Vegetius pointed out (although he was discussing the then anachronistic torsion-driven ballista), they needed to be "tuned" so as to deliver identical propulsive force on each side. Sinews could be hard to find and hard to treat; they could lose their torsion power if they became wet, or simply through ageing. The famous substitution of women's hair at the siege of Rome as described by Procopius (when animal sinews could not be found) was presumably not seen as providing a sufficiently reliable or regular alternative source of power. The engine delivered a mighty recoil shock as it released the stone shot (as vividly described by Ammianus Marcellinus) and, if large enough, could dislodge stones beneath it if it was stood on a tower. Anyone who has observed the replica constructed at Caerphilly castle in Wales (where it is wrongly described as a mangonel) will recognise its shortcomings as an effective artillery piece, as well as its "kick".

So while it is feasible that the *onager* continued to be used by armies in west and east throughout the medieval period, as so many historians continue to claim, it is more likely that the lever-action ("traction trebuchet") mangonel or petrary took over and may have done so from the seventh century.[77] Some historians assert that earlier descriptions of large stone throwers used during the Roman–Persian wars (for example, at Amida in 502) must signify that it was in use much earlier: there is simply no proof either way. The earliest definitive account of its appearance in Europe also gives insights into the role of the military engineer at this time. The story comes not from a military historian, but from the bishop (John) who compiled the *Miracles of Saint Demetrius* (probably about the year 620) and whose purpose was to explain how the patron saint of his bishopric based at Thessaloniki, the empire's second city, came (repeatedly) to its rescue. It was in the year 586 or 597 (this is not certain) that Avar and Slav raiders suddenly appeared outside and first attempted escalade of the (substantial) city walls, then tried with a battering

77 Purton, 'The myth of the mangonel: torsion artillery in the Middle Ages', *Arms and Armour* 3(1), 2006, 79–90, reviews the evidence for its survival into the Middle Ages and argues that it disappeared, with torsion artillery returning in the shape of the springald in the thirteenth century. Petersen has reached the same conclusion (*Siege warfare and military organisation in the successor states*, 274. Petersen argues for the much earlier diffusion of the stone thrower technology than is accepted here). D S Bachrach has argued that torsion weapons continued in England through the central middle ages: this is discussed below.

Fig. 3 Petrary or mangonel. Maciejowski Bible, Paris, France, c.1244–54. The Morgan Library and Museums, MS M.638, fol. 23v. Purchased by J P Morgan (1867–1943) in 1916. (The Pierpoint Morgan Library, New York), with permission

Fig. 4 Reconstructed petrary or mangonel at Mauvezin castle, Hautes Pyrénées. (Photo: author)

ram, which they abandoned under bombardment from the defenders, then the next day set up the first ever (in Europe) fully described mangonel/stone thrower. Their skill did not extend from constructing to effectively operating the weapons, as their shots failed to reach the target. Returning in 617 or 618, they attempted another attack using "massive" stone throwers, wheeled battering rams, rolling ladders and high wooden towers. Sailors operated the city's own stone throwers to some effect and the siege was abandoned after a month. The access of the Avars to such an apparently sophisticated arsenal of siege equipment has been ascribed to a fortuitous event that took place a few years before, in 587, when, according to the chronicler Theophylact Simocatta, a Byzantine engineer named Busas fell into Avar hands and in exchange for his life offered to help them construct a weapon for a "long-range assault", which has been interpreted as signifying the stone thrower. Another paragraph told of Slav fighters competing to offer their leaders advice on constructing siege engines, with the example given of someone "particularly experienced in the construction of machines" offering to construct an engine to be mounted on an "ingenious system" of wheels and cylinders, of three stages protected by fresh skins: a good description of the mobile siege tower. Had not the saint intervened to drive the builder mad, the team of carpenters and smiths assembled to build the device would have completed it, not suggesting any lack of skills among the besiegers.

There remain problems with this evidence but it is the first definitive description of the new weapon, and this must also mean that there existed, at least on the Byzantine side, sufficiently trained people to build the range of equipment reported, and to operate it successfully.[78] There is no question of the ability of the empire to deploy a major siege train for offensive purposes and artillery for both defensive and offensive purposes in these wars, and of the contemporary struggles with the Persian empire, of which the successful defence of Constantinople itself in 626 against both Avars and Persians represented a high point of crisis and of military action. The accounts tell of large-scale use of artillery (stone throwers), pre-fabricated siege towers and mobile wooden shelters by the Avars during their unsuccessful ten-day-long assault; they also preserve the instructions on strengthening the city's defences left by the emperor Heraclius (610–41) as he led

[78] P Lemerle, *Les plus anciens recueils des miracles de Saint Demetrius et le pénétration des Slaves dans les Balkans* (1979), I, 117, 122, 139–40, 146, 151–9, 206. The description of the stone thrower was also translated in K DeVries, R Brown, *Medieval military technology* (2012), 134. The complexities of the evidence are discussed in Purton, *HEMS*, 27–33. The Busas story is in Theophylact Simocatta, *The history*, trans. M & M Whitby (1986), book II, xvi (p. 66). The ability of the Avars to build a bridge over the Save (see above) suggests either they did not really need Busas, or else already had learnt Roman engineering skills.

an army outside the city.[79] The east Roman empire continued to function with a regular corps of military engineers and technicians during these years, even though their names are known only in the most exceptional circumstances, and the reference to the role of sailors as artillerists at Thessaloniki – the text says because of their familiarity with working ropes, which made them suitable candidates for the teams that pulled the ropes with sufficient coordination to enable the stone throwers to operate effectively – hints that a wider pool of people than a professional group of catapult operators could use the technology effectively.

It is equally clear that their new and most dangerous opponents took up the new artillery as the leaders of the Muslim armies began their conquests in the early seventh century, with similar issues around ability to operate the novel machinery properly. From early on, the equivalent Arabic term for the stone thrower was *manjanīq*, or some variant of the word, while it seems that another machine called the *'arrāda* was a smaller version of the same type. Both terms remained in common use for hundreds of year.

The growing capacity of the Arab armies of conquest under the Ummayad caliphs is captured in a jumbled and non-contemporary account of the siege of Caesarea in Palestine conducted by Mu'awiya (661–80) in which siege artillery was employed that failed to breach the walls, and entry was secured by simultaneous sapping of the wall, and escalade.[80] In 683, when the army of Syria had to storm the holy city of Mecca which was in the hands of rebels, their artillery, commanded by an unnamed Abyssinian, succeeded in smashing the columns of the mosque and burning it with naphtha bombs propelled over the wall.[81]

Evidence for what kind of weapons were being used is suggested by an account of the siege of the town of Synnada (modern Şuhut) near the key fortress of Armorion, which was occupied by an Arab army commanded by 'Abd al-Rahman ibn Khalid in 664. A master carpenter from Paphlagonia approached ibn Khalid and offered to build a stone thrower (original *manganike*) capable of taking the fortress. The commander ordered the collection of long logs and a machine was built "such as they had never seen before". It was set up in front of the gates, then the soldiers "drew back the catapult; a rock rose in the air and hit the gate … the second fell short … the third fell shorter. The defenders jeered and cried out 'pull your

79 Georgius Pisida, *Bellum Avaricum*, ll. 266–9 in Migne *PG* XCII; (Italian translation by A Pertusi, *Giorgio di Pisidia, Poemi: Panegirici epici*, 270–1); Sebeos, *Histoire d'Héraclius*; M & M Whitby (trans.), *Chronicon Paschale*, 173–9; Theodorus syncellus, *L'homélie sur la siège de Constantinople en 626*, trans. F Makk, xix–xxxv; J Howard-Johnston, 'The siege of Constantinople in 626', in C Mango et al. (eds), *Constantinople and its hinterland* (1993), 131–42.

80 The account is in the much later chronicle of Bar Hebraeus, discussed in Purton, *HEMS*, 51.

81 al-Ṭabarī, *History*, XX, 117.

weight, you are drawing badly'. They then used their own catapult to propel a 'huge' rock down onto ibn Khalid's catapult from above, wrecking it." This story is full of important information: this was evidently a manually operated stone thrower of the type in routine service. The defenders had their own. What is odd is that ibn Khalid was allegedly unfamiliar with it: perhaps this was no more than a reference to its size and capacity, rather than the design and shape? What is suggested is that the soldiers operating it were incapable of sustaining a consistent pulling force. The role of the master carpenter suggests that he belonged to a group of craftsmen whose skills now included building stone throwers. The form of the story – that of an engineer approaching an apparently ignorant commander with an offer he could not refuse – will be found repeated again and again across the centuries.[82]

Half a century later the Arab armies attempted another attack on Constantinople. The Byzantines took measures including refurbishing the engines; towers were provided with "catapults for darts and stones and other engines" (Theophanes) – suggesting ballistas and stone throwers. The large Arab army, led by caliph 'Abd al-Malik's son Maslama, including numerous artisans, arms, engines of war and naphtha, set out in August 716. But they were unable to dent the defences and were eventually forced to abandon the siege after thirteen months. Apart from the key problem of being unable to attack Constantinople by sea as well as land in order to achieve a complete blockade, the massive defences had proven again to be a match for the strongest siege machinery available in the early eighth century, and the triple walls and deep ditches no doubt kept the attackers too far off to cause any damage.[83]

One of the reasons for the failure of the Arab siege was the use by the Byzantine navy of "Greek Fire", the lethal flammable liquid that burnt in water and adhered to its target; it was first mentioned in the 670s, when its role in defeating an Arab fleet off Constantinople may have been decisive in thwarting that attack. Probably more significant than the composition of the liquid was the design of the siphon that allowed it to be propelled a sufficient distance to reach an enemy. Attempts at reconstruction have indicated that this was probably the decisive step, because incendiary devices themselves were not new – we have already mentioned naphtha bombs in regular use in the Middle East, taking advantage presumably of readily available oil, and propelled by hand, or by sling, or from siege engines to get them

82 Maronite Chronicle, trans. Palmer et al., 34–5, discussed in Purton, *HEMS*, 47. Petersen, *Siege warfare and military organisation in the successor states*, 672, provides a fuller translation of the original but the key points and conclusions are the same.

83 The sources include Theophanes, *Chronicle*, 534–46, Nicephorous, *Short history*, 105–23; Dionysius of Tell-Maḥrē, *Secular history*,212; Brooks, 'The campaign of 716–18 from Arabic sources', *Journal of Hellenic Studies* 19, 1899 (19–31); al-Tabarī, *History*, XXII, 198–9, XXXIV, 39–42. Discussed in Purton, *HEMS*, 55 and Petersen, *Siege warfare and military organisation in the successor states*, 707ff.

over walls. There is evidence for early use by Arab armies prior to this date.[84] It was no surprise that the Byzantine state sought to keep the secret of "Greek Fire" – and remarkably did – so well that in due course it was mislaid and no one was able to recreate the weapon. The engineering skill required to manufacture the siphons must have been considerable: the risks of handling the contents spoke powerfully against inadequate design or careless construction. Its inventor, according to the chronicler Theophanes, who is the main source for these events, was a chemist and master carpenter (or "artificer" according to Theophanes) called Kallinikos who had fled from Heliopolis in Syria.[85] If the name of the inventor of the device is known, the names of the people who had to turn the idea into practical reality are lost, but it is once again testimony to the existence of real engineering skill that this weapon could be built and replicated so as to provide sufficient numbers of siphons to equip the allocated naval warships when needed. The terminology itself becomes confusing because there are frequent references in chronicle accounts for events over many centuries to the use of "Greek Fire" when there is reason to think that this term was being used to describe any kind of incendiary weapon.

Another siege stands out for the detail provided of the attack: this was Kamakhon (Kamkh), on the Byzantine/Arab frontier in Asia Minor, in 766 (or 768). The besiegers brought craftsmen and wagons carrying large timber beams used to construct mangonels that were placed on a hillside to throw rocks into the fortress. The defenders also had stone throwers. At this siege, and at others, the sources refer to a machine called a *dabbaba* which was a mobile wooden device, probably a mantlet although possibly some type of siege tower. At this siege – which ended in a disastrous defeat for the Muslim army – local villagers were apparently employed as labourers on the siege works, and on the stone throwers, and were only replaced by regular soldiers when they suffered heavy casualties caused by the Byzantine artillery.[86] Assuming that the army was accompanied by the engineers who constructed or designed the artillery and siege equipment and the villagers' role was limited to labouring under instruction, and the statement that the matériel

84 H Kennedy, *The armies of the Caliphs* (2001), 176–7, discusses the evidence.

85 There is an enormous literature. Discussed in Purton, *HEMS*, 375–6. The technical problems associated with the weapon were highlighted dramatically in a reconstruction led by Byzantine scholar John Haldon broadcast on British Channel 4 TV on 28 July 2003. The invention by Kallinikos, the decisive impact of its use against Arab fleets in 673–4 and 717–18, and an important reference to the bronze tubes manufactured to propel Greek Fire when 36 of them fell into Bulgar hands when they captured Mesembria in 813 are all reported in Theophanes' *Chronographia* (*Chronicle*, trans. Mango and Scott, 354, 494, 528–9, 682). Theophanes relied on now lost sources for this period; he himself died around 818–20.

86 The sources, primarily the *Chronicle of Zuqnīn*, are printed and translated in Petersen, *Siege warfare and military organisation in the successor states*, 732–5.

was built on the spot using timber brought along, this interpretation is consistent with what we would expect at this time. All that is missing is any reference to the engineer himself.

The Christian and Muslim sources confirm the continuation of military engineering skills in the numerous conflicts between Byzantium and the Ummayad caliphate, although we remain largely ignorant of who was responsible for constructing or operating it. The story of Synnada suggests it was not enough to build the machine without having the skill to project the shot effectively. Such skills might be learned quickly, but still required someone to provide the training. For the Byzantines, usually on the defensive during this period, it was easier to store equipment inside fortresses than to have to construct it from scratch at a siege, and, if they indeed had engineers serving, it would explain why Byzantine cities and garrisons were able to mount artillery on their walls and – with the advantage of height – cause problems for the attacker. But there were also Byzantine offensive operations, and in their accounts of the attack on the Crimean city of Cherson in 711 the chronicles highlighted the role of a battering ram in breaching the defences, along with references to the mangonel and another commonly used term, *helepolis*, a word that may have been applied to any powerful siege weapon.[87] But the construction of battering rams large enough to generate sufficient impact to breach a wall was a significant piece of engineering design. It is possible that among the various "other types of engines" mentioned by chroniclers a ram was meant whenever the creation of a breach is implied, but the words rarely allow a precise interpretation.

The means by which the Arab conquerors succeeded in taking on the knowledge and technology of the lands they conquered remain opaque. A number of sources indicate that the first armies of conquest used non-Muslims in ancillary roles but also skilled craftsmen, without indicating what for. Recent studies emphasise the diverse and changing composition of the Arab-led armies of conquest.[88] The ninth-century writer al-Baladhuri recorded several examples of activities requiring advanced technological skills. The original conquest of Egypt had seen the use of artillery. Familiarity with hydraulic engineering was also rapidly acquired [Fig. 5]. During the reign of Mu'awiya (661–80), a large-scale land reclamation was carried out by the construction of dams and restoration of a canal led by a freedman of

87 Theophanes is the main source: cited and translated by Petersen, *Siege warfare and military organisation in the successor states*, 698. The use of *helepolis* as a term to describe the main weaponry was considered by G T Dennis, 'Byzantine heavy artillery: the Helepolis', *Greek, Roman and Byzantine Studies* 39/1, 1998 (99–115).

88 Wadād al-Qāḍī, 'Non-Muslims in the Muslim Conquest army in early Islam', in F Donner, A Borrut (eds), *Christians, Zoroastrians and Jews in the Umayyad period* (83–127), at 108–9, 112; R Hoyland, *In God's path: the Arab conquests and the creation of an Islamic empire* (2015), esp. 56–61, 131–3.

FIG. 5 Raising water from a well: reconstruction of a sixth-century *saqiya* at Qasr 'Amra, Jordan. Photo: author.

the caliph. In the early eighth century the city of Wâsit (modern Iraq) was built, supported by the construction of two canals and dams to ensure water supply. A stone bridge was erected across the river Tigris – a massive undertaking, albeit one that was soon washed away.[89] There is very little contemporary evidence, so the discovery of seventh-century papyri preserved in Egypt has offered vital clues about the day-to-day administration of the new rulers that confirm a rapid take-over of existing resources, including the tax and *munera* records of a village showing charges for the maintenance of skilled craftsmen working on the conquerors' projects including ships. Works commanded by the first Muslim governors in that country included dredging the canal linking Fustat to the Red Sea (done by Coptic Christians). The early Muslim armies swallowed up a large portion of the revenues of the state raised from land and poll taxes but records do not survive to confirm whether military engineers were part of the payroll.[90]

89 Abu-l Abbas Ahmed ibn Jabir al Baladhuri, *The origins of the Islamic state (Kitâb al-Buldân)*, trans. P Hitti (2002), 348, 450–2, 453–5.

90 P Sijpesteijn, 'The Arab conquest of Egypt and the beginning of Muslim rule', in R Bagnall (ed.), *Egypt in the Byzantine world* (2007), 437–59; K Morimoto, *The fiscal*

Faced with a Byzantine state which controlled the seas through its navy, Arab commanders had to adapt quickly to military activities in this new environment if their victories were to be sustained. Very soon after the first conquests, Arab fleets sailed south and eastwards on plundering missions. In 653 an Arab fleet was built at Alexandria to attack the Byzantines, equipped with engines of war including devices for throwing incendiaries, and stone throwers. The numbers given for the size of the armada are incredible, but its fate was not: it was wrecked in a storm. The work was done by skilled local (Greek and Coptic) shipbuilders inherited by the Arabs, along with the Byzantine shipyards at Alexandria and Clysma (Suez). A new arsenal was built in 672–73 on the island of Rawda. The papyri recorded the requisitioning of carpenters and caulkers, and other letters confirmed the recruitment of smiths and ironworkers and the purchase of nautical hardware.[91] It seems that the Byzantine empire had sustained a skilled workforce of shipwrights in Egypt (and probably the ports of the Palestinian coast) that was taken over and placed under Muslim overseers: shipbuilding was a skilled occupation that could not be mastered quickly.

The Muslims' conquest of the Visigothic kingdom at the other end of the Mediterranean had brought armies across the Pyrenees and into conflict with the Franks within ten years of crossing the straits of Gibraltar. The invaders probably brought some siege skills with them but the local population included skilled craftsmen and, as it appears from the accounts of the rebellion against King Wamba involving sieges of the cities of Narbonne and Nîmes in 673, possessed siege equipment and engineering capacity. The besiegers had engines that hurled "enormous" stones into the cities causing dismay among the populace, and Bishop Julian's account confirmed that ballistae were used.[92] In 721, Narbonne was captured by a Muslim army, and the invaders went on to attack Toulouse with "fundis et diversis generum machinis" – suggesting both stone throwers and a range of siege equipment.[93] The cities had substantial late Roman walls. Both the Visigoths and

administration of Egypt in the early Islamic period (1981), 105–7; Sijpesteijn, 'Army economics: an early papyrus letter related to 'atā payments', in R Margariti et al. (eds), *Histories of the Middle East* (2011), 245–67; H Kennedy, 'Military pay and the economy of the early Islamic state', *Historical Research*, 75 (no. 188), 2002 (155–69)

91 Sijpesteijn, 'Arab conquest', 447–8; A M Fahmy, *Muslim sea power in the eastern Mediterranean from the seventh to the tenth century AD*, 23–37; *Histoire d'Héraclius par l'evêque Sébeos, traduit de l'Arménien*, ed./trans. F Macler, 101, 110, 140.

92 Julian, 'Historia rebellionis Pauli', Migne *PL* XCVI, cols. 759–98; discussed in Petersen, *Siege warfare and military organisation in the successor states*, 678–82.

93 *Corpus scriptorum muzabaricorum*, ed. J Gil, I: *Chronica byzantina-Arabica, Continuatio byzantina arabica*, 358, identical words in *Continuatio hispana; Chronicon Moissiacense*, *MGH SS* I, 290. Discussed in Purton, *HEMS*, 59 and Petersen, *Siege warfare and military organisation in the successor states*, 709, with a translation.

the Muslim army had current types of siege equipment and the people who knew how to operate it effectively.

The progress of Muslim-led armies beyond the Pyrenees was halted – Charles Martel (d. 741) was the victorious leader of the Frankish army that turned back the Muslim incursions into Frankish territory. His son Pippin went on to remove and replace the last Merovingian king in 751 and to found the Carolingian dynasty. Charles' skill in warfare, inherited by Pippin and then Charlemagne, was a major factor in his rise, and this included victories both in the field and in sieges, where the sources give an indication of the presence of engineers. The source material is limited and is prejudiced by its overwhelming pro-Carolingian bias. The reports of some of his successful operations have no detail, but where they do, as at Charles' successful recapture of Avignon from the Muslims in 737, and then Narbonne (again) shortly afterwards, it sounds as if a full deployment of siege equipment took place. Charles' son Pippin (king 751–68) engaged in a number of operations against fortified places held by "internal" enemies and again, although many are recorded as being captured, rarely is detail offered in the chronicles. When it is, as with the capture of Clermont and Bourges in 761, we are told that the operations involved surrounding the places with fortifications and using "machines and weapons of all kinds". The same words tend to be repeated each time, although at Bourges it was said that the Franks had breached the wall, but not how this was done.[94] The chronicler noted that the ruler of Aquitaine decided that his fortified towns could not resist Pippin's siege train.

The military engineering skills available to the early Carolingians must have been a continuation of what already existed under the Merovingians, and the reality was that they were much more effective leaders. Stone throwing artillery was in use in the late seventh century at the latest. If, as identified, there had been a weakening of centralised resources, literate commanders and high-level engineering skills in the post-imperial times described by Gregory, the conclusion must be that the skills had been recovered and the practitioners continued to exist. We have already seen building skills being lent to Anglo-Saxon kings by this time.

(c) Mining and Miners

From ancient times to the modern age, those who earned their living by digging out metals (and later coal) from the ground have been called upon to use their skills in military settings. Sappers and miners have been integral parts of the engineering

94 *Annales Mettenses priores*, *MGH* SrG X, 51–4; Fredegar, ed. Wallace-Hadrill, *Chronicon*, 111–14. Discussed in Petersen, *Siege warfare and military organisation in the successor states*, 712–30.

establishments of armies for centuries. There are two distinct elements to the military aspects of the role:

- Hacking away at the base of an enemy's wall or tower, using picks or other tools, until the wall crumbles. This form of attack, while it might be carried out by professional miners, could also be executed by anyone able to wield the relevant tool after brief instruction. It could be a very dangerous occupation.
- Digging a tunnel underground from within one's own lines until the enemy's tower or wall was reached, whereupon the normal method was to excavate a chamber propped up with timbers, then fill the chamber with inflammable material, set it alight and retire to await the destruction of the props and the (hoped for) collapse of the masonry structure above.

A variant of the second method was to continue the underground tunnel past the wall and then exploit it by breaking through inside to allow soldiers to penetrate. Both variants required people to excavate the tunnels who knew what they were doing: the consequences of doing it wrong could be catastrophic.

Defence against underground attack also remained constant throughout history. Various methods were devised to listen out for the sound of people working below ground, because they created vibrations. The simplest was to keep watch – using the advantage of height provided by walls and towers – to see where the tunnel had begun. For the attacker, the equivalent advice was to start work outside the view of the defence, but this was not always achievable. A standard defence once a tunnel was identified was to dig a countermine from inside the defences to break into the mine, usually leading to a subterranean fight and the abandonment and destruction of the mine [see Fig. 6]. It is rare to find a form of warfare that lasted effectively unchanged for two millennia, but in mining and countermining the only significant change would take place around 1500 with the first use of gunpowder to replace burning the props, and the only alteration to the design of the operation was the need to introduce zig-zag tunnels to prevent blowback.

The Romans and Persians were masters of the use of mines for siege warfare. The excavations of the siege of Dura Europos (256 AD) revealed a countermine containing the human remains of Roman soldiers killed in the underground battle when the countermine met the mine. The whole had been immediately sealed up to prevent its reuse.[95] Zosimus and Ammianus Marcellinus recounted fourth-century sieges involving mines and countermines. Both gave detailed accounts of Emperor Julian's capture of a Persian fortress (identified as possibly Maiozamalcha) in 363. Two senior officers were put in charge of the mine which was dug by soldiers, but they were seen to be working "carelessly" so were dismissed and replaced. When the mine was ready, the emperor ordered an attack on the surface with a battering

95 Coulston, 'The archaeology of Roman conflict', in P Freeman, A Pollard (eds), *Fields of conflict: progress and prospect in battlefield archaeology*, BAR International Series 958 (2001), 23–50, here 37–9.

FIG. 6 (a) Descent into the countermine at St Andrews castle, Fife, Scotland (photo: author), and (b) how the countermine breaks into the mine itself as drawn on the display board at the castle. This mine dates from 1549 but the technology had remained unchanged throughout medieval times. (Courtesy of Historic Scotland)

ram as a diversion. This enabled his underground force to break through into a building inside the fortress, irrupt into the courtyard and bring about capture. The tunnels were lined with timber props and Ammianus added the detail that the simultaneous surface attack was designed to cover the noise of the iron tools being used by the tunnellers.[96]

Vegetius outlined the same approach, stressing the importance of starting to dig outside of the view of the besieged. Mines could be used either to collapse a wall or to pass under it and break into the fortress as happened at Maiozamalcha. He clarified that legionary armies included sappers under the authority of the prefect of engineers who were responsible for mining, and compared the latter to the Bessi, who were silver miners in Thrace, presumably cited as a means of illustrating the work of miners.[97] The conclusion from the historical accounts and Vegetius is that there were specialists attached to an imperial Roman army (and the histories as well as the excavations at Dura Europos confirm that the Persians were equally capable) who were responsible for organising mining, but that very likely, the physical work would be done by rank-and-file soldiers. Perhaps some of them also had relevant experience?

Working in mines in the ancient world was not something conducive to health or a long life, and miners were often slaves and criminals. Most of the mines, and miners, to be found in medieval Europe and the Middle East operated on a different basis. The expendability of the workers in Roman mines did not prevent technological progress permitting the creation of connecting galleries to exploit different levels, oil lamps to light the way, and drainage using Archimedean screws and water wheels. Propping with timber was only needed when the ground required it.

The ordinary mineworker's life counted for nothing, but skilled operatives were also needed. There were official mining experts, both administrators and technicians, in the later empire, necessary to ensure continued productivity and avoid premature abandonment of a seam of ore because the workers did not have the skill to work it. From the fourth century the state allowed "free mining", a precursor to much medieval practice whereby anyone could start their own mine subject to contributing a share to the state. Any link between Roman mine operatives and military miners remains, however, completely obscure. If there *was* a link between a particular mining area and military service, continued mining into the medieval period may provide some clues to recruitment by medieval commanders. Mines certainly did continue, as rulers needed silver for coins and warriors and farmers and everyone else needed iron. Roman iron mines are known from central and north-east Spain,

96 Ammianus Marcellinus, book XXIV, 4, 12–13 and 22–3 (vol. II, 436–43); Zosimus, *New history*, ed./trans. R Riley, book III, 20–2, pp. 61–2.

97 Vegetius, book II, c. 11, pp. 43–4, book IV, c. 24, p.128. The Latin term for miner in Vegetius and in the fourth century historians is *cunicularius*, derived from *cuniculus*, a hare's burrow.

the eastern Pyrenees, the Massif Central and parts of Aquitaine in Gaul, Trier and the Rhineland, Carinthia, and many locations in Britain (the Forest of Dean and the Weald in particular).[98] How much continued is the issue: even in the eastern empire, large-scale (capital-intensive) mining had collapsed by late antiquity, to be replaced with much smaller-scale operations, but there is little evidence.[99]

There was continuity of military mining practices in the east. Procopius described the role of mines in the Persian siege of Dara in 540: Chosroes' tunnels were dug from the only place where the soil made it possible, begun inside the Persian lines of circumvallation and therefore invisible to the defenders, but the Romans were advised by a man named Theodorus who was "learned in the science called mechanics", so they were able to dig a successful countermine. A similar attack by Chosroes against Petra (in the Caucasus) was more successful: the miners excavated the stones under one of the solid towers on the wall facing the only approachable side, replacing them with timber props, then set the props ablaze bringing down the tower and forcing surrender. There were also examples of the defending Romans themselves digging tunnels to undermine the great ramps that were a feature of Persian siege warfare, as at Amida in 502 and Edessa in 544,[100] and of underground mining by the Romans at Chlomaron in 578.[101] Mines featured as well in the Persian siege and sack of Jerusalem in 614.[102]

Mining continued to be a feature of Byzantine siegecraft into the seventh century and the Muslim caliphate very quickly acquired the ability to do the same, although it is not clear where they recruited the diggers. The earliest reference concerns the capture of Caesarea in (?) 640, although alternative interpretations suggest they gained access via a tunnel. There is a definite statement that mining was the means of victory in the capture of Baikand by the emir of Khorasan in 706: the later role of Khorasanis as mining specialists, and their possible service under the Sasanians, is known.[103] That the state had an interest in promoting the mining industry is confirmed by a reference to the appointment of a caliphal agent to oversee the iron, lead and silver mines at Tutis in northern Syria in 773–74.[104]

98 K White, *Greek and Roman technology*, 34–5, 238–40; O Davies, *Roman mines in Europe*, although elderly is still useful, particularly 4–5, 9–12, 20–32, and the inventory of locations, although there is sometimes rather a lot of guesswork about the dating of particular mines.

99 M Lewis, 'Antique engineering in the Byzantine world', in Lavan et al. (eds), *Technology in transition AD 300–650*, 372–3.

100 Procopius, *Wars*, book I, vii, 8–33, pp. 52–9 (Amida); book II, xii, 20–9, pp. 378–81 (Dara); book II, xxi, 21–4, pp. 446–7 (Petra); book II, xvii, 1–18, pp. 504–7 (Edessa).

101 Menander, frag. 23.7, p. 203.

102 Discussion of the various sources for this event in Purton, *HEMS*, 26–7.

103 Discussion and sources in Purton, *HEMS*, 51–3.

104 *Chronicle of Zuqnīn*, ed. Harrak, 300.

The picture in western Europe was different and there were no explicit references to a military role for mining after the end of the western Roman empire until the tenth century. It has been argued that Bishop Gregory's account of Clovis' capture of Angoulême in 508 meant it was undermined, a method also suggested at the capture of Clermont in 524 by his successor Theuderic (511–33). The reasons for doubting this are no less powerful than the reasons for accepting it. Gregory's account stated that Angoulême fell when the walls collapsed of their own accord as the king observed them.[105] The implication is that the writer was making (an oblique) comparison with the biblical account of Joshua's capture of Jericho. This is possible, but leaving aside divine intervention, which is the message Gregory was emphasising, there is no evidence that the biblical capture of Jericho was itself due to Israelite undermining.[106] Late Roman Angoulême was built on a hilltop and had substantial Roman walls. Maybe Clovis had enlisted local miners (where from?), or perhaps (more likely) his soldiers were able to bring a section down by surface sapping: there is no evidence for any interpretation. It is possible the walls built several hundred years before had not been maintained and were crumbling. Clovis' victories followed the defeat of the opposing Visigoths in battle (at Vouillé, 507) and many towns capitulated without a fight.

At Clermont, the language used by Gregory to describe the taking of the city – *muros urbem evertere* – may signify sapping on the surface.[107] At best, therefore, the case is not proven. The absence of evidence for Merovingian siege mining in later times is also telling. To mine by digging underground tunnels requires people capable of executing the work. Educated Frankish commanders may have been aware of the potential of mining but did not have access to skilled practitioners. Common soldiers with no previous experience could not undertake such a challenging task successfully, although they could execute surface sapping.

105 Gregory, book II, chap. 37 (ed. Thorpe), 154.

106 On the city of Jericho: A Ferrill, *The origins of war from the stone age to Alexander the Great* (1985), 28–31; Y Yadin, *The art of warfare in biblical lands in the light of archaeological study* (1963), 32–5, 99–100. The actual account (Joshua 6:3, 16, 20) says that the Israelites marched around the city seven times blowing trumpets, at which point they took it. Yadin compares this with a stratagem described by Frontinus at another siege, where the defenders had been duped into thinking that the circuits were a drill, so took no precautions and were overwhelmed. There is no reference to mining in any of this. Cf. P Kern, *Ancient siege warfare*, 30–1, 36.

107 There is no detail in Gregory's history but Petersen, *Siege warfare and military organisation in the successor states*, has analysed accounts from two saints' lives to decide there was undermining at Angoulême, 493, and sapping at Clermont (496–7), although nothing is explicit.

SHIPBUILDING

There is growing evidence for early medieval shipbuilding from painstaking analysis of remains by maritime archaeologists, which increases knowledge of skill levels. Distinct traditions continued in the Mediterranean and in northern and western Europe, determined by the different conditions prevailing there. In both areas it is apparent that there was no great transformation of the technologies in use when the Roman empire disappeared in the west. The few examples of ships recovered from this period that had cruised across the North Sea or traded along the coasts of northern Europe had a number of designs ranging from the formidable "Viking"-type longships to flat-bottomed barges used on rivers. The Germanic peoples who conquered eastern Britain used vessels powered by oars and sails, and technologically there was little advance between the possibly fourth-century "Nydam ship" (probably Anglian or Saxon) and the really large "Kongsgäde ship" of c. 600 or the famous Sutton Hoo buried ship (early seventh century), which had been 89 feet long with a beam of 14 feet and depth of more than 4 (27.4×4.3×1.2 metres). The ships were clinker built and represented highly effective vessels. But how much knowledge was needed to build one successfully? It has been argued that those responsible must have had a general idea of the shape they wanted, and specific knowledge of the shape of the bow and the stern, and that this would lead to a rough proportionality of the whole ship's hull, with the keel itself determining the final dimensions. "Rule of thumb" must have been the method, given general illiteracy in this region. Crumlin-Pedersen proposed "the boatbuilding master had fixed the whole concept of the ship ... before the keel was even laid" and hypothesised that the proportions were based on segments of circles. For all this, study of wrecks such as "Skuldev 3" shows "a refined construction with elegant lines and in carefully selected materials".[108] By the time the seaborne route of attack had become a serious threat again from the end of the eighth century, those at the receiving end found themselves having to pay attention again to their naval power, and the Capitularies of the Carolingian monarchs suggest reliance on existing maritime communities for the skills needed.[109]

The imperial navy of Byzantium was different. Using both galleys, which may have initially been smaller than their classical predecessors, and a variety of transports, it had been re-established at the end of the fifth century. Many wrecks

[108] O Crumlin-Pedersen, 'Aspects of wood technology in medieval ship-building', in the same, M Vinner (eds), *Sailing into the past. The international ship replica seminar Roskilde 1984* (1986), 138–49; S McGrail, 'How were vessels designed before the late medieval period?', in C Beltrane (ed.), *Boats, ships and shipyards*, 125–31; J Haywood, *Dark Age naval power* (1991), 45–8, 65–8.

[109] The Capitularies of 802, 808 and 811 are in *MGH* Legum Section II, Capitularia Regum Francorum I/1, nos 34 (p. 100), 50 (p. 139), and 74 (p. 167).

have been discovered, and the dating of timbers has enabled trends to be identified, such as a slow move from shell to skeleton construction techniques (by the ninth century), and the change from square to lateen sails. War galleys were called *dromons* from the sixth century (as Procopius described), and were driven by two banks of twenty-five oarsmen. Some of them would soon be adapted to mount the siphons needed for "Greek Fire". Enough is known of these vessels to confirm that they were sophisticated constructions. Illiterate craftsmen managed to create fleets that helped the empire survive and eventually, after two centuries, to return to expansion.[110]

THE START OF A NEW AGE?

In the mid-eighth century, new and powerful dynasties took political control in the western European world (in the shape of the Carolingians) and in the Muslim caliphate (the 'Abbasids); in between them, the Byzantine empire continued to exist, if not prosper, in the face of intermittent but dangerous threats.[111] In traditional historiography, we are now leaving the 'dark ages'. Leaving aside for the moment the broader social and economic developments of the preceding three centuries, it has become clear that there were times when progress in science, learning, technology – and therefore also technicians – either halted or was lost, but that this happened at different times in different places. The eastern Roman empire survived the dramatic explosion of the Arab empire, while the Ummayad caliphate almost effortlessly absorbed the best of classical civilisation and, under the following 'Abbasid dynasty, achieved new heights. In western and central Europe, the dominant Frankish kingdom was the only one of the western successor states to survive the eighth century. The world inherited by the Carolingians was inevitably different from that taken over by the first Merovingians. However, in both the wider picture and in the details of the subject under study, there was no sudden break or revolutionary change in the eighth century. Were there any connections between the undoubted new progress in science and technology, and engineering and its military applications?

110 J Pryor, 'Ships and seafaring', E Jeffreys et al. (eds), *Oxford Handbook of Byzantine Studies* (2008), 482–91; F van Doorninck, 'Byzantium: mistress of the sea 330–641', in G Bass (ed.), *A history of seafaring based on underwater archaeology* (1972), 133–58; U Kocabas, 'The Yenikapi Byzantine ship wrecks, Istanbul, Turkey: a preliminary report and inventory', *International Journal of Nautical Archaeology* 44/1, 2015 (5–38) for analysis and approximate dating of 37 shipwrecks found in a harbour at Constantinople abandoned in the twelfth century. It included the first evidence of an actual Byzantine galley, a light scouting *galea* with a single bank of oars; and see below, pp. 103–04.

111 W Treadgold, *The Byzantine revival 780–842* (1988).

THE REBIRTH OF EDUCATION IN WESTERN EUROPE

Charles the Great (Charlemagne) earned his title not only for his decades of often brutal wars of conquest that extended the Frankish empire eastwards to the Elbe and the borders of modern Hungary, and southwards to the borders of Byzantine southern Italy and across the Pyrenees, but also for the accompanying administrative reforms (vital to governing an enormous territory) and for the attention he paid to raising the level of education. Charlemagne was intelligent enough to see that he could not hope to run his empire without a large cohort of educated people. Such a group seems not to have existed when he assumed the crown in 768.

Historians generally agree that the result of his instructions on education (issued first in 789) was a transformation. Evidence shows a remarkable change, with not only the clergy (vital for running the empire) but also lay people acquiring at least basic skills – defined as literacy in Latin and simple arithmetic. Copies of texts found their way to Anglo-Saxon libraries, and new monastic foundations also copied them in the territories conquered in Germany. Texts that had been common in late antiquity became the staple of Carolingian education, including Boethius, Cassiodorus, Isidore, Pliny, and Vitruvius (*De architectura*, now rediscovered) amongst others.

Numeracy was a central component of Carolingian education and was essential for both ecclesiastical and lay purposes, including architecture. An increase in the provision of such texts is confirmed by the catalogues of monastic libraries known from the era, which contained Bede and the pseudo-Boethius, the *corpus agrimensorum*, Vegetius and classical works of history. It is likely that the copying was slow and the numbers who had access to these works must have been limited, but over the years the copies multiplied and more people will have had the opportunity to study them.[112]

BUILDING AND ENGINEERING UNDER THE CAROLINGIANS

If direct evidence for the practical application of learning is absent, hypotheses can be proposed about the level of knowledge of those who designed buildings, or

[112] J Contreni, 'The pursuit of knowledge in Carolingian Europe' in R Sullivan (ed.), *The gentle voice of teachers: aspects of learning in the Carolingian age* (1995), (106–41); the same, 'Counting, calendars and cosmology: numeracy in the early middle ages', in Contreni, Casciani (eds), *Word, image and number*, 43–83; B Bischoff, 'Manuscripts in the early middle ages', in M Gorman (ed.), *Manuscripts and libraries in the age of Charlemagne* (1994), 1–19; the same, 'The court library of Charlemagne', in the same, 56–75; the same, 'Libraries and schools in the Carolingian revival of learning', in the same, 93–114; the same, 'Benedictine monasteries and the survival of classical learning', in the same, 134–60.

built bridges, or constructed military equipment. Charlemagne was determined that his military commanders had been to school – and that such education would include not only the Roman histories but also Vegetius and Rabanus Maurus. So the aristocratic generals were familiar with ancient as well as having been trained in current practice. Incessant military activity would also have provided numerous opportunities for hands-on practical experience.

The master builders' and carpenters' knowledge can only be guessed at: builders must have had at least a little mathematical knowledge and been able to interpret the drawings that almost certainly must have been provided for them by the patron. The people who built the famous octagonal chapel at Aachen must have had sufficient knowledge to execute the sophisticated design, consciously modelled on the proportions of the Holy Sepulchre in Jerusalem,[113] the now lost audience hall apparently aping Roman imperial plans likewise, and the builders of Charlemagne's palaces similarly. It is known that the builders of Aachen (his biographer Einhard, and Odo of Metz) were part of the emperor's court circle, and that circle of scholars had been trained in the classics. The schools set up under Charlemagne flourished under his son Louis the Pious, and it is to the first decades of the ninth century that can be attributed important mathematical texts.[114] The famous plan of an ideal monastic complex preserved in the library of St Gall, and believed to have been drawn for the abbot of Reichenau around the year 820, was a scale drawing (using a Roman foot) and used geometric progressions. It also laid out a plan that included provision for a large amount of lay activity around the abbey church and its associated buildings, and this involved industrial workshops such as arms manufacturers. Plans must have been in use throughout, but the one at St Gall was innovatory in being a scale drawing of an entire complex.[115] This monastery, had it ever been constructed, would have reflected the model being used by hundreds of such establishments across the Carolingian empire, that formed a keystone of the empire's social, economic, political and military functioning. Changes in the status assigned to those actually doing the physical work also took place over these centuries, although this can be hard to verify: nonetheless recognition of the role in society of skilled labour can be picked up from monastic rules, images of people working and other sources.[116]

Monasteries took on the role traditionally assigned to towns – as centres of population, manufacture and trade – during this period. If the term "dark ages"

113 Stalley, *Early medieval architecture*, 71–3, 97 (audience hall).

114 Butzer, 'Mathematics in west and east', in P Butzer, D Lohrmann (eds), *Science in western and eastern civilization in Carolingian times* (1993), 452–7.

115 W Horn, E Born, *The plan of St Gall* (1979), I, xxi–xxiv, 89–95, 341–7; II, 189–95; III, 103, 132–7. Kleinbauer, 'Pre-Carolingian concepts of architectural planning', 74.

116 Particularly useful in summarising the evidence is J Le Goff, *Time, work and culture in the middle ages*, trans. Goldhammer (1980), esp. 71–86.

can be applied to anything, it is to the town in Europe between the seventh and tenth centuries, with the exception of a few metropolises used as centres of government and a number of important trading entrepôts such as Dorestad and Hamwic. Instead, 300 episcopal sees and 700 monasteries in the Frankish empire were the centres of these functions under Charlemagne and his successors. The building complexes themselves represented a massive investment in labour and materials that employed large numbers of craftsmen and technicians alongside the local labour force. Archaeological investigation of monastic sites from this time demonstrates that the set up of an "ideal" Carolingian monastery, with its large provision for lay occupations and workshops, was also a practical reality. The excavators of San Vincenzo Maggiore described the construction as "a massive technical operation" involving a range of highly skilled craftsmen around the year 800 and identified a vaulted tunnel to the workshop as "a major engineering feat". Similar evidence has been uncovered at other major monastic sites and there is also documentary evidence to confirm this picture.[117]

The crafts evidenced at San Vincenzo and the technologies that had been described in Theophilus' *De diversibus artibus* confirm that high levels of technical knowledge and skill were present in sufficient quantities to supply the needs of hundreds of major institutions. Artisan skills required to supply Charlemagne's large military expeditions were present, and often based in monastic workshops: Frankish laws (capitularies) included royal instructions to abbots to manufacture, store, supply and then return to storage after use a range of weaponry and materials for the army. Contingents of *milites* and other soldiers who held their land from monasteries were a significant part of Carolingian armies.[118]

However, all the items listed were hand weapons or armour for the individual warrior, and (with the exception of the instruction to prepare carts for the army, for the transport of provisions) the only reference to these workshops or artisans being the source of any engineering expertise was (importantly enough) to prefabricated boats. Crossing rivers remained vital. There was no mention anywhere of the manufacture of siege equipment. That rams, ballistae and stone throwers were in regular use is certain. Chronicle accounts made frequent references to rams and stone throwers in particular in conflicts involving Lombards in Italy, Muslim rulers in Iberia, Avars and Moravians in central Europe, and Saxons and Slavs in (modern)

117 R Hodges et al., 'San Vincenzo Maggiore and its workshops', *British School at Rome* Archaeological monograph 17 (2011), 129–30, 157, 432–6 (on evidence from other sites and sources); Hodges, *Dark Age economics*, 69–81; F Schwind, 'Karolingerzeitlicke Klöster als Wirtschaftorganismen', in L Fenske et al. (eds), *Institutionen, Kultur und Gesellschaft* (1984), 101–23; A Verhulst, *The Carolingian economy* (2002), 72–84.

118 'Capitulare de Villis', *MGH* Legum II, Capitularia regum Francorum, I, I, ch. 42, 45 (p.87); 'Capitulare Mantuanum', same vol., ch. 7 (p.190); 'De villis', trans. in H Loyn, J Percival, *The reign of Charlemagne*. Documents of medieval history, 2 (1975), 70–2.

Germany. Archaeology has uncovered and dendrochronological techniques have permitted the accurate dating of Carolingian-era works of fortification, too, to confirm that the Frankish kings had sustained and developed technical and engineering skills in the arena of warfare: the people with engineering skills who designed, constructed and operated the equipment required or who laid out and oversaw the construction of the new works of fortification ordered by the rulers existed, but their identities are unknown.[119]

An area of engineering skill with military relevance involved water. There are chronicle references to bridge-building. In 789, Charlemagne ordered two bridges across the river Elbe, on one of which earth and timber fortifications were constructed at each end, for a campaign against a Slav king, reflecting both Vegetius and the practice enjoined on Byzantine commanders for bridges in hostile territory. Maintenance of existing bridges may well have been a burden placed on local communities but construction of temporary military bridges must have been carried out by the soldiers. Whether there was an engineer who decided how many pontoons were needed and of what dimensions remains unreported. As Bachrach points out, a number of other Carolingian military operations involved crossing large rivers and therefore must have required bridge-building[120] even when this is not recorded. A bridge with a more civic purpose was the large-scale construction across the Rhine at Mainz, noted by Einhard as being 500 feet long. The work caught fire in 813 and plans to rebuild it in stone were apparently shelved when the emperor died in the following year.[121] It has been pointed out that this undertaking took a decade to achieve and the pontoon-based design[122] apparent here and in the other bridges where there is evidence did not suggest an acquaintance with Roman bridge-building technology – but perhaps it is more likely that they did not have people with the skills to reproduce the technology that required stone piers embedded in the river to cope with the flow of the water. Evidence from Anglo-Saxon England similarly suggests that only timber pontoons were ever attempted and that many former Roman bridges (such as the bridge across the Thames at

119 For a summary and analysis of the chronicle accounts and reference to archaeological investigations of Carolingian fortifications, see Purton, *HEMS*, 65–80, and for discussion of the organisation and campaigns of the armies, B S Bachrach, *Early Carolingian warfare* (2001) (throughout).

120 Although the possibility that even large rivers might be forded should not be ruled out.

121 Einhard, 'Life' (ed. Thorpe), ch. 17, p. 71; 'Royal Frankish Annals', in B Scholz (ed.), *Carolingian chronicles*, 68; Bachrach, *Early Carolingian warfare*, 220–1, 237, 254–5. In the other examples, discussing campaigns by Charles Martel and Pepin, no source is offered for the assertion that the crossing was by bridge, or might have involved crossing in boats.

122 That is, the bridge comprised wooden planking laid across a number of small boats or otherwise watertight vessels (pontoons) floated in the waterway between the two banks.

London) did not survive and were not replaced, even when London was refortified under King Alfred after 886. The only exception was the Roman bridge across the Medway at Rochester (Kent), where the timber superstructure seems to have been maintained throughout.[123]

That Carolingian and Anglo-Saxon engineers did not attempt to follow the bridge-building advice presented by Vitruvius is not to argue that they possessed no skill, as the practical result of using pontoons was sufficient for their purpose. Reproducing ancient Rome's ability to create stone piers deep under a fast-flowing river using coffer dams was a serious challenge and it would be safe to assume that after several centuries when no such engineering had been attempted, it would take some time before people were able to replicate the feats of their predecessors.

That Merovingian and Carolingian engineers did not take up Roman methods indicates the limits of available technicians, but other evidence demonstrates a willingness to undertake massive hydraulic engineering. Similar timber bridge building existed in many other parts of Europe, including dense networks in tenth-century Sweden and long causeways in Denmark and Mecklenburg dendrochronologically dated from even earlier.[124] Another major project of Charlemagne's helps assess engineering skills: the construction of a canal known subsequently as the *Fossa Carolina* (Karlsgraben) to create a navigable link between the Rhine (via the Main) and the Danube via tributaries of both. Historians have sometimes relied on the rather dismissive report in the Royal Frankish Annals that the idea could not be turned into reality by reason of natural causes (geology, weather) and the statement that the emperor (whose wisdom would otherwise have been faulty) had been misled by "experts". In fact, more recent archaeological and scientific analysis has restored the canal as a highly significant piece of early medieval engineering. The canal was built between Weissenberg and Treuchtlingen and connected the rivers Rezat and Altmühl, on either side of the watersheds of the Danube and Main, a distance of more than 2000 metres. It was 300 feet (90 metres) wide. The project involved digging retaining embankments each side of the new canal (of which little remains visible at ground level) and dams and a reservoir to sustain the supply of water to the canal, the level of which changed by some 10 metres over its course. Modern analysis suggests that the engineer who decided on the start and end points and the route had succeeded in finding the optimum solution. The canal did fail: the labour required for its continual maintenance (and problems of silting) and the probability that the cost of portage of boats across the dams made it uncompetitive for merchants led to a

123 N Brooks, 'Medieval European bridges', in the same, *Communities and warfare 700–1400* (2000), 10–13; B Watson, *Old London Bridge* (2004), 12–20. See also the discussion by Haywood, *Dark Age naval power*, 97–103. On Rochester: N Brooks, 'Rochester Bridge AD43–1381', reprinted in the same, *Communities and warfare*, 219–65, here 229–41.

124 Brooks, 'Medieval European bridges', 15–19.

rapid fall into disuse and the progressive collapse of its earthworks. Nonetheless, its inspiration may have been military (it was first proposed while the campaigns against the Avars were continuing), or commercial, or it may have been a prestige project in which Charlemagne decided to emulate the canal-building of the ancient Romans. There are two other significant points to note. First, this was a project calling for knowledge that Charlemagne himself cannot possibly have possessed. The "experts" cited by Einhard must really have known what they were about. This was not an exercise where a mistake could be covered up. Secondly, only a very powerful ruler could have mobilised the resources needed. It has been calculated that thousands of labourers were required to excavate the 126,000 cubic metres of soil over two short autumn campaigns, an immense logistical challenge.[125]

Placed alongside the evidence for large-scale building works at palaces, cathedrals and monasteries and significant construction of new fortifications, the *Fossa Carolina* confirms that, with the support of a vast royal patronage, wide-scale engineering work took place throughout the Carolingian realm requiring large numbers of skilled masons, carpenters and other craftsmen. Such works also required the existence of masters who could design and oversee the construction. It is true that the only references there are to named individuals fitting this job description are in a much later troubadour's *chanson de geste*,[126] but the reality of the existence of such persons must be beyond question, even if they were not known by the song's name of Maurin.

The next major construction work involving water and with an unquestionably military function was the effort of Charlemagne's grandson Charles the Bald (840–77) to protect his kingdom from Viking attacks up the country's great rivers. The effort involved was similarly enormous and the outcome was disappointing.

125 R Koch, 'Fossa Carolina', in K Elmshäuser (ed.), *Häfen, Schiffe, Wasserwege* (2002), 54–70; the same, 'Fossa Carolina. An early medieval canal near a watershed', in C Westerdahl (ed.), *The significance of portages. Proceedings of the first international conference on the significance of portages, 2004*. BAR International Series 1499 (2006), 259; H Hofmann, *Kaiser Karls Kanalbau* (1969), 24–5, 43–9; K Spindler, 'Der Kanalbau Karls des Grossen', in K Spindler (ed.), *Mensch und Natur in mittelalterlichen Europa* (1998), 47–100. There is now a local museum dedicated to the Fossa Carolina in Treuchtlingen and a stretch has been restored.

126 P Taylor (ed.), *Gerbert de Metz*. Gerbert's (twelfth-century) *Geste des Loherains* describes Charlemagne's engineer (engigneor) Maurin constructing a definitely twelfth-century siege engine for the emperor (lines 2720, 2738ff, 2752ff), while "Ogier the Dane" had the misfortune to face the work of Maurin, "most skilful of his engineers", who organised hundreds of carpenters and labourers to build a seven-storey siege tower – the defenders, however, sortied and managed to kill Maurin (*Champion of Charlemagne*, ed./trans. M Butt, R Linker (1967), containing 'Les infortunes d'Ogier le Danois', 72–3). It is highly unlikely that the name refers to a historical person but there must have been real people with similar skills and roles.

The king, having succeeded in beating off the Norsemen and cutting their retreat by erecting a bridge across the river Marne in 862, opted to develop a longer-term defence by constructing a fortified bridge across the Seine where it was joined by the Andelle and the Eure, the site of a Roman bridge. Archaeological investigation suggests the structure might have been at Pont de l'Arche, a few kilometres from the royal town of Pîtres, whence the king's orders for the work were issued. The width of the rivers (the Seine and the here parallel Eure) was some 300 metres, and the bridge was protected by earth and timber forts erected at each end, the clay banks strengthened with the insertion of logs and stone facing on the outside. The collection of 1500 oaks cut to a length of 12 metres suggests that the bridge itself was of timber trestles and was planned to block passage to boats. Despite a number of repeated references in royal records to the obligations on lords and locals to supply materials and labour, the whole project may have taken fifteen years to complete. It may have achieved part of its objective, but it did not stop river-borne raids as far as Melun in 866, when the fort was not complete but what about the bridge? Certainly, it did not stop the Vikings from getting to Paris for the siege of that city in 885.[127] While there were detailed instructions about the obligations designed to secure the labour and materials, there was no reference to anyone being responsible for undertaking the planning, layout or design of any of these substantial works.

Elsewhere in Europe, there are a few large-scale building projects involving knowledge of engineering for water. The new palace at Ingelheim (Germany) begun around 800 obtained its water supply from an 8-kilometre-long aqueduct that ran along the contours of the ground, a massive piece of engineering. In Italy, there were a few examples of water works to control flooding by the river Po by both Byzantine and then Lombard rulers, and references to popes such as Hadrian I (772–95) organising work on the Tiber, but little else.[128]

127 See C Gilmor, 'Logistics of fortified bridge-building on the Seine under Charles the Bald', *Anglo-Norman Studies*, XI, 1989 (87–106); B Dearden, A Clark, 'Pont de l'Arche or Pîtres? A location and archaeomagnetic dating for Charles the Bald's fortifications on the Seine', *Antiquity* 64, 1990, esp. 568–9. The royal plans are recorded in J Nelson (ed.), *The annals of St Bertin*, 98, 100 (862: "Charles caused all the leading men of his realm to assemble about 1 June, with many workmen and carts, at ... Pîtres, where the Andelle from one side and the Eure from the other flow into the Seine. By constructing fortifications on the Seine, he closed it off to ships ... This was done because of the Northmen."); 118 (864, the fortification order repeated); 127 (repeat of the order in 865); 129 (Melun, 866), 131 ("Charles marched to ... Pîtres with workmen and carts to complete the fortifications" (866)); 151–4 (arrangements for garrison duties of the fort, 868–9).

128 K Grewe, 'Water technology in medieval Germany', in P Squatriti (ed.), *Working with water in medieval Europe* (2000), 128–60, at 136; R Magnusson, P Squatriti, 'Technologies of water in medieval Italy', same vol., 216–66, at 221–2.

Viking attacks led to another chronicle reference that has given rise to speculation: in 873, according to Regino of Prüm, Charles the Bald besieged the invaders in Angers and created *nova et exquisita machina* to press home his attack. It is to be noted that the attack failed. Were these machines siege towers or battering rams? Who knows?[129]

Viking attacks also had a major impact on early medieval England, which at this time was divided into several kingdoms and only (under the prolonged pressure of Viking raids and then Danish invasion) finally consolidated into a single realm under the rule of the kings of Wessex in the tenth century. There has been a long debate about the origin of the duty of everyone in the kingdom to carry out repairs to royal fortifications, to bridges and to military service summarised in the much later term *trinoda necessitas*, the earliest (but not undisputed) reference often claimed to date from a charter of 770. The concept of the populace being liable to various duties in the service of the crown was deep-rooted, but what it signifies is that local communities (unless exempted) had a role in maintaining structures deemed important, here fortresses and bridges.[130] It does not suggest who was responsible for the design and oversight of new work. In Anglo-Saxon England, the largest building project over a short period of time was the construction at the command of King Alfred (871–99) of the system of more than thirty fortifications known as burhs (and new ones added subsequently under his son Edward the Elder (899–924)), as part of the measures against Danish invaders. The survival of numerous earthworks and the existence of historical and documentary evidence reveal much about the skills at Alfred's disposal, including officials able to calculate the distribution of duties of labour, defence and taxation according to land assessments, the enormous requirements imposed on the population, but also the skill to design and oversee the construction of large modern fortifications. I have shown above that surveying expertise was part of the skill set available in early and middle Anglo-Saxon England. These same skills were called upon both to lay out new towns and to measure out the burhs.

A recent study of the surviving earthworks of Wallingford reveals the level of engineering skill employed to construct the burh there. The site was planned

129 *Reginonis chronica*, ed. Dümmler et al., 240.

130 Modern debate was initiated by W Stevenson, 'Trinoda necessitas', *English Historical Review* XXIX, 1914 (689–703). For later views, see R Abels, 'Bookland and Fyrdland', *Anglo-Norman Studies* 7, 1984 (1–25) and N Brooks, 'The administrative background to the Burghal hidage' in D Hill, A Rumble (eds), *The defence of Wessex*, 128–50, and B S Bachrach, R Aris, 'Military technology and garrison organization: some observations on Anglo-Saxon military thinking in light of the Burghal Hidage', *Technology and Culture* 31, 1990 (1–17). For a broader perspective on England under Alfred and the reforms he initiated, R Abels, *Alfred the Great. War, kingship and culture in Anglo-Saxon England* (1998), esp. 95, 195–9, 206–34.

and measured, and it involved massive ditches, ramparts and stockades, gateways, designed road layouts, and internal structures. It also involved engineering a water system identifying springs and both an internal water supply and a means of sustaining water levels in the (wet) ditch. The excavators suggested that "achieving … the rapid creation … of some 30 forts and fortresses in a coherent and interconnected system demonstrates that the ninth- and tenth-century Wessex royalty was able to muster both engineers of high ability and substantial workforces …". The massive remains at Wallingford (partly covered by the later Norman castle) offer powerful visual testimony to the skills of the nameless experts who designed and directed the construction.[131]

This review of engineering – there are no grounds for a distinction between military and civil – indicates a considerable level of technical expertise, awareness of the qualities of various materials (in particular, of course, timber) and design skill. Once a burh, for example, had been first designed, others could follow the pattern, but somebody was responsible for the original design, and local topographical variation meant that this could not be copied in every situation. Any link with scholastic learning is completely impossible to demonstrate: only the rare rulers like Alfred who had read the available learning and seemed able to grasp what it meant are known to have been in a position to apply their theoretical knowledge. But there is nothing to suggest that Alfred did sketch out plans for his burhs or could anyway have been in thirty places at once. If he had, his admiring biographer would presumably have told us. Therefore, in the absence of other possibilities, one can only conclude that the engineers in practical charge of Charlemagne's and Alfred's projects did have the requisite knowledge, some of it the same as the learning that the scholars were busy copying and teaching in the monasteries and schools.

IN THE CALIPHATE

The same can be said with even more direct evidence for the 'Abbasid caliphate. One of their first acts was to remove the capital from Ummayad Damascus and begin the construction of a new one at Baghdad on the river Tigris and linked by canal to the Euphrates, at the nub of many trade routes, in a fertile region benefitting from historic irrigation systems. The change was also symbolic, visible

131 N Christie et al., *Transforming townscapes. From burh to borough: the archaeology of Wallingford AD800—1400* (2013), 68–80 quotation at 72. There is a substantial literature on the burhs and historical references in the Anglo-Saxon Chronicle. On the apparently rectilinear plans of the new towns built over the Roman site at Winchester, M Biddle, D Hill, 'Late Saxon planned towns', *Antiquaries Journal* LI, 1971 (1), 70–85.

evidence of a new regime in charge, and it entailed a shift from Arab to Persian as the dominant culture in a number of ways.

Central to the foundation was the construction of the famous "round city" at its heart by the caliph 'al Manşūr around the year 760; and it was ready for use within seven years. Made largely of mud brick, but with stone foundations for its inner core containing the caliphal palace and central mosque, Baghdad's construction was evidence of high levels of planning, design and construction skills and must have involved the work of many engineers and architects as well as the labour of thousands of ordinary people: al-Ya'qūbī described how the ruler assembled "engineers, men with a reputation in the art of construction and surveyors skilful in measuring lengths and surfaces". Construction also involved establishing bridges (of which five were part of the original plan – all of them apparently built on pontoons provided by boats), waterways and mills.[132] Despite the large discrepancies between the dimensions offered in several of these early descriptions, the scale and complexity of the work are clear – not to mention the debt to those aware of Euclidean geometry – and testify to the availability of a substantial number of architects and master craftsmen. The first caliphs of the new dynasty also established the foundations of a centre of scientific excellence, attracting and supporting scholars from far and wide. The same question arises as it does for the Carolingian court: how far did the knowledge gathered in the capital of the caliphate influence the practical engineering applied in the empire and the military activities of the caliph's generals? The flowering of the sciences under the early caliphs of the new dynasty included the creation of the "House of Wisdom" under 'al Manşūr as a central library for the collection and translation of the scientific and philosophical works of the ancient and contemporary worlds, wherever they originated,[133] and, taking advantage of having learnt how to make paper (from the Chinese, in 750), it became much cheaper to circulate that knowledge. This was not just an exercise in copying: new generations of scholars built on the classical Greek, Persian and Indian foundations to take theoretical knowledge to new levels, and it would be the transfer of these ideas, as well as the translations, to Europe that underlay the rebirth of scientific knowledge there from the eleventh century on.

There is sufficient evidence to show that there was a connection between science and practice. A text by al-Baġdādī giving advice to those holding office as government secretaries, written around 870, said, "The Persians say ... 'who doesn't know

[132] G Wiet, *Baghdad: metropolis of the Abbasid caliphate* (1971); the Arabic sources are translated in G Le Strange, *Baghdad during the Abbasid caliphate* (1900), 15–26, 42–5; G Salmon (ed.), 'L'introduction topographique à l'histoire de Baghâdh d'Abou Bakr Ahmad ibn Thōbit al-Khatīb al-Baydādī', *Bibliothêque de l'école des hautes études*, 148, 1904, 151–60.

[133] For example, J Lyons, *The House of Wisdom. How the Arabs transformed western civilization* (2009).

the art of irrigating, of digging canals and making barrages, the way (waters) rise and fall with the days ... use of scales, the way to measure triangle, square, different angles, to construct arches, floating bridges, raisers of water, does not know the artisans' instruments ... doesn't have the competence necessary to be a secretary."[134] Two generations later, al-Fārābī (c.870–950) in his work on the divisions of the sciences allowed seven subdivisions of mathematics including mechanical artifices, itself further divided into "applied or practical" and theoretical. While the master not surprisingly lauded the theoretical as the superior form (just as western monastic scholars did), the list included the practical as well, and commentaries on it specified, for example, that engineering covered both "ingenious devices" and the "practical arts", which included weaponry.[135] The indication that knowledge was not only required to carry out the functions of the state, but that the scientific knowledge developed and handed down through the schools also (in part) included access to practical knowledge strengthens the likelihood that people applying for official positions could be expected to demonstrate command of a wide range of tasks. It is no surprise to find that heading the list of knowledge expected of a government secretary was a detailed knowledge of irrigation technology.[136]

The range of skills necessary to implement effective irrigation of the soil for agriculture and to ensure the supply of fresh water for the cities in which Islamic civilisation flourished were sustained and developed across the medieval centuries, had been continued unbroken from their Roman or Persian predecessors and were reimported into conquered territory such as al-Andalus. An immense network of canals brought water to the new city of Baghdad; when caliph Mutawakil (847–66) tried to establish a new capital at Samarra, this required hundreds of miles of new *qanat* (underground conduits). Treatises explaining how to carry out such works were regularly composed. As early as the seventh century, when the Arab conquerors of Egypt created a new city at Fustat in 643, it had been necessary to supply water to the new city not directly from the Nile, because of its regular floods, but to devise a system of dams and canals taking water to artificial reservoirs then into cisterns from which it would be distributed by water wheel to

134 D Sourdel, 'Le Livre des secrétaires de 'Abdallāh al-Bāġdādī', *Bulletin d'Etudes Orientales* XIV, 1952 (115–53), here 121.

135 I Netton, *Al-Fārābī and his school* (1992), 38–9; O Bakr, *Classification of knowledge in Islam*, 112, 123; J Jolivet, 'Classification of the sciences', in R Rashed (ed.), *Encyclopedia of the history of Arabic science* (1996), 3, 1008–25, makes the same judgements on the relationship between theoretical knowledge and practical application.

136 That the tradition continued is demonstrated by the account of a (Jewish) bureaucrat, Samaw'al-Maghribi, serving the Seljuk sultan in the twelfth century, who listed his education as including Indian mathematics and equations, accounting, the 'science of surveying' and the geometry of Euclid: translated in A Peacock, *The great Seljuk empire* (2015), 76–9.

a network of local distribution points. Even more complex arrangements (though on the same principles) were called for when the new Fatimid dynasty created its capital close by at al-Qāhira (Cairo) in 969. These works required large-scale investment available only to a centralised state apparatus, but also a large number of skilled and experienced engineers familiar with surveying and construction and the properties of water.[137]

The House of Wisdom flourished and saw a vast range of theoretical and practical work produced. The sons of Musā b. Shākir produced works on a wide range of topics including engineering. The manufacture of devices such as the astrolabe, which was built for astronomy but had a wide range of additional uses, was a result, and if many of the mechanical devices created were intended to be no more than fascinating toys, suitable to provide gifts to impress foreigners, the engineering required to make them was equally applicable to more mundane applications. In fact, the distinction may seem sharper today than it did to contemporaries, as the reach of such knowledge was considerable among the elite of society who included, of course, the people who led the caliph's armed forces into battle and siege.[138]

CONCLUSIONS

The Islamic world reveals a history of uninterrupted continuity between the classical and the medieval, confirmed in the many areas in which the skills of the engineer were required, from building palaces and fortresses to creating and using the weapons with which to capture them, from surveying the land to create efficient irrigation and water supply to being able to measure distances for military purposes or the planning and laying-out of fortifications. This was achieved through a readiness to employ local, knowledgeable inhabitants, and if this caused arguments about the rightful place for non-Muslims, in most cases it seems that practical need for their skills trumped religious doubts, and that it was debated confirms it was happening. In due course, many would convert to the new dominant religion. The introduction of paper and the expansion of education increased the number of

137 D R Hill, A Y al-Hasan, 'Engineering in Arabic-Islamic civilisation' (2011), 3–7, 10–11, http://www.history-science-technology.com/articles/articles%2011.html; H Wulff, *The traditional crafts of Persia* (1966), 244–51; A Levanoni, 'Water supply in medieval middle eastern cities: the case of Cairo', *al Masaq*, 20/2, 2008, 179–205.

138 Hill, 'Science and technology in ninth-century Baghdad', in Butzer, Lohrmann (eds), *Science in western and eastern civilization in Carolingian times*, 485–502; D R Hill, *Islamic science and engineering* (1993), 11–14; G Saliba, 'The function of mechanical devices in medieval Islamic society', in P Long (ed.), *Science and technology in medieval society*, 141–9.

people able to share the knowledge. If the great works of this age are all credited to the ruler who ordered them rather than the people who executed their commands, there can be no doubt of the existence of a substantial cadre of skilled engineers and technicians.

The catastrophes of the sixth and seventh centuries severely tested the fabric of the Byzantine state and society and wrought big change in the economy and population levels. But the empire succeeded in maintaining a centralised military system with engineering capacity and was by the ninth century in a more stable condition and able to consider once more expanding its frontiers.

Conclusions about western and central Europe need to be more nuanced. The question here is the extent to which the major changes that did occur (varying in degree, place and time) interrupted the continuity that is now generally accepted. What is known – chiefly from archaeology – about broader economic and social developments suggests different effects on the levels of technology and engineering surviving from the imperial Roman age during the political upheavals of the fifth century. How far did the change from urban to rural centres of power, the gradual replacement of an educated (administrative) Roman ruling class with a militarised and probably illiterate aristocracy, and changes in how state resources were mobilised impact on knowledge and skills (outside of clerical circles)? Did fewer large-scale building projects, or bridge construction, lead to a drop in the call for particular skills? The notion of a "Carolingian renaissance" only makes sense if such a falling away had taken place, but in the area of military engineering, this raises some problems that cannot easily be resolved. Charles Martel and Pippin had access to significant engineering abilities: unless it is to be argued that these were suddenly imported, they must already have been there. Diffusion of knowledge certainly took place with the widespread take up of the mangonel, for example. So it is safe to conclude that skill and knowledge transfers between Islam, Byzantium and the Franks did happen, even if the originators, routes and dates remain unknown. But for it to work there had to be skilled engineers able to make use of the knowledge. That has to be the conclusion of this study of military engineering in practice.

There were a number of preconditions for the continuation of military engineering skills: the preservation of the knowledge itself; the people able to put this into practice in the first place; and the presence of rulers who could provide the opportunities to use the skills, without which the specialisms would wither. Once lost, the knowledge might be re-acquired from book-learning, but even for the tiny minority who could access and read (say) a copy of Vegetius, turning that learning to practical use could only be something achieved under the eye of an experienced master carpenter or mason. With a few exceptions (such as – perhaps – underground mining in the west), the conclusion has to be that far from being an age of darkness, in terms of military engineering there was sufficient exchange of knowledge and sufficient skilled engineers to keep alive the tradition in the early medieval world.

One feature of this period has been the remarkable absence of reference to any of these engineers. The tiny number of names handed on reflects the prejudices of the clerical writers, and the disappearance of almost all administrative records means there is no record of these people being paid for their services. The situation begins to change only slowly in the following centuries, but evidence of engineering for warlike purposes becomes more consistently available, as will be shown in the next chapter.

3

Anonymous But Effective

The Engineers and Technicians of the Ninth to Eleventh Centuries

THE PREVIOUS CHAPTER ENDED with the re-established Roman empire of western and central Europe having to learn how to cope with the prolonged attacks of Scandinavian sea-borne assailants, while the eastern Roman empire had recovered from its own darkest days to be able to confront more effectively than for many centuries the power of the Muslim rulers of the Middle East and North Africa. During the period comprising (roughly speaking) the ninth to eleventh centuries of the Christian era, a number of new features developed that make a study of military technology and engineering particularly fascinating. These included the appearance in the Byzantine empire of a number of treatises that provide evidence of what Byzantine commanders were expected to do. The historical accounts suggest that the texts were not abstract ideals, but offered practical guidance largely followed by the generations of successful generals who flourished particularly in the tenth century. Contemporary evidence also supports the case that in the world of the caliphs (a title to which there were three simultaneous claimants by the end of the tenth century), despite political turmoil, the wealth of the rulers contributed to continuity in the level of theoretical and practical knowledge and technical skill on which military rulers could draw.

The empire established by Charlemagne was sub-divided into what would become France (ruled by Charles the Bald, 840–77) and Germany (ruled by Louis the German, 840–75), either side of a "middle kingdom" (under Lothair, who inherited the title of emperor, 840–55) originally stretching from the North Sea to Provence and encompassing northern and central Italy (subsequently divided again, with the northern parts absorbed by their eastern and western neighbours thirty years later). The Carolingian dynasties reigning in all of these territories gave way to new families, the Capetians from 987 in "France" and the Saxon dynasty from 919 in "Germany". There were simultaneously major social changes, most rapid in "France" but spreading everywhere in time, commonly associated with the term "feudalism".

One element of these new regimes was the disappearance of central authority: those who had formerly served Charlemagne as royal local officials (the counts) established their own dynasties and in turn created followings of local land-holders, who were sometimes only theoretically but sometimes also practically obliged to serve them. In due course, central power in the French realm all but disappeared and in its place the titular kings had little more power than the regional magnates who fought one another while their own social inferiors, local knights, set themselves up in small castles. Their conflicts offer what evidence there is for the level of military technology, while archaeology has helped to identify – although with difficulties in accurate dating – the skills involved in the construction of churches, cathedrals and castles by these generations of feudal lords and their dependents. Underpinning these developments was the beginning of a long period of economic growth that over time closed the gap between the more prosperous European lands and their eastern neighbours and eventually saw the creation of equally wealthy and well-populated states in the following centuries.

BYZANTINE THEORY AND PRACTICE

The military engineering employed by Byzantine armies, and for which the right practice had been spelt out in historic texts such as *Strategikon*, was rehearsed and developed to a new offensive approach in new texts that now appeared. Analysis has suggested that in starting with ancient practice, these writers updated it and advised commanders to rely on their own experience and the guidance of experienced engineers.[1] Earliest among these texts was the *Taktika*, attributed to the emperor Leo VI (886–912). Leo's work repeated the formulas set out by Maurice in describing how to bridge a river in enemy territory, using either wooden beams to create the roadway or small boats. He also recommended erecting towers to protect the end of the bridge, using timber, earth or dry stone work. When it came to the role of the surveyors, once again Maurice's advice was repeated about how to lay out and build a camp for the army on campaign.[2] A highly detailed specification for the work of the surveyors, with precise measurements for each section of a camp and instructions for how to dig the surrounding ditch and erect a rampart, how to assign the different parts of the camp to each part of an expeditionary army, and how to design the entrances, along with detailed drawings to illustrate what was required, was covered in a text written about a century later, author unknown, which focussed on warfare in the Balkans, where the Byzantine empire now faced

1 E.g. D Sullivan, 'Tenth-century Byzantine offensive siege warfare: instructional prescriptions and historical practice', in J Haldon (ed.), *Byzantine warfare*, 497–518 (originally published 1997); and E McGeer, 'Byzantine siege warfare in theory and practice', in I Curtis, M Wolfe (eds), *The medieval city under siege* (1995), 123–8.

2 *The Taktika of Emperor Leo VI, text, translation and commentary*, ed. G T Dennis (2010), 395 (bridges), 55, 243 (surveying).

a neighbour carved out of former Roman territory by the Bulgars.[3] That surveying remained part of the skills widely required across the empire was demonstrated by its use in measuring land for taxation purposes, the subject of texts containing treatises on how to achieve this, which implied again that surveying practice was a continuing skill and that it was linked to an understanding of at least basic geometry.[4]

Another anonymous text, the purpose of which was to teach how to deal with the fluid and smaller-scale warfare that became endemic on the frontier between largely Byzantine Anatolia and largely Muslim-controlled Syria and Mesopotamia during the mid tenth century, incidentally offered confirmation that the Muslim armies followed the same standard practices as the Byzantine in their operations. In the book now called *On Skirmishing* is the sentence "the enemy usually sends an advance party of troops, whom the Romans ... call mensuratores ... ahead to the site to get the camp arranged for them".[5]

The best surviving manuscript collection of Byzantine military treatises was prepared on the orders of Emperor Constantine IX Porphyrogennetos (913–59) in the middle of the tenth century. The fact that the emperor felt it was important to gather together the substantial body of such texts, and to state that his generals were expected to use them, is powerful evidence that they were seen as important. These works were practical training manuals that required study by those entrusted with command of the empire's armed forces. They suggest a different truth far from the populist image of the medieval soldier: and it will be shown that the training was followed in the field. The texts themselves lay out a whole range of practical engineering tasks without specifying – with the exception only of the surveyors (using the Latin *mensuratores*) – who was responsible for executing them. If the commanding generals were themselves familiar with the principles, they relied on the engineers knowing exactly how to construct a bridge, or of what size to build a piece of artillery, or to give instructions as to how long a beam should be prepared to construct a battering ram.

Two other texts also give important information about the state of knowledge then available, as well as clues about current Byzantine practice. The name of the

3 Anon, 'Campaign organisation and tactics', in G T Dennis (ed./trans.), *Three Byzantine military treatises*, 247–61. The text is sometimes ascribed to Nikephorous Ouranos, a contemporary of Emperor Basil II, and dated to the 990s.

4 J Lefort et al., *Géométries du fisc byzantin* (1991), reproduced a probably tenth-century text entitled "Method of Geometry" describing how to measure triangles, circles and irregular shapes using two instruments of ancient origin and emphasising the importance of accurate measurements. However, the editors point out that the arithmetic needed was elementary (38–9, 218–19, 243–6).

5 Anon, 'On Skirmishing', in G T Dennis (ed./trans.) *Three Byzantine military treatises* (1985), 189.

author of these two further works, the *Parangelmata Poliorcetica* and the *Geodosia*, dated to 938 on the basis of the astronomical data used in the text, has been lost, but because he based much of his writing on the work of a classical writer of the same name, he is now known as Heron of Byzantium. He may or may not have had proven experience like other named authors, but overall his manuals suggest he knew the subject well.[6]

"Heron" claimed to be reproducing the advice of the ancient authorities Apollodorus of Damascus, Athanaeus Mechanicus, Philo and Biton, some dating back to the third century BC, but he had recast the language so that it was accessible to modern military commanders who were not "engineers" – the language reflected that used to describe the "architects" and "master builders" last seen in the building work under Justinian rather than the modern sense of the term, but his suggestion that the construction methods explained by "Heron" could have been carried out by "anyone" was surely rhetorical. The texts contained drawings that had been altered from those of the originals "Heron" used; they were designed to be more "useful for construction and carpentry".[7] The word this author used for contemporary craftsmen was the same as will be seen employed by other writers to describe the people who actually built siege engines (*tekhnitai*).

The first text explained the construction and use of a wide range of equipment for siege warfare, with particular attention paid to various kinds of "tortoises" (*chelonen*) – for covering excavation work, for holding battering rams, as protection for soldiers advancing towards the defences, for covering the filling of ditches and so on. Also covered were the construction of siege towers, rafts designed to cross defended water, "drop-bridges" and various types of ladder both for scaling and for erecting an observation position to allow occupants to see over walls. The method of construction was laid out with detailed information on dimensions for each part, with a clear understanding of how to secure the stability of the resulting structure and its resistance to enemy attempts to destroy or burn it, while ensuring it was not so heavy as to be immobile. Particularly intriguing were the instructions for building a siege tower of multiple storeys, with successive storeys above the first decreasing in height and width proportionately. Here, the author did not simply reproduce the words of the ancient forebear (in this case Apollodorus) but, as he stated, reduced the overall height from the original to a mere 20.5 metres on a base 5 metres square (16 *podes* – the Byzantine foot of 31.23 centimetres).[8] The instructions for constructing battering rams were equally specific, and the resulting weapon would have stood 24 *podes* long and created a protected space

6 "The work of the Byzantine 'Heron' ... presents a clearly 10th-century approach to poliorcetic instruction": Sullivan, 'Tenth century Byzantine offensive siege warfare', 512.

7 D Sullivan (ed./trans.), *Siegecraft. Two tenth-century instructional manuals by "Heron of Byzantium"* (2000), c. 1, 26–7.

8 *Siegecraft*, cc. 30–7, 75–81.

using inward-facing beams of the same length to enable crews to operate freely and to swing the ram, which "gathers more momentum … (and) renders a strong blow" (c. 22).[9] The text was summed up with the words "and it is clear that the ancient engineers (*mechanikoi*) and the very learned master builders (*architektones*) carried out the construction of machines scientifically and rationally" (c. 38).[10] The practical applicability of this scientific knowledge possessed by the ancients had been asserted early on when "Heron" wrote praising "… mechanics, which surpasses teaching by argument through actions", with reference here to the construction of artillery, although this work spent little time on that subject and it is open to question whether its descriptions of torsion-powered bolt throwers were nothing more than a reproduction here of the classical examples (in this case from Heron of Alexandria of the first century AD) rather than contemporary practice.[11]

The second text was devoted to the use of the ancient version of the theodolite, the dioptra.[12] *Geodesia* began with the words that anyone intending to begin a siege "must not miscalculate in the measurement of walls from afar, of intervals from distances, and of widths of rivers … must make their investigation … trained in expertise of mathematical figures". "Heron" pointed out that "many have got it wrong, led astray by perception and estimation" (c. 1). In the same way that the *Parangelmata Poliorcetica* began, the author explained that he would simplify the theory presented by the ancient author (again, Heron of Alexandria) and make it easy to understand for "shrewd men", "especially for those who have in any way studied geometry".[13] "Heron" stated that generals of his time were not expected to match the theoretical understanding of mathematics possessed by engineers of previous ages, but they *were* expected to have a considerable range of practical skills based on formal education in geometry and measurement rather than knowledge acquired "on the job".

The texts were clear that the work was to be carried out by craftsmen who might be described in modern usage as technicians, but what is missing is the link: was the general, who was advised here that he should possess the knowledge of how tall each storey of a siege tower should be, really the person who stood over the carpenters while they measured out the timbers for each construction? Had he personally set up the dioptra to calculate the height of the town wall and thus decided how tall his towers needed to be? Leo's *Taktika* specified that the commander should defer to the advice of the engineers (*manganarioi*).

9 *Siegecraft*, 57–61.

10 *Siegecraft*, 83.

11 *Siegecraft*, c. 3, 33; c. 45, 93.

12 The dioptra was an ancient Greek device for measuring precisely the angular distance between two points. There are many descriptions taken from Greek texts (such as that of the original Heron) on the internet.

13 *Siegecraft*, 115–17.

The late seventh century had seen the emergence of what became the first Bulgarian empire in lands south of the Danube, previously part of the Byzantine realm. From modest beginnings it consolidated and expanded across the Balkans, and subsequently its ruler, Krum (803–14), was crowned as a rival emperor. In 808–09, Krum captured the Byzantine city of Sardica (modern Sofia), and the name of Eumathios was recorded, "who had experience with engines". Eumathios fled to the Bulgarian ruler and provided him with the apparently lacking expertise. The historian whose account provided the information was hostile to the emperor Nikephoros I (802–11), and blamed him for enabling the Bulgarian to humiliate east Rome. One way to present his culpability while preserving the idea of the cultural superiority of Constantinople was to suggest that the enemy could only have captured imperial territory with the help of a turncoat driven out by imperial stupidity. The story does not end there: in 811, having famously ambushed and slaughtered Nikephoros and his army as they withdrew from an attack on his capital (Pliska), Krum assaulted the important Byzantine Black Sea fortress city of Mesembria "with siege engines and *helepoleis* of which he had learned thanks to the folly of Nikephoros". The Byzantine emperor had enlisted a "certain Arab" who was "very experienced with engines", but had then mistreated him so that he fled and taught the Bulgars "the requisite skills for siege engines" with which they captured the city and – to compound the offence – found in its arsenal (it was a base for the fleet) thirty-six bronze siphons and "the liquid fire", which, it seems, they did not know how to use.

It is possible that Eumathios and the "Arab" were the same person, or else it might be observed that to lose one engineer to a dangerous enemy was careless, but to lose two was indeed folly. Scholars of Bulgarian history believe that contrary to the images presented by Byzantine writers, who were propagandists and invariably members of, or attached to, the ruling elite, substantial numbers of Byzantine subjects may actually have chosen to live under Krum's rule and might have helped him develop a technical resource previously lacking.[14] In this scenario, Krum and his successors would not have needed to rely for long on enlisting "traitorous" experts. In addition to the two (?) highly skilled military engineers, other sources refer to a whole group of Byzantine subjects who had been captured and had chosen to remain rather than be ransomed, including high-ranking officers as well as craftsmen.

These events – as well as offering a rare name – confirm that there was a cohort of experts in siege artillery serving the empire, and that it was drawn from diverse ethnic origins. Nothing explains how Eumathios became such an expert, or his

14 Theophanes, *Chronicle*, ed. Mango and Scott, 682. See also Treadgold, *Byzantine revival*, 184–5, and P Sophoulis, *Byzantium and Bulgaria 775–831* (2012), 221–3; R Browning, 'Byzantines in Bulgaria – late 8th to early 9th centuries', *Studia Slavico-Byzantina et medievalia Europensia. In memoriam Ivan Dujčev*, I, 1988 (32–6).

social origins, but there is already hinted at here the emergence of what would become a common theme, of the experts whose skills were sufficiently in demand to recommend them to rulers anywhere. Whether they did acquire the knowledge that "Heron" said was required or managed without it remains unknown.

No engineers' names were given in other accounts of operations where Byzantine armies engaged in siege warfare of Muslim or Bulgarian-held cities and fortresses, but the frequent references to the use of siege engines suggests it is unlikely that Eumathios was alone during these centuries. These events were numerous, with varying amounts of detail recording attacks with siege engines including rams and stone-throwing artillery, the construction and destruction of temporary forts and permanent city defences, operations on land and sea, defeats such as the capture of Crete (824) by Muslim forces fleeing the new regime in al-Andalus and Sicily (by 878) by forces from North Africa, and victories such as the capture of frontier cities in the Middle East.[15] For our purpose, it is more fruitful to examine related documents and a detailed history from the middle of the tenth century, from which date lists of the forces and materials assigned to various expeditions, including one intended to recover Crete and preserved in Constantine Porphyrogennetos's book *On Ceremonies*, and the unusually detailed accounts of the successful reconquest carried out by future emperor Nikephoros Phokas in 960–61.

On Ceremonies has been the subject of extensive study because, alongside detailed information about the working of the Byzantine court and government, and possibly the result of later interpolation, are documents that provide, in a level of detail unrivalled for centuries, accounts of preparations for an expedition (which may have targeted Crete) planned under Leo VI in 911, a campaign in Lombardy organised in 935, when Romanus Lecapenus (920–44) held the throne, and an attempt to recover Crete in 949, when Constantine IX was back in charge.[16] This ended in catastrophic failure due to the carelessness of its commander, Gongyles, who failed to protect his encampment and was the victim of a successful surprise onslaught by the Cretans: a story that gave powerful support to what the textbooks said he should have done to protect his camp.

Detailed instructions were given to commanders across the empire to bring together not just soldiers and sailors (the latter required in large numbers) but also immense – and precisely specified – quantities of provisions and hardware, down to the numbers of nails (of different sizes), hammers and crowbars that were to be provided. A large fleet was required and it was clear that workers able to

15 See the descriptions and analysis in Purton, *HEMS*, 93–104.

16 For the text in English: Constantine Porphyrogennetos, *The book of ceremonies*, trans. A Moffat, M Tall; for the text of the parts discussed here and detailed commentary, J F Haldon, 'Theory and practice in tenth century military administration. Chapters II, 44 and 45 of the Book of Ceremonies', *Travaux et mémoires* 13, 2000 (201–352). Both present the Greek original alongside a translation.

prepare large numbers of ships of specified size and type were essential, along with an even longer list of the materials needed to equip each one of the ship types.[17] The number of craftsmen called upon in every province that contributed must have been substantial. However, there was no separate listing for anyone who might be seen as an engineer among the more than 40,000 men recorded for the 911 expedition. But the list for the Italian campaign of 935 included the pay for the "Armenians who volunteered with the 36 engineers", where the term used was translated as "mechanics". The same word appeared in the 949 list where it was included alongside the Rus' (the Varangians serving the empire) and the Churchmen. The wording and location in the list therefore suggested that engineers were regarded as a distinct element from the soldiers of the regular army, though not of high status.

This text also included a tantalising section for students of medieval artillery: the names of three (or four – one of the names might have been generic) types of weapon and the equipment for them. The most likely explanation of what they were is that they represented different types of stone thrower, in which the name reflected a weapon's appearance: for example, it is probable that the *labdaria* used an L-shaped frame (the name suggests the letter *lambda*), while on a similar interpretation the *tetrarea* was a stone thrower using a foursquare wooden framework. One of them was possibly a large ballista-type bolt thrower. Alongside the artillery were instructions to convey a wooden siege tower (a *xylopyrgos*), a "tortoise" (*testudo*, the Latin form) with fifteen large iron rams, and a large amount of material needed to make the siege equipment function, such as pulleys, ropes, strings and nails. There is no doubt that the people for whom the instruction was compiled knew what was required, but what was not revealed was whether the weapons – only four of each of the artillery were specified, so it was not a very large battery – had been stored in arsenals or constructed anew. It may be more probable that they were already in storage: the quantity of arms called for in this (as well as the previous) operations implies strongly that there was continuing arms manufacture taking place in the empire, and that the weapons were centrally stored. Evidence for the continued functioning of an arsenal at Constantinople (last noticed three centuries before) under the direction of an important official suggests the continuation of a skilled group of artisans throughout the empire's history to this point.

Finally, these lists also confirmed the central manufacture for the navy of Greek Fire, as they identified the number of siphons needed for each of the ships as well as the materials required to service them (10,000 pounds of pitch and 200 pounds of tin provided for the "metal caster Michael for soldering the [parts of the] siphons together"[18]). It can also be deduced that there were specialists operating

17 See below for more on Byzantine shipbuilding.

18 Moffat and Tall edition, 651–59, 664–75. Haldon edition, 210–15, 222–7 (text); 268–78 (commentary, weapons), 291–2 (weapons production).

the equipment: the list included corselets for the "siphonators". Otherwise, there is no evidence to confirm the skills needed to pre-heat the fuel for these deadly but short-range weapons (if that process was necessary) and to aim and shoot the device. Greek Fire operators therefore join the list of experts whose professional expertise proved vital to the success of Byzantine military (and naval) operations. It is likely that both the constructors of the devices and their operators in battle must have received training, but there is nothing to tell us how. The method of manufacture of the siphon and/or of the mixture of fuel, and/or of the method of projection and/or ignition, seem to have been maintained as a state secret, quite remarkably, for centuries, and the logical conclusion is that a small number of experienced operatives were responsible for training their successors.[19] Incendiary weapons were common already in countries where oil was readily available on the surface, but none of the various incendiary grenades very often described as Greek Fire across many centuries were the same as the original.

BYZANTINE RECOVERY OF CRETE 960–61 AND THE EVIDENCE OF SUBSEQUENT OPERATIONS

In 960, a large Byzantine fleet landed an army led by a senior military commander, Nikephoros Phokas, on Crete and in March 961 the main city of Chandax (Candia, Heraklion, now Irakleio) was captured and the island returned to the empire's control.[20] The campaign was reported in many histories and in the most detail by Leo the Deacon, whose work was composed perhaps forty years on. The author would have been about ten years old when the news of the conquest was brought to Constantinople.

This time, the errors of the previous expedition were not repeated. The fleet – which had included "fire-bearing" *dromons* (warships armed with weapons for Greek Fire) – was found a secure anchorage and the army, having dispersed Cretan opposition on landing (or landing unopposed, according to other accounts),

19 J Haldon, M Byrne, 'A possible solution to the problem of Greek Fire', *Byzantinische Zeitschrift* 70, 1977, 91–9, rehearsed in J Haldon, A Lacey, C Hewes, 'Greek Fire revisited: recent and current research', in E Jeffreys (ed.), *Byzantine style, religion and civilization* (2006), 290–325 (here 316). "Heron" made reference in the *Parangelmata Poliorcetika* to the use of Greek Fire in siege warfare, projected from siphons mounted on ladders or siege towers, although described simply as incendiaries, and the work illustrated one being hand held on top of a scaling ladder, which must have been a terrifyingly dangerous assignment, but without indicating who may have been in charge of these operations: c. 49, *Siegecraft*, 99 and figure 22.

20 There is an authoritative modern account in V Christides, *Conquest of Crete by the Arabs (ca. 824)*, describing the reconquest at pp. 175–85.

created a well-defended siege camp around the city at a distance, it was incidentally reported, of three stadia (around 550 metres). The camp was protected by an earthwork topped by a palisade. Leo reported that siege engines (*helepolei*) were constructed by "selected craftsmen". Excavation of the old city indicated it was well fortified, with walls around 2.5 metres thick, less than Leo claimed. Artillery was used on both sides – in one case, a stone propelled from the walls fell harmlessly among Phokas and his commanders and in the other, Byzantine weapons (*petrobola*) hurled the heads of enemy dead (following the defeat of an attempt at relief) over the walls into the city to the intense dismay of the inhabitants. From both cases (if they recorded actual events), it is necessary to conclude that there must have been forward positions much closer to the city walls as no pre-gunpowder artillery was able to project missiles as far as the half a kilometre reported for the distance to the siege camp. A second position must have been constructed considerably closer to the walls in which the artillery was mounted, and the Byzantine chiefs had gathered there when they were targeted by the remarkably accurate, or extremely unlucky, Cretan artillerist. Phokas' decision to remain over the winter must have created a major challenge of provisioning, but not apparently for the inhabitants, because Leo reported that the siege ended not through hunger, but when the Byzantine army successfully sapped through the foundations of the wall, set up wooden props, heaped in kindling, then set this ablaze and brought down a section of wall between two towers at the same time that an assault was underway using the rams. The army then charged the breach, overcame the defenders massed behind, and captured the city, with Phokas intervening to prevent a total massacre.[21]

Many aspects of Leo's history seem to confirm the major part played by military engineering in the success of the campaign: the successful landing of a large army from ships on a beach (there was no available harbour), the laying out and construction of a large fortified camp, the building and operating of battering rams and stone-throwing artillery, and the undermining of a section of wall leading directly to a successful storming. However, Leo's description of the apparently critical mining copied the words used by Agathias in his description of the sieges of Gothic Cumae and other towns by the Byzantine general Narses in Italy in 552. So did Leo just make it up? The main outlines of the siege were confirmed by others: Nikephoros Phokas landed successfully on a hostile shore, besieged the major city of Crete, including staying over a winter, and conquered it. Other Byzantine historians repeat several of the details: we owe the "three stadia" to Michael Attaleiotes and the failed stone shot from Chandax to Skylitzes (both a century later). Rams and artillery featured in the account of Theodosius the Deacon; the defended siege camp was emphasised in the histories of Symeon and the continuator of Theophanes, who also indicated that the walls were

21 A Talbot, D Sullivan, *The History of Leo the Deacon* (2005), book I, cc. 3–5, book II, c. 6: pp. 59–62, 65–9, 76–8.

brought down. So whether or not Phokas did order the mining of Chandax as the prelude to victory – and in Agathias' history, the attempt failed and the siege was abandoned, unlike here – the rest of the story appears to be corroborated by other accounts, and if this is so, the various important roles of military engineers stand validated.[22] However, it is also possible that the corroborators themselves were just copying.[23]

Confirmation of the claim that the general was familiar with well-established siege techniques and commanded experts able to make them happen is found a few years later when Nikephoros was emperor, and led to the capture of Mopsuestia (on the frontier of Anatolia – now Misis or Yakapinar). He identified a weak spot in the defences and ordered his men to mine, using the proximity of the river to drown out the sound of digging and carry away the spoil. The method was the same as in Agathias' account, but the circumstances of the siege were different and were described differently.[24] There was confirmation of the technology and the engineering skills available to Byzantine commanders in the Arabic account of Ibn Miskawaih, who reported from distant Baghdad that Phokas used "moving towers" against Anarzaba (in 962), was accompanied by a large number of artisans (his 30,000 is certainly a big exaggeration) in his successful attempt on Aleppo, and had thousands of mules to carry the caltrops that were scattered around to protect the army's overnight camps. In 964, he tells us, the Byzantines drove "60 tunnels" under the walls of Massisah, although unsuccessfully on this occasion – one may suspect that the (far-away) writer was exaggerating for effect.[25]

Successful tenth-century Byzantine commanders were evidently familiar with both the theory and the practice of siege warfare and knew how and where to direct the efforts of their miners, artillerists and engineers to maximum effect. A few examples from the many other military operations conducted by this succession of warrior-emperors or their equally experienced commanders confirm this assertion. In 968, a siege of the great city of Antioch, begun by

22 Agathias, *Histories*, trans. Frendo, book I, pp. 16–18; M Attaleiotes, *The history*, trans. A Kaldellis, D Krallis (2012), ch. 28, pp. 406–15; J Skylitzes, *Empereurs de Constantinople*, trans. B Flusin, ed. J-C Cheynes (2003), 207, 210; Theodosio humili diacono, *Expugnatio Cretae*, Migne *PG* CXIII, cols. 1018–21; Symeon Magister, *Annales*, ed. Bekker (1838), 758–9; Theophanes Continuatus, ed. Bekker (1838), 475–8, 480–1. Advice on sapping and mining was also given in the *Parangelmata Poliorcetica, Siegecraft* cc. 13–17, 47–53, which reflects what the historians described.

23 On the reliability of medieval accounts, see the vigorous debate between Abels and Morillo ('A lying legacy?' (2005)) and B S Bachrach ('A "lying legacy" revisited' (2007)) (see bibliography). The debate focussed on Anglo-Saxon England and Francia but has wider significance for this era.

24 Talbot, Sullivan, *Leo the Deacon*, book III, cc. 10–11, pp. 101–2.

25 Miskawaih, *The concluding portion of the Experiences of the Nations*, trans. Margoliouth (1921), IV, 206–7, 210, 216.

Nikephoros but concluded by one of his generals, was described again by Leo the Deacon: having reconnoitred the wall for where it was low, the Byzantine worked out the exact height and constructed scaling ladders which were transported for a successful night-time assault: someone in the army was capable of making what was a critical calculation at a distance.[26] In the same year, the eastern Anatolian city of Manzikert was besieged and captured by undermining the walls.[27] The military successes of the Byzantine empire continued under Nikephoros' successors, with Basil II (inappropriately nicknamed the "Bulgar-slayer") being the leader who conquered the Bulgars as well as waging victorious war against the caliphs: descriptions rarely identify independently the role of engineering skills but one chronicler confirms their continuation, as at Moglena in 1016, where the foundations were undermined, caves dug and filled with inflammable material, and the walls brought down.[28]

THE WORLDS OF THE CALIPHS

One reason for the long run of Byzantine military success was the dissolution of effective central power in the Muslim caliphate. The 'Abbasid caliphs had been unable to prevent autonomous rule in many of the more distant regions and over a period, the caliphs themselves had replaced their native army with foreign mercenary forces, including many Turks from central Asia. This began in the form of slave armies, but before long their leaders were the de facto rulers of the lands and took control of the major cities. Subsequent struggles were as often between these new local rulers as against outside opponents.

In al-Andalus, the dynasty created by the last of the Ummayads had long ruled their own realm as a rival caliphate, while a new dynasty, the Fātimids, seized power in Egypt and North Africa during the tenth century and established a third caliphate. Numerous chronicle accounts confirm that engineering skills both in building and in military operations survived without any loss of continuity, even if the details are often lacking, and record mining as a common means of attack where circumstances called for it. Certainly, it was evident that access to great wealth went hand-in-hand with being able to command the necessary resources for building projects and military success, including being able to employ engineering and building experts. There is a rare detail provided by one chronicler about the resources available to the caliph 'Abd al-Rahmān III of al Andalus (912–61) in 936, when he sent his leading architect with 30 masons, 10

26 Talbot, Sullivan, *Leo the Deacon*, book V, c. 4, p.133.

27 Des Stephanos von Taron, *Armenische Geschichte*, trans. Gelzer, book 3, c. 8, p.134.

28 Cedrenus, 'Historiarum Compendium', Migne *PG* 122, col. 195.

carpenters and 15 diggers in response to a letter from a local commander asking for expert help in constructing a fort.[29]

At the eastern end of the Muslim world, meanwhile, a new dynasty had established itself under one of the Turkish newcomers, Sebüktegin, in what is now Afghanistan, and his son Maḥmūd of Ghazni (977–1030) extended his domains into northern India. Great skills in siege warfare were necessary, effective artillery and a role for mining featuring in accounts of his campaigns. A fortress in Herat was captured following deployment of "great and small slinging machines", and in 1037, Sultan Mas'ūd I (1030–40) captured Hánsi (Haryana, north-west of Delhi), bringing down the walls in five places with successful undermining. This sultan seems to have been particularly qualified in geometry, if his admirers are credible, drawing the plans for his buildings "with his own exalted hand".[30] An analysis of the Ghaznavid army suggests that in addition to its main component – mounted warriors – it comprised elements drawn from each of its subject peoples, that Persian bureaucrats provided logistical organisation and that siege experts from Ghur were in charge of artillery. A contemporary description of India confirmed that a high level of culture and science continued to flourish, including knowledge of "mathematics, geodesy, geometry, astronomy", and it had inventive designers of machines and weapons. So, in the skills available from all the lands between Iran and Delhi, the rulers of this realm had access to expertise, although how far this could be recruited for military purposes is unknown.[31] One of the scholars who flourished at the Ghaznavid court was the court astrologer and counsellor to the sultan known as al-Bīrūnī. His text (c.1020) on the accurate measurement of distances between cities, using astronomical measurements, demonstrated again how practical use could be made of theoretical learning.[32]

Meanwhile, returning to caliphal Baghdad, incidental evidence of the continuation of engineering skills exists. When in 961, under Caliph Mu'izz, a new palace was built in eastern Baghdad, "masons famed for their skill [were] fetched from far and wide". On an internal expedition downriver towards Basra in 964,

29 Ibn Hayyan, *Cronica del Califa 'Abdarrahman III An-Nasir 912–942 (al-Mutabis V)*, trans. Viguera, Corriente (1987), 290.

30 Quoted in G Necipoğlu, *The Topkapi scroll: geometry and ornament in Islamic architecture* (1995), 4.

31 On the Ghaznavids, chronicles by Al-Utbī, *Kitab-i-Yamini. Historical memoirs of the Amir Subaktagin and the Sultan Mahmud of Ghazna*, trans Reynolds, 56, 89, 96, 384; and Abu'l Fazl al Baihaki, 'Tarikhu's Subuktegin', in Dawson, *History of India*, II, 140 (Hánsi); C Bosworth, 'Ghaznevid military organisation', *Der Islam* 36, 1960 (37–69). On Indian knowledge: V Minorsky, 'Gardizi on India', *Bulletin of the School of Oriental and African Studies* XII, 1948, 628–9.

32 *The determination of coordinates of positions for the correction of distances between cities. Translation of al-Bīrūnī's Kitāb taḥdīd Nihāyāt al-Amākin Litaṣḥīḥ Masāfāt al-Masālin*, trans. Ali (1967), with a description of the measuring tools at 37–41.

the bridge at the capital was physically moved and relocated so the army could cross to the west bank, and in 979, following revolts and riots that had wrecked the city, repair and restoration included widening the main bridge by building new barges to provide the pontoons. The army decided in 971 to dam canals and construct a long dyke towards an opposing fortress which was protected by marshes (albeit unsuccessfully on this occasion because of unexpected flooding), while an incidental reference in the account of a conspiracy aimed at the ruler of Kirman identified one of the conspirators as "an engineer who had been in ['Ali ibn Alyas'] employ, named Mazuban".

Eventually, after a period of tumult, the emir 'Adud al-Daulah restored order; he reinstated the scientific and philosophical facilities in the capital and these included "ample allowances given to ... arithmeticians and engineers". One large dam constructed under his rule survived to modern times over the river Kurr near Istakr, 30 feet (9 metres) high and 250 feet (77 metres) long, built of masonry with a rubble core.[33] Among those who had flourished despite the rapid changes of political fortune was the great mathematician Ibrāhīm Ibn Sinān (909–46), who established a school of applied maths which had many practical uses. His now lost instruction manual on the construction of instruments was emblematic of the continued tradition of exploiting theoretical learning for practical purposes.[34] The surviving practical manuals across different disciplines included a treatise of the eleventh century written for government administrators explaining assessments of taxation, designs for machines to lift water for irrigation (such as the Noria and the Shadūf), methods for surveying the route of new canals, including descriptions of the measuring tools – the alidade and the astrolabe among them – and the dimensions of dykes, and other easily transferable skills. The book of Mohammad al Karagi, written in 1017, was a guide to how to find water underground in the first place, requiring knowledge of triangulation, algebra and the construction of aqueducts. The astrolabe was, of course, of ancient (Hellenic) origin, but in Muslim hands it was perfected [Fig. 7].[35] Rulers were expected to concern themselves with such matters (even if they paid others to execute them), and advice offered to Seljuk sultans pointed out their responsibility for "preparing underground watercourses for irrigation, digging canals, building bridges over great rivers and so on, including

33 Miskaiwah, *The concluding portion of the Experiences of the Nations*, IV, 198–9, 219, 268, 317–18, 445–7; Hall and al-Hassan, *Engineering in Arabic-Islamic civilisation*, 10–11.

34 R Rashed, H Bellosta, *Ibrāhīm Ibn Sinān et géometrie au Xe siècle* (2000), 10–11, 245, 291.

35 Hill and al-Hasan, *Engineering*, 18–20; Wulff, *Traditional crafts of Persia*, 246–8; C Cahen, 'Le service de l'irrigation en Iraq au début du XIe siècle', *Bulletin des études orientales* XIII, 1949–51, 117–43 for a French translation of the Kītab al-Hāwī li'l-a'māl as-Sultāniya wa rusūm al-hisāb ad-dīuāniya; M al Karagi, *La civilisation des eaux cachés*, ed./trans. A Mazaheri (1973), was rooted in ancient Persian experience.

FIG. 7 Thirteenth-century precision engineering: astrolabe with *qibla*, c.1230, made in Damascus, at the National Maritime Museum, Greenwich, London. (Image: National Maritime Museum, Greenwich, London)

building fortifications".[36] Skills in astronomy and measurement also found practical application in cartography and navigation, as this period witnessed the first Arab journeys to the Far East and the production of maps and advice for the travellers[37].

THE SITUATION IN CENTRAL AND WESTERN EUROPE

How does this picture compare with what was happening in Latin Christendom? King Charles the Bald's defences on the river Seine did not prevent a Danish army from reaching, and besieging, Paris in 885. An eye-witness account was written a decade later by Abbo, although considerable caution needs to be used in interpreting it. The siege appears to have been of the Roman-walled Île de la Cité (although there were already suburbs beyond the river) and of two bridges connecting it to each bank, with defences consisting of towers at the ends,

36 Niẓām al-Mulk, 'Sigāsat-nāma', in B Lewis, *Islam from the Prophet Muhammad to the fall of Constantinople* (1987), I, 181–3.

37 S Nasr, *Science and civilisation in Islam* (1968), 99–100.

probably constructed of timber on stone bases. The defenders deployed liquefied incendiary devices to set fire to the Danes as they attacked the tower at the end of the Grand Pont and there are several references to their use of ballistas. The Danes, having failed to win the city by an initial assault, settled in for a longer haul and began to construct giant roofed battering rams on sixteen wheels. Shots from the defenders' ballistas killed the two people in charge of the construction, indicating that the Danes were using engineers, since their deaths apparently ended this form of attack. This attack on the tower did not succeed, and the defenders deployed a piece of artillery called a *mangana* to throw "enormous stones" causing mayhem among the assailants, who abandoned their rams. Eventually, after a siege that had lasted from late November to March, Paris was relieved. This summary highlights the military engineering skills on both sides, and while the stone thrower has been endlessly analysed, let it suffice here to say that the description provided by Abbo suggests strongly that this was a lever-type manually propelled mangonel.[38]

A politically confused period, when the crown oscillated between rivals, marked the high politics of western Francia in the decades that followed. Carolingian kings were restored, and in 938 Louis IV (936–54) besieged Count Heribert in his newly constructed citadel (*arx*) at Laon. "The wall was undermined and broken by many machines and the citadel was taken by great effort", according to Flodoard of Reims, and this terse description suggested a role for engineers in the use of siege engines (it is not clear whether this meant artillery or rams) and by mining. The later account by Richer suggested the engines included mantlets constructed of timber mounted on four wheels, with sloping roofs of wicker, which were pushed up against the walls and, under the protective shooting of archers, enabled the wall to be sapped. The significance of this report is that it represented the first unequivocal reference to the use of mining or sapping for military purposes in western Europe since late antiquity.[39] Many other attacks on and defences of fortifications were taking place every year but little detail was recorded. It is therefore not possible to affirm what level of skill was deployed by the princes, bishops and lords who were involved and what resource in engineering skill they could draw upon, except that expertise seems to have been available: in a rare further example providing detail, Richer noted that another royal attack on Laon ten years later was abandoned because it was winter, and there was no time left to construct siege engines. This also confirmed that engines were built according to need rather than stored in an arsenal.[40]

38 Abbo, *Le siège de Paris par les Normands*, ed./trans. H Waquet (1942), 20–2, 42. See Purton, *HEMS*, 83–5.

39 Flodoard, *Annals 919–966*, ed. S Fanning, trans. B Bachrach (2004), ch. 20, p. 30; Richer of Saint-Rémi, *Histories*, ed./trans. J Lake (2011), book. II, c. 10, p. 184.

40 Richer, *Histories*, book II. c. 84, p. 348.

The next surviving detailed account of siege equipment is from a half-century later. This related to the siege of Verdun in 984 when Louis's son Lothair (954–86) was king. The city surrendered to his array of siege engines. Duke Thierry of Upper Lorraine having reoccupied it, Lothair returned to besiege it again the following year. This time, we are told that he cut down large oaks to construct machinery. Four beams 30 feet (9.2 metres) long were placed on the ground to form a lattice, and at each corner pulleys were used to erect a frame 40 feet high, secured together by 10-foot (3-metre) logs. Diagonal bracing was applied and planks were laid across the lattice. Logs were sunk deep in the ground with their tops protruding to enable the installation of pulleys, so that oxen could propel the tower forwards while they were driven in the opposite direction (meaning they were less vulnerable to being killed by the enemy shooting at them). Progress was achieved by placing rollers in front. The enemy constructed a smaller engine and a fierce fight ensued, in which grappling hooks enabled the king's men to pull over the other tower and achieve capture. The chronicler also provided details of the construction work carried out by Hugh, first of the Capetian kings (987–96), at his siege of Laon in 988. A ram (*aries*) was constructed of four beams laid out as a rectangle with cross-beams. Ropes were fastened to the tops of poles and hung down to support the ram, which was fitted with an iron-plated head. Ropes were also suspended at the middle and the other end of the beam, enabling a large team of men to work it. It was placed on three wheels. The triangular-shaped ram was described by the chronicler as ideally suited for breaking down walls. Unfortunately, the attackers had failed to take account of the fact that the hilltop location of the citadel made it impossible to bring the engine up close, so it and the siege were abandoned.[41] This story is strange: Laon had often been besieged before (see p. 91) so its site would have been familiar. The truth cannot now be recovered.

The dimensions of King Lothair's tower demonstrate that very considerable skills were required to achieve the result, which seems to have actually worked to great effect. The dimensions given by Richer were not the same as those "Heron" had derived from classical proportions for Byzantine commanders to follow, ruling this out as a source. The question of who the builders might have been remains without a definitive answer, despite some much later naming of possible candidates. The twelfth-century chronicler-monk Orderic Vitalis named Ivo de Creuil as the *regis balistarium* of King Louis IV of France in 943, a name transposed to Yvo de Bellême in additions to another later chronicle. The family of Bellême would become important in Normandy in the eleventh century, which explains the repeating of the citation, but leaves unanswered what the title means. There are two obvious translations: master of the crossbows or of the royal artillery, and no

41 Richer, *Histories*, book III, c. 101, p. 174; book III, cc. 103–7, pp. 178–88 (Verdun); book IV, c. 22, pp. 240–2 (Laon).

way now to decide between them. But it is clear that the king did have artillery, and it is unlikely that there was not a master in charge of it.[42]

Meanwhile, elsewhere in the former Carolingian empire, rivals sought to win the crowns of its former parts. The capture of the strong city of Bergamo by the German king Arnulf with a south German army and in alliance with the future king of Italy, Berengar, in 894 had two contemporary descriptions that led to the conclusion that the attackers organised to scale the walls but in fact brought them down by sapping away the foundations from the town ditch, resulting in a successful storming inside two days.[43]

The first half of the tenth century in (modern western) Germany and other distant targets was marked by a long period of raiding by the nomadic Magyars, leading to much fortress-building by local potentates urged on by the kings Henry I (919–36) and then the first of three named Otto, who became in turn Holy Roman Emperor. At least some of these monarchs had benefitted from the kind of education urged by Charlemagne and Louis the Pious. It is no surprise that there are references to a range of engineering skills being put to use. In 955, for example, when Otto I (936–73) was campaigning against a Slav prince, his army was transferred across the river by the rapid construction of three bridges. The implication is that they must have been prefabricated as laid down in previous (Carolingian) regulations, because of the speed of the action in the face of an enemy. This army had people with bridging skills.[44]

The chronicle accounts of operations where siege machinery was deployed confirm the presence of those able to build, maintain and operate stone throwers and other constructions. Emperor Otto III (983–1002) led an army to Rome and in 998 besieged his opponent Crescentius in the Castel Sant'Angelo, built on Hadrian's mausoleum on the banks of the Tiber. Otto ordered the construction of a siege tower made of fir, which was pushed up against the walls at the same time as soldiers attempted to break in through the doors.[45] In 1001, he was attacking Tivoli, which was well protected by wet ditches, and prepared all kinds of siege engines. This siege also stood out as an example of prelates contributing their learned skills to the cause of the secular power – here it was Bishop Bernward of Hildesheim, one of a considerable number of multi-talented clerics. Bernward's biographer

42 M Chibnall (ed./trans), *The ecclesiastical history of Orderic Vitalis* (1969–90), book VI, vol. III, 306–7; discussed in G White, 'The first house of Bellême', *Transactions Royal History Society* XXII, 1940, 67–99, at 68.

43 *The annals of Fulda*, trans. T Reuter (1992), 126–7; 'Gesta Berengarii Imperatoris', *MGH* Poeta Latinorum, IV (1), book III, lines 94–103; J Jarnut, 'La conquista di Bergamo nell'894', *Archivio storico Bergamesco* I, 1981, 26–7.

44 Widukind, *Res Gestae Saxonicae*, book III, chap. 54.

45 'Annales Quedlinburgenses', *MGH* SS III, p.74; Glaber, *Historiarum libri quinque*, ed. France, book I, 12, 25–7;

praised his study of things mechanical and his building skills[46] (see below). A few years later, under Henry II (1002–24), there was conflict in the archbishopric of Trier, a city with standing Roman defences and an important bridge over the river Mosel. The candidate for the bishopric supported by Henry had to besiege his rival (who erected towers, presumably timber defences, on the bridge) inside the episcopal palace from 4 April to 1 September 1008. Henry ordered the construction of engines made from the material of local houses (presumably not concerned at the response of their occupants), but the defenders were able to burn them; engineering skills were thus deployed on both sides.[47] Henry had little better success with siege weapons in Italy in 1022: although the Byzantine garrison of Troia at first succeeded in making a successful sortie leading to the destruction of his engines, he had them rebuilt stronger and covered with rawhide, suggesting in this case rams or siege towers.[48]

It seems that access to engineering skills was not limited to the highest ranks of society. In their feud Wichman and Balderich, rival noblemen with estates in the Rhineland near the (present) border with the Netherlands at the end of the tenth century, engaged in violent attacks, abductions, sieges and rapid (small) castle-building that required the intervention of the emperor. His foe having erected a threatening new earth-and-timber fortress, Balderich gathered his "friends" and vassals to attack, and to reach the target had to use *experti* to build a bridge across the waters – a project that failed because of the "immensity" of the water. He then constructed engines to be placed aboard boats, the defenders abandoning their fort. The implication was that these noblemen cannot have been alone in being able to recruit people with military engineering skills.[49] The subject of earth-and-timber private castles will also require further comment.

This brief overview confirms that the eastern successors of the Carolingian empire continued the traditions either established or sustained since the early middle ages.[50] The great majority of the copies of Vegetius were preserved in

46 'Thangmari Vita Bernwardi episcopi Hildesheimensis', *MGH* SS IV, 758, 769–70.

47 'Gesta Treverorum', *MGH* SS VIII, ch. 30, pp. 171–2.

48 Glaber, *Historiarum*, book III, ch. 4, pp. 101–3.

49 Alpertus, 'de diversitate temporum libri III', *MGH* SS IV, book II, p. 710.

50 D S Bachrach, *Warfare in tenth-century Germany* (2012) continues the Bachrach tradition of proposing unbroken continuity with skill and aplomb, but from time to time proposes details that are simply not sustained by any evidence: for example, jumping from proposing that special texts were needed by engineers and architects (citing Cassiodorus' *Variae*) to the claim that these manuscripts had been copied and in some cases updated, and implying they must have been used by the engineers, and arguing that advanced mathematics was required for surveyors (p. 107). Evidence of surveying techniques discussed in chapter two would question this conclusion. While it is evident that skills were needed to build the kind of siege tower and rams discussed here, the assertion that special training was needed is justified only by a reference to

monastic libraries, but it is recorded that copies were made for Charles the Bald (in 840), Louis the German and a few other secular lords. It is claimed that the founder of Angevin power, Count Fulk Nerra (987–1040), made use of the advice in his operations.[51] The methods and the technology available *appear* not to have changed since the days of Charlemagne, although the sources do not allow any conclusions to be drawn about the detail. Apart from the apparent precision of the description of the weapons built to capture Verdun, there is little that allows one to conclude what the engines so frequently mentioned even were, let alone work out their characteristics, how effective they were, or how or by whom they were built and operated. None of the detail offered by the Byzantine manuals is on offer – even if what was offered there was what actually happened. However, there is nothing to suggest that the level of technical skills developed by craft workers and artisans differed between societies in western or eastern Christendom or in the Muslim world. It is safe to conclude that the people needed to manufacture equipment could be identified by rulers anywhere. What cannot be known is where the engineer was found who could instruct the carpenters or masons or metalworkers about what was required. The conclusion is that such people existed, still largely anonymous but definitely regularly employed by those with the wealth to sustain expenditure on warfare.

Scientific learning had not disappeared despite the Viking, Magyar and Muslim onslaughts and indeed witnessed new progress. One leading scientist's work was not purely academic. Gerbert of Aurillac was taken up by the count of Barcelona in 967, sent by the pope to educate the future Otto II and became Pope Sylvester II (999–1003). He became an expert in many areas including the abacus, a method of calculation of massive significance for many engineering needs which had previously been little known or understood in the west. Among his letters are several to pupils and clerics dating from the 970s to 999 describing how to use the abacus to multiply and divide, how to make measurements, and other skills with many practical applications.[52] It is possible that a link to practicality was

the words of Rabanus Maurus; there is no reference to where this training could have been provided from (pp. 166–7). Similarly, he concluded that the reference to artillery at Trier must signify that the weapon in question was the *onager*: there is no basis for this conclusion (p. 161).

51 Allmand, *The* de re militari *of Vegetius*, 64–7; B S Bachrach, 'The practical use of Vegetius' *De re militari* during the early Middle Ages', *The Historian* XLVII/2, 1985, 239–55, argues that Fulk Nerra's known military practices seem to be based on Vegetius' teaching, but accepts this cannot be proven (250).

52 H Pratt Lattin (ed./trans.), *The letters of Gerbert with his papal privileges as Sylvester II* (1961), 2–6, 44–5, 64, 301; F Yeldham, *The story of reckoning in the Middle Ages* (1926), 27–37; P Riché, 'L'enseignement de Gerbert à Reims dans le contexte européen', in Riché, *Education et culture dans l'occident medieval* (1993), 51–68.

created in the transmission of ideas from educated princes to those responsible for engineering their instructions.[53]

The exception to the general anonymity of engineers and builders over the tenth and eleventh centuries appears with the arrival in the written evidence of clerics with an affinity for building not only churches and cathedrals but also castles and city defences, including some who could also turn a hand to siege machinery – Bernward of Hildesheim has already come to our attention. Bishops and abbots were familiar with the liberal arts because they were part of their education, and that they should have been responsible for the diffusion of knowledge and skills first observed in one location to other places was a natural corollary of their unusual opportunity to travel. That they should have been engaged in commissioning ecclesiastical projects was the result of their place in society. That they were able to undertake what amounted to a massive expansion of building work was down, ultimately, to a gradual increase in the wealth of the society in which they were dominant figures. There is universal acceptance among historians that the tenth century witnessed the beginning of a prolonged period of economic growth in Europe, represented by the expansion of areas of settlement and clearances of the vast forests, combined with population growth, in turn reflected in a shift from subsistence agriculture to the production of surpluses that were to be appropriated and spent by the landowners: the nobility and the church. A substantial proportion seems to have been expended on new building. If there had been less large-scale new building in the ninth century, this had been largely because there had not been the wealth available to pay for it, itself partly the result of long periods of external raiding that targeted centres of accumulated wealth but also inhibited long-term construction work – even the defences urgently required could not be completed in timely fashion, if Charles the Bald's activities were representative.

Two further aspects were new. One, a resurrection of an old tradition, was the extent to which the patron became the architect or master-builder for new ecclesiastical construction. It has been proposed that this happened because the number of architects (such as those responsible for Charlemagne's great construction work) had dwindled, so the clerics had to take over that role if they were to spend their new wealth on emblematic new churches, cathedrals and monasteries.[54] It has been argued that the turn of the eleventh century witnessed a step forward in building skills and architectural planning. This suggests the development of progressively better skilled craftsmen, in particular masons, able to construct the new buildings and to understand (however empirically) the technical

53 See for example T McCarthy, *Chronicles of the investiture contest* (2014), for the interest in the Quadrivium at the monastery of St Emmeram, Regensburg, over several generations of abbots (p. 15).

54 A Erlande-Brandenburg, *The cathedral builders of the Middle Ages*, trans. R Stonehewer (1995), 37–42.

demands of the stone vault. The study of ecclesiastical structures of this time seems to confirm that new engineering skills were developing from the early eleventh century.[55] The few surviving contemporary illustrations show the existing range of equipment and tools for building – ladders, scaffolding, cutting tools, scales and other weighing devices, masons' tools, smith's forges and furnaces – and images of substantial stone buildings.[56] The emergence of masons' marks at this time may be indicative perhaps of greater recognition of their role, as well as confirming their mobility between all kinds of building projects.

The second new aspect was the application of these skills to more secular ends, at the direction of the secular ruler, or the ecclesiastical in their secular role. The responsibility of bishops and abbots for the defence of their local communities was a central feature of the Ottonian and later ages in the German kingdom and owed much to the urgency of protecting the people of the realm from Magyar devastation, or for backing the drive eastwards to conquer Slav territories, as well as providing one of the few reliable resources for the crown, as the kingdom was already subdivided into large and powerful duchies with their own systems of patronage. Once the monarchs had decided that building powerful walls was a central element in their strategy, the question became who was to do it? Evidently, while the secular dukes and counts had a necessary part in building defences, on their own this was not sufficient. There was no relaxation in the secular building operations undertaken by clerics once the Magyar threat had been terminated after the battle of the Lech (955).

Traditionally, this phenomenon has been approached from the viewpoint of the bishop as architect and as part of the debate as to what that word then meant. What were the bishops described as doing? An early reference involved Bishop Theolach who from 873 restored the walls of his city, including the outworks (*propugnaculis*). In 954, Bishop Ulrich of Augsburg (lived 890–973) was credited with the construction of a *castellum* at Menchingen as well as the restoration of churches destroyed by the Magyars. Bernward of Hildesheim was responsible for a small fortress (*munitiunculam*) at the junction of the rivers Ocker and Aller, the garrison of which was to protect the local people against the ravages of pirates. Other bishops whose skills included major building projects included Burchard at Worms

55 Stalley, *Early medieval architecture*, 106–7, 130–3, 191–4; O Lehmann-Brockhaus, *Schriftquellen zur Kunstgeschichte des 11. und 12. Jahrhunderts für Deutschland, Lothringen und Italien* (1938), I, 41, 59, 136, for references in contemporary sources to the work of the 'architectus' in new churches, and under German king Henry IV to a summons for *omnes sapientes et industrios architectos, fabros et cementarios aliosque opifices regni sui …* (I, 272).

56 Such as the 1023 manuscript of Rabanus Maurus' encyclopedia published by A M Amelli (ed.), *Miniature sacre e profane dell'anno 1023 illustrati l'enciclopedia mediovale di Rabani Mauro* (1896).

(lived c.965–1025), Konrad at Constance (934–75), Meinwerk at Paderborn (1009–36) and Heribert at Cologne (999–1021). In northern Germany, Alebrand (bishop 1035–43), having reconstructed the monastery at Bremen in stone, moved on to complete the wall around the city, including a "strong tower" over the gate "fortified in the Italian way". The same source also recorded the ecclesiastical and urban defence projects undertaken jointly by the archbishop and duke in reconstructing Hamburg following its destruction in a Slav attack.[57] The most renowned came a couple of generations later: Benno (II of Osnabrück, bishop 1068–88), who served King and Emperor Henry IV (1056–1106) in his Saxon wars. His building work included churches, cathedrals, palaces and royal castles.[58]

Their "lives" record that they were well educated, including in the liberal arts, and it seems certain from its availability in their libraries that they had access to textbooks including Vitruvius' *De architectura*. Copying of manuscripts continued in the monastic schools in the German half of the empire encouraged by the kings, who were keen to secure the service of educated people. Pupils of schools at Trier, Cologne, Metz, Liège and Verdun included those who went on to become bishops and were the builders recorded in history. Whether they had responsibility for the overall concept of their building, or for overseeing its construction by master masons and carpenters, or for actually directing the work is now impossible to recover. A combination of all three is probable. Certainly, that they could turn their hand to the erection of fortresses as well as churches indicates an understanding of what a defensive structure required.[59]

Assessing these skills is hard as so little of the work remains standing, but skill there must have been, although stone walls dating from this time offer little to suggest any great level of quality. The new wall erected to protect papal Rome by Pope Leo IV in 847 following a devastating Arab raid, which still stands in part, largely reused material from other buildings and its 3-kilometre length took five years to complete, but it was of a substantially lesser quality than the city's existing Aurelian walls of the third century.[60] This suggests a shortage both of resources

57 Adam of Bremen, *History of the archbishops of Hamburg-Bremen*, trans. F Tschan (2002), book II, lxix, pp. 103–5 and book II, lxx, p. 105. On the various bishops: 'Gerhardi Vita S. Oudalrici episcopi', *MGH* SS 4, 381–428; 'Thangmari Vita Bernwardi episcopi', same vol., 754–82; 'Annales Wormatienses', *MGH* SS 17 (19), 37; G Bindung, *Der früh- und hochmittelalterliche Bauherr als sapiens archictectus* (1996), summarises what is known of the bishops' lives, education and building works.

58 G Bindung, 'Bischoff Benno II von Osnabrück als "architecto et dispositor caementarii operis, architectoriae artis valde peritus"', *Zeitschrift des deutschen Vereins für Kunstwissenschaft* 44/1, 1990 (53–66);

59 Riché, 'L'enseignement de Gerbert à Reims', as note 52. Koch, 'Vom Nachleben des Vitruv', *Deutsche Beiträge zur Altertumwissenschaft* 1, 1951, 11–14.

60 S Gibson, B Ward-Perkins, 'The surviving remains of the Leonine wall', *Papers of the British School at Rome* 47, 1979, 30–57; 51, 1983, 222–39.

and of skilled masons in central Italy in the mid ninth century, though buildings further north offer a different picture.[61]

In the following century there were many references to fortifications being constructed, destroyed or repaired. A major scholarly debate surrounds the so-called *Burgenordnung* issued (according to written sources) by Henry I in 925 calling for the fortification of numerous places to protect against the Magyars, but no one has yet been able to identify any fortifications specifically connected to the order. Where archaeologists have excavated dateable sites of the tenth century, they were largely small earthwork-and-timber defences, with royal palaces being prominent among the few constructed with stone buildings. Numerous charters and surviving letters refer to orders to secular and ecclesiastical lords to erect *castra* or *castella* or to fortify towns. Little remains of these today. More durable stone walls and towers are known from a number of sites, such as the stone wall around the monastery of St Gall erected by 950 and the royal palace at Werla in the Harz mountains. Excavations suggest that while this place was a significantly superior building to what was there already, it did not compare with the rich quality of building work being carried out by the Byzantines and the 'Abbasids at this time. Nor was there anything to distinguish most of the defences thrown up by German nobles and bishops. The Magyars, once they had settled down to impose their rule on the indigenous population in (modern) Hungary, rapidly acquired a range of engineering skills of their own, exemplified by the discovery of an extensive network of canals for irrigation in the Carpathian basin, described by its excavators as "uniformly controlled with a management system covering the whole region".[62]

The picture in western Europe was not much different. The realm now known as France was divided into major territorial units such as Flanders, the Vermandois, Anjou, Aquitaine, Burgundy, and from 911 when it was granted to Rollo the Viking by the king, Normandy, along with independent Brittany, and in the south-east the lands belonged to rival rulers. Within these territories local nobles took control of former counties and vied for regional domination. Some of them were in the lead in constructing massive stone castles as bases of their power and lordship, such as Fulk Nerra in Anjou, working in a region where the very first rectangular stone donjons were built (around the year 1000). Below them, their vassals and retainers, whose loyalty or otherwise depended on their being provided with sufficient land to support their families, began during the tenth century to consolidate their own local power bases and central to this process was the emergence of the private

61 G Brogolio, 'Architetture in pietra lungo l'Adige tra X e XI secolo', *Archeologica dell'architettura* XVII, 2012 (80–93).

62 For a summary of the evidence and examples of excavations of tenth-century central European fortifications, see Purton, *HEMS*, 133–45; Magyar canals: K Takács, Gy Füleky, 'Remains of the medieval canal systems in Carpathian basin', BAR International Series 1163 (2003), 65–77, quotation at 75.

castle. This development has long been the subject of detailed historical study and more recently archaeological investigation. But it is often very hard to date many of the thousands of candidates to be castle sites, of which the more successful went on to be rebuilt and modernised time and again.[63] Towards the end of the century, the first mottes appeared in northern France, particularly familiar to students of castles in Britain because their introduction by the Normans (including a small number of castles built before 1066 by Normans invited in by King Edward the Confessor) have a definite "not earlier than" attached to their construction. What do they reveal about engineering skills? What do we know of the method of construction?

Historians owe a debt to the labours of Victor Mortet who, in 1911, published a collection from thousands of sources of all the references to be found to building in France after the year 1000: the overwhelming majority are references to ecclesiastical building but they are a treasure trove confirming the skills, the ideas, the knowledge and the techniques of medieval builders in general.[64] The earliest references report building activities of the early eleventh century, but there is no reason to doubt that the methods and skills did not appear from nowhere, even if there was a big step forward in both the quantity and quality of building around this time. The various chronicles, biographies, charters and other surviving texts present a picture of large and frequent building works involving teams of masons, carpenters, smiths and labourers, with occasional references to the (high) level of knowledge possessed by (usually) the patron, who could be the abbot or bishop, and there are references to monk-builders as well. These reports suggest the availability of large numbers of craftsmen across the territories now making up France, while the length of time it took to complete a major building project meant that the workers there might be tied up for some years, and therefore not able to accept other work during that time.

The evidence also supports the continuation of other construction work involving particular skills, such as bridges. During the years 1033–37, Eudes, count of Blois and Tours, commissioned the construction of a bridge across the river Loire at Tours at his expense to deal with the consequences of a flood; at the same time, further south, the lord of Albi and the local bishops ordered the construction of a bridge over the Tarn at Albi "for the benefit of the inhabitants". In Languedoc, at the same time, local abbots jointly agreed to have built a bridge of stone and timber across the Herault near Montpellier.[65] Constructing even timber bridges across large and fast-flowing rivers such as those named here was not a task for someone with no experience in the operation.

63 There are numerous accounts: for a review, see Purton, *HEMS*, 145–63.

64 V Mortet, P Deschamps, *Recueil de textes rélatifs à l'histoire de l'architecture et à la condition des architectes en France au moyen âge* (1929).

65 Mortet, *Recueil*, XXVI (102), XXVII (103), XXX (109).

At this date, many of the castles mentioned were constructed of earthworks and timber, occasionally with stone buildings inside, although there are references also to stone walls and towers and also to town defences of stone. The main difficulty in analysing the level of skill required to construct castles of this period is that very few have been subject to archaeological study. Earthworks that have been identified as castles exist in their thousands across western Europe in particular. In some cases, they may never have been castles at all: perhaps a prehistoric mound, or the base of a windmill, or a natural feature that happens to resemble a "traditional" motte. Sometimes it is hard (now) to establish whether a basic enclosure inside a low bank was a farm or a castle, where the original banks would have been higher and steeper and defended with a timber palisade. Nor did even so apparently easily identifiable an erection as a motte spring fully formed into being in its now traditional form everywhere at the same time, and many different forms of "early" mottes of different shapes and different relationships with an external or attached enclosure at a lower level are known. These include the earthwork castles involved in the feud between Balderich and Wichman.[66] Study of other sites region by region confirms how little is actually known of the original history of many of them, or of the construction techniques.[67]

There are a number of technical skills that would have been as necessary for the construction of a castle as for that of a stone church. The site must have been measured or marked out before work could begin, and this may have required a surveyor to mark out the enclosure boundaries, whether this was to be rectangular or, as was usual in the case of a mound, circular. From the tenth century there are no surviving references to individual surveyors being used on the building of fortifications, although there are from earlier centuries and from later. It is evident that if a space on the ground was to have a regular shape, however, and particularly if it was to be circular, someone would need to mark out the shape before digging could begin. That surveyors carried out this work in Byzantium and in the Muslim realms, and that the many and sophisticated stone churches of this time must have been mapped out first on the ground suggests strongly that surveyors were available, although the skill required to use a rope attached to a central stick to trace a circle was minimal enough.

66 These have been closely studied by Bas Aarts: 'Motte and bailey castles of Europe', *Virtus* 14, 2007, 37–56; and 'The origin of castles in the eastern part of the delta region (NL/D) and the rise of the principalities of Guelders and Cleves', *Château Gaillard* 25, 2012, 3–16.

67 Among a large number of studies of early earthwork castles, M-P Baudry, *Châteaux 'romans' en Poitou-Charentes X–XII siècles* (2011), esp. 63–95, is a modern review of what is known (and not known) of such sites in an area where there are many potential and actual early castle sites.

Whereas the throwing up of earth banks to demarcate a space would have been commonplace in agricultural communities, with the spoil from a ditch being thrown inside to provide the bank, where this bank was to be of some height and solid enough to mount a palisade additional skills would be necessary to provide a firm base. This was even more critical if a mound was to assume the proportions normally associated with a fully fledged motte and be tall enough to stand out in the landscape (whether as a symbol of lordship and dominion or for defensive purposes or both is a debate not relevant to this point), and in addition be strong enough to sustain the weight of a timber building on top – palisades, halls, towers and gate towers have been identified in archaeological investigations as well as contemporary manuscript illustrations:[68] then the builder needed special skill. To pile up a large heap of earth and then put a heavy structure on top would be a recipe for disaster, as a result of the soil settling unevenly and the structure standing on it collapsing. There have been attempts to quantify such issues as how much earth was required to build a motte, how the layers were put down and how long it took for each one to settle sufficiently before the height could be increased, but this aspect of construction (unlike studies of stone buildings) remains largely unexplored. Knowledge of the type of soil must have been important, and of course the many early castles taking advantage of natural features reflect the possibility of avoiding such questions altogether.

Where a low-lying site was to benefit from wet ditches (moats), the design had to include calculating how to divert local flowing water into the ditch and how to sustain its level once there. The skill to do this was available: in northern France, for example, defensive hydraulics were common over this period – a canal to protect Senlis (1068), moats to protect St Denis, a new moat created by diverting water from the river Verse to protect Noyon at the end of the ninth century, draining marshes and creating a wet moat at Beauvais (1015 onwards) and new canals to feed the moat at Étampes at the same time. The works at Beauvais required extremely good knowledge of levelling and surveying and made use of "technical mastery that still astounds engineers", while the diversion of the rivers and creation of a harbour at Caen on William the Conqueror's orders between 1066 and 1083 was a large-scale engineering exercise in water management.[69]

That these engineering skills were exploited in work in cities or on large ecclesiastical sites is known, because the owners or their subordinates would often record the facts. The local noble follower of a great lord rich enough to construct a new castle to become (he hoped) the seat of a great family would rarely have an

68 The Bayeux Tapestry offers several illustrations of motte castles including showing the construction of one after William landed, at plates 49–50.

69 A Guillerme, *The age of water* (1988), 25–38, quotation at 38. The evidence for Caen is contained in charters of William I: D Bates (ed.), *Regesta regum Anglo-Normannorum. The Acta of William I (1066–87)* (1998), 217 (no. 45), 239–40 (no. 53).

educated scribe to write down his achievement or how it was executed, but it is likely that his expenses must have included recruiting people with the technical skills suggested here, to direct his peasant labour in the work.

SHIPBUILDING

Shipbuilding is another area where ancient skills, particularly in carpentry, had been sustained, and where rulers would have had to access large workforces to carry out their projects. The importance of the navy to the survival and then revival of Byzantine power has already been suggested. The records cited included references to very large-scale mobilisation of fleets of ships to different designs depending on their purpose, ranging from swift scouting ships, to war vessels carrying soldiers and equipped with Greek Fire, to transports for bulky equipment and special cargo such as large numbers of horses. Throughout this period, the Byzantine empire managed to organise successful maritime expeditions involving the transport of armies, horses and equipment long distances across the sea.

Over several centuries, a small number of ship types made up the navy. Vessels for fighting, such as the *dromon*, appear to have had a standard design to carry marines and archers and to mount siphons on a structure in the bow. A larger and slower vessel – called a *chelandion* – provided the transport. At some point ship design was changed to enable horse transport, a challenging task because of the size of the animal, which not only required a method to allow loading and unloading but also safe stabling on board, as well as a way to cater for the great increase in need for water and food during transport. It is not certain that there were technical improvements during the long period separating the description of sixth-century warships by Procopius from the vessels described under Constantine Porphyrogennetos. At some point the superior navigation achieved by lateen sails was introduced (and immediately copied across the Mediterranean). Sometime around the eleventh century (probably) a new, lighter and more seaworthy oared ship (the *galea*) became the mainstay of navies, a trend that continued in these waters. The complexity of the combined naval and military operations undertaken by the empire during these years testifies to the level of skills of the commanders. What skills were needed to build the ships is harder to know. One of the rare texts discussing naval warfare, the *Naumachika* that concluded the *Taktika* of Emperor Leo, did not answer this question in any usable way. Nonetheless, it was clearly the case that a cohort of shipwrights existed (and renewed itself) to create the Byzantine navy and to continue building reliable vessels, deploying skills particularly in carpentry but also other crafts specific to marine transport.[70]

70 J Pryor, 'Shipping and shipbuilding' in Jeffreys et al., *Oxford handbook of Byzantine studies* (2008), II, 8, 9, 483–7; Pryor, Jeffreys, *The age of the ΔΡΟΜΩΝ*, 175–8 (Leo),

Little information survives on the Muslim navies. The Fātimid dynasty had regular recourse to naval operations, although it faced major problems in accessing timber supplies, and repeated chronicle references confirm that incendiaries were used from their ships just as on the Byzantine fleets. Based at arsenals established at al-Mahdiyya (Tunisia), Cairo and Alexandria, shipbuilding and its associated crafts (as well as the sailors themselves) were reliant on the pressed work of reluctant subjects, which may explain the unimpressive record of this navy once the conquest of Egypt had been accomplished.[71]

The Merovingian Franks had made use of ships for campaigning purposes in the Channel and in the Low Countries, as had the Anglo-Saxon kings: the attribution to Alfred the Great of the title of founder of the British navy, following his successful creation of a fleet to combat Viking invaders, is testimony enough.[72] Analysis of the preserved remains of Norse and Baltic ships has confirmed the skills used, albeit definitely in a rule-of-thumb, handed-down skill set, and discoveries of trading ships, such as the tenth-century vessel found at Graveney in Kent, confirm their standard of workmanship.[73]

One of the largest and most famous naval operations was the Norman Duke William's invasion of England in 1066, although the intricacies of the naval operation have tended to be lost in the attention given to the subsequent battle. Nonetheless, analysis of the evidence of the chronicles and comparison with contemporary practice – particularly the very similar operations carried out by the Byzantine empire – confirm the ability of eleventh-century rulers to carry out massive logistical and building operations when necessary. In this case, William's requirement was probably for around 700 ships, which had to be constructed anew, involving the recruitment of shipwrights from across Europe. Woodworkers of all kinds were vital, but the role of those in charge must have been critical to ensure that designs were followed. The proof of the existence of standard designs for ships covering a wide range of functions is in the success of the expedition

449–53; van Doorninck, 'Byzantium, mistress of the sea', 145–6. The analysis of wrecks found in a Byzantine harbour in Istanbul has provided real evidence of the building methods and materials used for a wide range of vessels, including single-banked light galleys dated to between 690 and 890 and seen as the forerunner of the *galea*: U Kocabas, 'The Yenikapi Byzantine era shipwrecks', *International journal of nautical archaeology* 44/1, 2015, (5–38) at 26–33.

71 Y Lev, 'The Fātimid Navy, Byzantium and the Mediterranean Sea', *Byzantion* LIV, 1984, 220–52.

72 On the subject in general, Haywood, *Dark Age naval power*. He stresses (93) the reliance of the crown on those involved in trade, fishing and piracy to provide the skill and manpower for their naval operations.

73 V Fenwick (ed.), *The Graveney boat: a tenth-century find from Kent*, BAR British Series 53, 1978, 93–8.

and is suggested too by contemporary illustrations of shipbuilding, such as in the Bayeux Tapestry.

By this time, there were regular contacts between the Normans of Normandy and their cousins now based in southern Italy, who were carving out their own lordships at the expense both of Muslim Sicily and Byzantine southern Italy. The Norman invasion of Sicily attempted in 1061 must have relied exclusively on Byzantine ships and crews, so the mechanism for a transfer of knowledge was there, and the Normans proved themselves adept at learning from other cultures.[74] Shipbuilding skills were in constant demand in any case, but it would have been meaningless to distinguish between the different purposes for which ships were required. William's hired craftsmen were constructing these ships for military purposes, but no doubt then resumed their normal function of building vessels for trade goods, seafarers (and pirates).[75]

BEGINNING OF A NEW PERIOD

I have suggested that new developments occurred in the areas of theoretical and practical knowledge and that evidence for the application of engineering skills in military operations became available from the tenth century in Europe, while continuity of these attributes in a resurgent Byzantium and in the now fractured Muslim worlds can be identified. In the European context, military technology itself scarcely advanced at all before the eleventh century, but perhaps the number of technicians expanded with the growing population, growing wealth and increasing demand for the services of people who could erect new buildings and construct ships and bridges. Apart from the builder-bishops, little is known about the people who constructed fortifications or built siege equipment, except that they must have existed. The rare exceptions to this anonymity were in the eastern Christian empire, and these examples were only recorded because of the extraordinary circumstances involved and because their names could be cited for polemical purposes. The continuity of engineering skill and therefore of engineers has been shown as well for the 'Abbasid, Fātimid and Cordoban caliphates, without the benefit of knowing who the people were. Throughout the eastern and Middle Eastern worlds, there have also been continuous references to the military uses of mining skills, all the time with no indication as to who provided them.

74 G Theotokis, 'The Norman invasion of Sicily 1061–72', *War in history* 17/4, 2010, 381–402, here 391.

75 C Gillmor, 'Naval logistics of the cross-Channel operation 1066', *Anglo-Norman Studies* 57, 1984 (105–31); B S Bachrach, 'On the origins of William the Conqueror's horse transports', *Technology and culture* 26/3, 1985, 505–31; *The Bayeux Tapestry*, ed. D M Wilson, plates 35–7.

In the next period, marked by accelerated economic growth and another "renaissance" of knowledge in Europe, the possible "trickle-down" consequences of which require examination, this anonymity disappears and there appear the first definitive references to engineers employed by rulers for military and other purposes. In this, the start of the age of crusades generated a much closer relationship and interchange between Christian and Muslim societies from Portugal to Armenia and Syria, which may well have further accelerated the diffusion of techniques and skills.

4

The Engineer Recognised

FROM A POSITION OF almost total anonymity in the tenth century, there are records of the names of engineers, builders, master craftsmen and master miners in the twelfth. There is still no evidence that a separate category of military engineer was coming into existence, and there remain numerous accounts of events requiring engineering in which no engineer was named, but over the course of around 150 years between the middle of the eleventh and the end of the twelfth centuries a trend towards both the identification of individuals and a certain level of specialisation can be suggested that is not entirely due to the accidental survival of the records. In this chapter I will present the economic background, the possible relevance of advances in science and the evidence for technological improvement. I will then review what happened in actual campaigns and the role of the engineer within these contexts.

AN AGE OF ECONOMIC GROWTH – AND SOME OF ITS CONSEQUENCES

Underpinning these and many other important historical developments was a long period of economic growth rooted in a substantial increase in population, leading to extensive clearance of lands as well as the settlement of new territories – one feature of this period was a powerful drive eastwards by German rulers to take over the lands and peoples of the Slavs in what is now eastern Germany, as well as the establishment of new monarchical states in largely Slav lands such as Bohemia (which was part of the empire), Poland, and the western Balkans, and Magyar Hungary (which were not). It is not possible to more than guess at the size of medieval populations, based on very tenuous evidence, but it is generally accepted that in the tenth century Europe was a very empty continent by comparison with later ages and there was plenty of uncultivated land to spare. It has been argued that the return of (relative) stability following the ending of the Scandinavian, Magyar and North African raids that had caused disruption and insecurity in many areas, and the establishment of new political entities (which sometimes included the former invaders, as in Normandy and England) exercising control over their states, based in agriculturally fertile areas, encouraged the production of an economic surplus beyond what was required for subsistence. There is much to suggest that the great spurt in church building that started during the eleventh century was the result of the investment of this surplus in stone. Economic recovery seems to have begun in Italy, spread to (parts of) "France", then the Rhineland, then (slowly) to the rest of "Germany" before eventually reaching further east and north (into Scandinavia) and south-east to the Balkans. This process continued into the fourteenth century.

This trend included the growth of existing towns and the foundation of new ones, as it was now possible for peasant surpluses to be traded to feed their occupants. Some townspeople engaged in trade, others in crafts, which led to a gradual process of specialisation among the artisans, later reflected in the growth of craft guilds. It was to be expected that the growing number of craftspeople contributed to the expansion of ecclesiastical and secular building, since the availability of those able to build church, palace, castle, town hall or town wall, along with the greater surplus wealth of bishops, abbots, kings, princes, dukes and counts, made the work possible. This combination of greater activity and the larger number of people with technical skills also made viable the emergence of a larger group of engineers and architects. The evidence that this urban development began in Italy and that the coastal cities (first Amalfi, then Pisa, then Genoa and Venice) were the first to establish substantial trading empires also explains largely why during the early crusades and the twelfth century these same cities were to be found providing many of the military engineers for whom there are historical references.

SCIENCE AND TECHNOLOGY: ANOTHER RENAISSANCE AND ITS PRACTICAL IMPACTS

Concurrent with this growing wealth (in money and in people) was a period that witnessed new scientific advances. The twelfth century in particular witnessed such intellectual progress in Europe that this period has been dubbed a renaissance in its own right. The place of Aristotle was firmly established as the greatest philosopher and the foundation for understanding the natural world,[1] and it would be a renewed discovery of the works of the Greek transmitted through translation of Arabic versions of many of these texts, previously lost in western and central Europe, that would help drive forward this medieval renaissance. Some of these

1 Otto of Freising, *The Two Cities*, trans. Mierow, book 2, c. 8 (161–3), writing in 1146, described Aristotle as "the most renowned of all the philosophers", praising in particular his explanation of logic, using the syllogism properly and as an effective tool for understanding.

writers produced texts that were relevant to the questions that might be of interest to those responsible for practical tasks. Modern historians of science debated for decades whether there was any connection between the theoretical and the practical. Just as in late antiquity, theoretical knowledge in its own right, as *scientia*, was seen by scholars as a form of human endeavour that aimed better to understand God's creation, not assist with the mundane. The physical work of human beings occupied an entirely different world: the mechanical arts necessarily were the responsibility of ignorant commoners, and the people who practised them were merely tolerated because their labour – drudgery, of low status, mindless: something vulgar – was necessary to society. Many of the attempts to classify the sciences since the days of the Roman empire maintained this distinction, originating in the approach of the Greek philosophers, and excluded the mechanical arts altogether from their lists. But some did not, and mechanical arts were included not just in the lists but sometimes in the practical application of the teaching of the "liberal arts".[2]

The many works of Adelard of Bath (c.1080–1152) included a text of around 1116 describing geometry as the science of measuring land and going on to present rules on measuring heights, using the same methods as the Roman *agrimensores*.[3] Adelard was among the first to translate into Latin from Arabic books IV to VIII of Euclid's *Elements*, which enabled a far more elaborate grasp of the principles of geometry than had hitherto been possible – although some claim that a better job was done by Gerard of Cremona, who lived in Toledo and died in 1187[4] – as well as al-Khwarizmi's astronomical tables with their practical and intellectual implications.

Unusually, Adelard was not a cleric. Hugh of St Victor (1096–1141) was, however, and two of his works are of particular interest here. His *Didascalion* contained a new classification of the sciences which included the mechanical arts given that name because they were "adulterate, because their concern is the artificer's product" (book II, c. 20), but despite his traditional dismissal of such commonness he went on to detail how men make buildings and armaments and the other forms of

2 An excellent summary of the debate to that point and presenting a well-argued and convincing case for the twelfth century having witnessed a considerable flexibility on the importance attached to the mechanical arts is E Whitney, 'Paradise restored. The mechanical arts from antiquity through the thirteenth century', *Transactions of the American Philosophical Society*, 80/1, 1990, 1–149. For the views of the scholars discussed, see entries for Alessio, Allard, Beaujouan, Crombie, Le Goff and L White in the bibliography.

3 A Drew, 'The *De eodem et diverso*', in C Burnett (ed.), *Adelard of Bath. An English scientist and Arabist of the early twelfth century*, Warburg Institute Surveys and Texts XIV, 1987, 17–24.

4 W Stevens, 'Euclidean geometry in the early middle ages: a preliminary reassessment', in M-T Zenner (ed.), *Villard's legacy. Studies in medieval technology, science and art* (2004), 229–63, here 260.

manufacture: this was not therefore simply mindless work and had a place among the liberal arts. The same author's (attributed) *Practica geometriae* contained advice on how to measure areas and heights, including the comment that "less skilful surveyors, with no equipment or training for better methods, judge a height by shadows". Quadrants and the astrolabe featured among the instruments covered by the text, which while they *could* reflect a purely theoretical intent, may also imply some practical purpose consistent with the rest of the contents.[5] The work of Gundisalvo (1120–84) of Toledo, who translated the Arabic work of al-Fārābī, recognised that the "science of devices" (*scientia de ingeniis*) was the science of devising ways "to make all the things happen whose modes were stated and demonstrated in the sciences (*de doctrinis*)". He referred to the devices needed for geometric measurements, including for stonemasons and for making armaments, among many other practical applications.[6]

I have argued that the twelfth century saw a partial but progressive realignment of scientific thought with its potential practical applications in areas where the knowledge might have been useful to builders and engineers. It is interesting that this period did demonstrate some progress in technological development to accompany the knowledge, as will be shown. A basic knowledge of geometry had always been important if surveying was to be accurate enough for the purpose, and for the design and layout of buildings. It would be valuable in assessing how to set about crossing a river the width of which was not already known. Measurement of distance was necessary to site siege artillery and of height to construct ladders or a siege tower. Other calculations made by modern engineers were not so easy for the medieval engineer, or scientist, because the science of motion had not yet progressed beyond Aristotle's theories presented in his *Physics*, which argued that a projectile was moved by its propellant and kept going by the air surrounding it.[7] Nothing more sophisticated would be developed in Christian Europe until the

5 *The Didascalion of Hugh of St Victor*, ed./trans. J Taylor (1961, 1991), 75, 76; Hugh of St Victor (attrib.), *Practical geometry (Practica geometriae)*, ed./trans. F Homann, SJ (1991); S Victor (ed./trans.), 'Practical geometry in the high Middle Ages. *Artis cuiuslibet consummatio and Praktike de geometrie*', *Memoirs, American Philosophical Society*, 134, 1979, 1–297; see also R Baron, 'Sur l'introduction en occident des termes "geometria theorica et practica"', *Revue des sciences et leurs applications* 8/4, 1955 (298–302); L Shelby, 'Geometry', in D Wagner (ed.), *The seven liberal arts in the Middle Ages* (1983), 196–217.

6 Original Latin available in D Gundisalvo, *De scientiis*, ed. P Manuel Alonso Alonso (1954) (quotations at 108, 110, 111); in English, D Gundisalvo, 'Classification of the sciences', trans. M Clagett, E Grant, in E Grant (ed.), *A sourcebook in medieval science* (1974), 59–76, at 76.

7 J Murdoch, E Sylla, 'The science of motion', in D Lindberg (ed.), *Science in the Middle Ages* (1978), 206–44, here 206–10. Deeper theories were developed in the Muslim world by Ibn Bājjah (d. 1138): see Nasr, *Science and civilisation in Islam*, 138.

fourteenth century. The reality was that people observed projectiles in motion and knew about momentum, even if they did not have the right words to describe it.

Did non-scholars have access to the new learning in Europe? Among rulers, the German emperor Henry IV (1056–1106) had, "like his father" (Henry III, 1039–56), spent time with "learned" clerks and discussed both the scriptures and the liberal arts.[8] Henry II, ruler of England and much of western France (1154–89), was described as knowing many languages. But Henry the Lion, duke of Saxony and Bavaria, was illiterate (signifying that he could not read or write Latin) and Emperor Frederick Barbarossa (1152–89) "can understand Latin more readily than he can speak it".[9] The warrior class was schooled in warfare rather than intellectual pursuits.[10] But to what extent this training now involved studying ancient texts like Vegetius or the Roman histories (or, for Byzantine commanders, recent military treatises) is rarely known, and the cases where the particular skills of individual nobles and princes were highlighted in the written evidence might be interpreted as indicating their rarity. The university would only make its appearance at the end of this period, with Bologna, Paris and Oxford (in that order) being among the earliest to be recognised.

Who else was able to read? Attempting to grasp the extent of lay literacy in the middle ages is a challenge attempted by many but perhaps inevitably doomed to failure. Even the most positive about Charlemagne's reforms believe that the impact among those of lower rank was minimal, and the collapse of the empire during the ninth century will have reinforced a decline.[11] Outside of the ruling class, evidence is rarer still. There is a logical argument that traders must have been able to (ac)count, and sometimes to record information, and that the bailiffs and reeves who ran estates needed a form of record-keeping. In the late twelfth century, Walter Map (famously, although for polemical effect) bewailed the fact that mere peasants were seeking education for their children in order to achieve riches rather than to get rid of their vices, even though only

8 'Chronicle of Frutolf of Michelsberg', interest in the quadrivium at St Emmeram (Regensburg), 15; and 'Anonymous Imperial Chronicle', 205, both in T McCarthy (ed.), *Chronicles of the investiture contest.*

9 W Map, *De nugis curialium*, ed./trans. M James, revised N Brooke, R Mynors (1983), 476–7 (Henry II); K Jordan, *Henry the Lion*, 23; Otto of Freising, *The deeds of Frederick Barbarossa*, trans. C Mierow, 333 (book IV, c. lxxxvi).

10 Not just men, either. Matilda of Canossa was a long-standing enemy of Emperor Henry IV and her armies successfully resisted his efforts on many occasions, reflecting a skill in warfare (and this was largely siege warfare) matching any male ruler: D Hay, *The military leadership of Matilda of Canossa 1046–1115*, 7–9.

11 See, for example, F Ganshof, 'The Carolingians and the Frankish monarchy', *Studies in Carolingian History*, ed. J Sondheimer (1971), 129–34; J Nelson, 'Literacy in Carolingian government', in R McKitterick (ed.), *The uses of literacy in early medieval Europe* (1990), 258–96, passim.

"free men" were "of right" allowed to learn the arts.[12] In the flourishing city of Milan during the twelfth century, a patriotic native recorded the presence among the citizens of more than fifteen *doctores artis gramaticae* and seventy *magistri* teaching *initiales litteras* to boys.[13] England saw a growth from the twelfth century of cathedral schools, with there being at least thirty in major towns by 1100. Government under the Anglo-Norman and then the Angevin kings witnessed a massive increase in administration reliant on the written word.[14] This period also witnessed the writing of textbooks for the schools, to teach the liberal arts, including the basics of arithmetic and geometry and the use of the abacus for calculation. Hugh of St Victor's book on education specified that geometry was for practical purposes and that it was for measurement, which contrasted it with the theoretical version of the subject.[15] It is a safe conclusion that some lay people acquired a level of education, even if it was just what was necessary to carry out their functions. Where they did, it is at least possible that they included some master craftsmen. But it remains most likely that a large majority of master craftsmen acquired their skill through apprenticeship to an existing master. Some of these crafts began to specialise and then adopt the structure of the guild in order to protect their interests – although from the viewpoint of skills with military application, it should be noted that miners and masons remained outside these structures. The engineers who will have begun as master craftsmen themselves may have learnt their trade at the feet of existing engineers. By learning on the job during military operations – there would have been no shortage of opportunities – perhaps when they became masons or carpenters they engaged in the construction of engines, observing close up and sufficiently sharp-witted and intelligent to remember what they saw, and sometimes to suggest ways of making the engine more effective, sometimes to help in the diffusion of knowledge and experience from one region to another.

12 Map, *Nugis curialium*, 12–13.

13 G Fiamma, *Chronicon Extravagans et Chronicon Maius*, ed. A Ceruti (1869), 51 (Chron. Extravagans).

14 R Swanson, *The twelfth-century renaissance* (1999), 23–30; J Thompson, 'The literacy of the laity in the Middle Ages', *University of California Publications in Education* 9, 1939, 196–7; M Parkes, 'The literacy of the laity', in D Daichen, A Thorlby (eds.), *Literature and western civilisation, III: The medieval world* (1973), 555–78; M Clanchy, *From memory to written record. England 1066–1307*, 2nd edn (1993), 26–32, 328–33. Insufficient evidence survives of the practices in England under the Anglo-Saxon kings to reach any safe conclusions about the reach of King Alfred's educational reforms. R V Turner, 'The miles literatus in twelfth-century England: how rare a phenomenon?', *American Historical Review* 83/4, 1978, 928–45.

15 G Evans, 'The development of some textbooks in the "useful arts" c.1000–c.1250', *History of Education* 7/2, 1978, 85–94, here 89–93.

BUILDERS AND BUILDING CRAFTS

Experience of building sites would have been common for masons and carpenters during the great expansion in the construction of stone churches, cathedrals, castles and bridges starting in the eleventh century. The two following centuries would see, in western Europe, the glorious culmination of what is now called Romanesque architecture and then, from the twelfth century under the influence of the work of Abbot Suger of St Denis, the first Gothic-style churches. These developments and the many surviving buildings have been studied exhaustively by architectural and art historians and my purpose here is to see if it is possible to identify any of the skills in arithmetic or geometry that might have been learned from the new educational culture. It has been argued that what became known as the Gothic style emerged from experimental church-building in the Île de France and that observation of the results led to refinements in the design.[16] This may have been *how* it evolved, but someone must have decided to abandon the massive columns of the Romanesque for the entirely different support offered by flying buttresses, and while this was almost certainly not based on engineering calculations, what kind of thinking lay behind the decision? Furthermore, experts have identified that the change of concept involved significant changes in building techniques, not just in the masonry, where it was necessary to deal with novel factors such as the "sail" effect generated by the opposite pressures of wind and suction on very high stonework, but also in timberwork for the enormous spans that now became commonplace and are reflected most visibly in surviving examples of Gothic great barns [Fig. 8].[17]

How much mathematical skill must have been deployed to create a large stone building like a cathedral or large castle is debatable. Using the present dimensions to demonstrate the use of mathematical proportions incurs the risk that the result was a matter of chance (especially if the figures are not exact, which is often the case), or had been copied from another site without awareness of the theory behind the design. No building plans survive in Europe from this period (which does not prove they did not exist). The progress of education increased the likelihood that clerical patrons were aware of some of the theory – and in some cases, we know, as in the case of Bishop Gundulf, who was responsible not only for Rochester cathedral but the new castle there and also the Tower of London.[18] But just how

16 R Mark (ed.), *Architectural technology up to the scientific revolution* (1993), 232–3.

17 L Courtenay, 'Scale and scantling: technological issues in large-scale timberwork of the High Middle Ages', E Bradford-Smith, M Wolfe (eds), *Technology and resource use in medieval Europe: cathedrals, mills and mines* (1997), 42–75; R Mark, 'Technological innovation in High Gothic Architecture', same volume, 11–26.

18 *Vita Gundulfi: the life of Gundulf Bishop of Rochester*, ed. R Thomson, Toronto Medieval Latin Texts (1977): "arti studeret grammaticae, coepit … multo magis arti

Fig. 8 High quality timberwork in a fifteenth-century barn, Harmondsworth, Middlesex. (Photo: author)

theoretical this knowledge was is not possible to establish. Gervase of Canterbury's well-known contemporary description of the rebuilding of the cathedral carried out after destruction wrought by a fire in 1174 recounts the recruitment of William of Sens, "a workman most skilful in wood and stone", who surveyed the walls and planned the new work, creating "ingenious machines" that sound like cranes (for loading and unloading stones from ships) and moulds so that the sculptors could shape the stones. The new work was of a "different fashion" from what it replaced. In 1178, while preparing machines for the turning of the great vault, he fell 50 feet when the scaffolding collapsed and the work had to be continued by another "ingenious" monk who directed the masons.[19]

utiliori studere", 26; and Textus Roffensis (same volume), bishop "in opere cementarii plurimum sciens et efficax erat" (c. 4), 78–9.

19 R Willis, *The architectural history of Canterbury cathedral* (1845), 32–6, 50–1 (being a translation of the Latin account by Gervase).

It is clear that skills based on practical knowledge were necessary for the construction of buildings or military machinery. These are a requirement in any age: sufficient awareness of how to set about the task, the right tools to achieve it, and the right materials. Scholars have identified improvements across them all during this period, ranging from the developing skills of stonemasons already discussed (chapter three), to the existence of machines vital to enabling the construction of a large building. Collections of the illustrations of building construction from medieval manuscripts demonstrate that building workers (covering by this term everyone from the master builder to the newest apprentice and simple labourer) were acquainted not just with appropriate tools, but also with equipment to allow them to use their skills. Hammers, saws, trowels, axes – all are repeatedly shown. Scaffolding – its use still identifiable from the putlog holes that climb the walls of countless still standing masonry-built castles – and lifting devices such as cranes driven by pulleys, windlasses or treadmills were clearly ubiquitous [Fig. 9]. Implements such as plumb lines and measuring staffs confirm the means whereby buildings were kept vertical and the correct dimensions identified (incidentally confirming again the evident necessity for someone to have decided what the dimensions were to be).[20]

Improvements in timber working have been shown by archaeological investigation of woodwork on London's waterfront, including the rediscovered timbers of parts of the abutment for old London bridge, study of which identified signs of improved technology and carpentry skills. These included the introduction of base plates and the adoption of diagonal braces integrated into framed structures, well-squared and precisely cut timbers, sawing of timber length-wise to produce planks and the use of prefabricated components. Studies of standing buildings confirm that the change from Romanesque to Gothic was also marked in carpentry from the mid twelfth century on, with greater use of triangulation and the deployment of mortice and tenon joints, also found at London bridge. In turn, this called for exploitation of younger timber and different ways of managing the supply.[21]

20 Erlande-Brandenburg, *Cathedral builders*, 97–8, 118–20; G Binding, *Medieval Building Techniques*, trans. A Cameron (2004), no. 238, 266, 268a, 295b, 13a, 431, 541 are all illustrations from the tenth or eleventh centuries of tools and machinery.

21 G Milne et al., 'Timber building techniques in London c.900–1400. An archaeological study of waterfront installations and related material', *London and Middlesex Archaeological Society* Special Paper 15 (1992), 131–5; R Darrah, 'Working unseasoned oak', in S McGrail (ed.), *Woodworking techniques before AD 1500*, BAR International Series 129 (1982), 219–29, at 229; R Hill, 'Tenth-century woodworking in Coppergate, York', same volume, 231–44, esp. 232–4; F Épaud, *De la charpente romane à la charpente gothique en Normandie. Évolution des techniques et des structures de charpenterie aux XIIe–XIIIe siècles* (2007), 581–3 (summary); reviewed by N Alcock, *Medieval Archaeology*

Fig. 9 Treadwheel-operated crane at height, reconstructed at Guédelon, Yonne. (Photo: author)

Did those putting these skills into practice understand that their calculations (or estimates) were rooted in Euclidean geometry (now that it was better understood)? The appearance of the fuller text meant that its readers, familiar with the basics, were able to apply this to new needs – another way of arguing that building techniques had become more sophisticated.[22] The evidence that they did understand proportions continues to be argued on the basis that standard ratios inherited from antiquity, such as the proportion of $1{:}\sqrt{2}$, can be found in new constructions. Two examples in England may offer confirmation: the new cathedral at Norwich, begun at the end of the late eleventh century, has been measured closely and seems to confirm a universal application of this proportion in its layout and design, while several generations later and also in East Anglia, Orford castle [Fig. 10] built at the orders of Henry II (1154–89) has been analysed and the proportions of this magnificent and ornate great tower have been found to comply with proportions of $1{:}\sqrt{2}$ and $1{:}\sqrt{3}$ throughout, surely not by chance.[23] The plans of many other new castles built during the twelfth century reveal innovation and experimentation with new designs, including many great towers evidently built using geometrical principles, combining circles with squares but polygonal shapes, all of which required accurate measurement. But although the resulting proportions were deliberately chosen, how they were planned on site is another matter. This could well have been achieved with a rope and pegs or by what might best be described as "rule of thumb". Such is the conclusion of modern students of the subject.[24]

In Byzantium, new building was no longer the responsibility of the named architects (*mechanikoi*) who had erected Hagia Sofia, but of the anonymous *oikodomos*, a term used equally for master masons and carpenters, whose training was solely through the workshop. The masons and carpenters were of

---

52, 2008, 448–9. On the significance of the discoveries relating to London bridge for understanding timber working changes, B Watson et al., *London bridge. 2000 years of a river crossing* (2001), 190–1.

22 See, for example, Stevens, 'Euclidean geometry in the early middle ages', here at 262.

23 C Radding, W Clark, *Medieval architecture, medieval learning: builders and masters in the age of Romanesque and Gothic* (1992), 11–17, 33–6, arguing for the education of master builders of this age, and eleventh-century builders represented "a decisive break with the past" (54); M-T Zenner, 'Imaging a building', in Contreni, Casciani (eds), *Word, Image, Number*, 219–46, 228, 243–4; E Fernie, 'The ground plan of Norwich cathedral and the square root of two', in L Courtenay (ed.), *The engineering of medieval cathedrals*, 107–16 (originally published in 1976); T Heslop, 'Orford castle, nostalgia and sophisticated living', in R Liddiard (ed.), *Anglo-Norman castles* (2003), 273–58, at 284–7.

24 P Kidson, 'Architectural proportions', J Turner et al. (eds), *The dictionary of art* (1996), II, 343–52. On geometric castle designs, see most recently M Hislop, *Castle builders. Approaches to castle design and construction in the Middle Ages* (2016), 8–9, 110–16.

Fig. 10 Orford castle, Suffolk. The twelfth-century great tower was designed according to geometric principles. (Photo: Neil Guy)

low status and the building trade, like the others, was under strict government control, although the rules applying to all trades and enforced by the prefect of Constantinople indicate that a significant level of what would now be called quality control was expected of them.[25] This situation applied to both ecclesiastical and secular building projects – Anna Comnena reported how Emperor Alexios I Comnenos (1081–1118) sent "all the construction materials needed ... together with the architects (*oikodomoi*)" on a fleet to build a new fortress at Kibotos on the Sea of Marmara, led by Eustathios, commander of the fleet.[26] This officer was probably, then, the director of operations, but the practical work was executed by the professional, but probably illiterate, master craftsmen. It has been argued that Byzantine building of these centuries was not based on architectural drawings, that the use of proportions such as the golden mean was rare, and that instead simple measurement would have been undertaken using stones, ropes and measuring sticks, whether to design a church or lay out a military camp. There is some confirmation of this in the mid-eleventh-century life of the tenth-century saint Nikon among whose miraculous accomplishments in the Peloponnese was building a new church following a vision: "he began the building, having earlier delineated it with a rope". The account went on to describe how he completed it with "a technical skill even more ambitious than his physical skill".[27]

Although the intellectual progress of the rest of Europe seems not to have happened in the Byzantine empire, the fourth-century scholar Pappus of Alexandria maintained a continuous place in the curriculum. His works on mathematics include the words "the man who has trained from his youth in the aforesaid sciences as well as practised in the aforesaid arts ... will be the best builder and inventor of mechanical devices". It should therefore be expected that those ruling and governing the empire continued to be educated in the same works as their predecessors – but that the people responsible for executing their orders relied on an inherently conservative training in the workshop or on the job. In the non-military context, there is a complete absence of information concerning numerous aspects of the way that engineering or building projects were managed or executed, and no evidence to suggest what overlap existed between civilian and military technology.[28] What must remain, however, is the probability that there was in practice no difference between these spheres.

25 J Nicole (trans.), *Le livre du préfet, ou l'Edit de l'empereur Léon le Sage* (1894), ch. XXII, section 4, pp. 80–2; modern edition, J Koder (trans./ed.), *Das Eparchenbuch Leons des Weisen* (1991), 138–42.

26 *The Alexiad of Anna Comnena*, trans. E Sewter (1969), 204.

27 R Ousterhout, *Master builders of Byzantium* (1999), 7–11, 42–52, 58–64; D Sullivan (ed./trans.), *The Life of Saint Nikon* (1987), 117–19. The church no longer exists although two possible sites have been identified: 286.

28 A Kazhdan, A Wharton Epstein, *Changes in Byzantine culture in the eleventh and twelfth centuries* (1985), 27–8, 121–5, 148–9; K Vogel, 'Byzantine science', *Cambridge medieval history* IV pt. 2 (1967), 264–306, at 264, 271–4, offers the traditional view. This has been partly challenged by later writers including: C Niarchos, 'The philosophical

There was also technological progress in metalworking during this period. Contemporary evidence comes from the fascinating text written by Theophilus, apparently a Benedictine monk, possibly from north-west Germany and probably himself a metalworker, whose *De diversibus artis*, the oldest manuscript of which is dated to before 1150 and which may have been written by 1100, provided instruction in a range of manufacturing techniques and described the process of creating iron from the ore and how to make steel. The intended audience was presumably fellow monks rather than the village smith. However, the improvement in technique was represented by the manufacture of cast iron with the invention of the blast furnace, assigned to Sweden in the twelfth century and confirmatory evidence has been found in Swabia in Germany. Over the same period, there is evidence that a metalworking industry was developing fast in Lombardy (a massive increase from the middle of the twelfth century), and slowly the social status of metalworkers in the centres of the iron industry there improved.[29] The practical usefulness of better iron for military purposes was obvious, and if the manufacture of armour for the knight was central, the improved technology would have significant benefits in other forms of engineering and construction as well.

This information is background to reviewing developments in military engineering. The rulers who were patrons for new churches and castles were also the military commanders – including bishops – who led armies across the landscape, crossed rivers, laid out siege camps, ordered the construction of equipment for capturing or defending a fortification, and decided whether or not to try to undermine those defences. The craftsmen who had to construct siege towers and artillery or temporary defence works were the same masons and carpenters who had previously worked on the cathedral and learned new or

background of the eleventh-century revival of learning in Byzantium', M Mullett, R Scott (eds), *Byzantium and the classical tradition* (1981), 127–35; and N Wilson, *Scholars of Byzantium* (1983), 148–66. D Geanakoplos, *Byzantium, church, society and civilization seen through contemporary eyes* (1984), 431–2 for Pappus on mechanics and mathematics. C Bouras, 'Master craftsmen, craftsmen and building activities in Byzantium', in A Laiou (ed.), *The economic history of Byzantium from the 7th through the 15th centuries*, Dumbarton Oaks Studies XXXIX, vol. II (2002), 539–54. The continuity in use of longstanding techniques of land measurement for taxation was suggested by N Svoronos, 'Recherches sur le cadastre Byzantin et la fiscalité aux XIe et XIIe siècles: le cadastre de Thébes', *Bulletin de Correspondence Hellénique* LXXXIII, 1959/1 (1–145), at 144–5.

29 Theophilus, *De diversibus artis*, ed./trans. C Dodwell (1961), book III (metals); A Williams, *The Knight and the blast furnace. A history of the metallurgy of armour in the Middle Ages and Early Modern period* (2003), 881–3; F Menant, 'Metallurgie lombarde au moyen âge', in P Benoit, D Cailleux (eds.), *Hommes et travail du métal dans les villes médiévales. Actes de la table ronde 'La metallurgie dans La France médiévale', université de Paris 1, 1984* (1988), 127–61, at 131–2.

existing skills through their service-based training; the engineers who designed them had worked in both arenas.

WORKS WITH WATER AND SURVEYING IN EUROPE

Archaeological finds continue to demonstrate the skills needed by carpenters and other craftsmen to construct efficient watermills. If the technology was known in pre-Carolingian Europe (see chapter three), a growing population and the expanding use of water power to provide energy for industry were features that drove the construction of more mills driven by water and improvements in their design and technology, including the use of the camshaft to enable a watermill to drive hammers, with significant improvements resulting in metalworking (possibly from the tenth century) [see Fig. 11]. Cistercian abbeys provided a major impetus. But even before then Domesday Book, for example, had recorded 5,264 mills in England. Many have described the watermill as the most complex piece of medieval technical equipment. The skills required to build and maintain the mills and their accompanying dams, sluices and channels also meant a growing demand for skilled craftsmen – master carpenters, other carpenters, ditchers and dikers whose talents might well be taken up in other fields.[30]

A growing population and demand for land also gave impetus to new projects for draining marshes and digging canals. In Hungary, a massive network of canals for both drainage and irrigation, which must have been managed across the whole region, was created under the new state established by the Arpad dynasty at the end of the tenth century.[31] In the Anglo-Norman kingdom, work to drain the fens had to wait longer and there were few examples of new work. One such began in 1121, when Henry I "cut a large canal from Torksey to Lincoln, and by causing the river Trent to flow into it, he made it navigable for vessels". The most substantial examples of new works were in northern Italy, where Milan (a city that was at the

30 Clay, Salisbury, 'A Norman mill and other sites at Hemington Fields, Castle Donington, Leicestershire', *Antiquaries Journal* 147, 1990, 276–307. For a general history of its development, T Reynolds, *Stronger than a hundred men. A history of the vertical water wheel* (1983) – more modern discoveries suggest that the choice of vertical or horizontal wheels was not technologically significant. See also B Blaine, 'The enigmatic water mill', in B Hall, D West (eds), *On pre-modern technology and science* (1976), 163–76, and Graham-Taylor, Valor (eds), *The archaeology of medieval Europe*, I, 214–17. Contemporary manuscript illustrations of watermills at work are found in L Verriest (ed.), *Le polyptique illustré dit 'veil rentier' de messier Jehan de Pamele-Audenarde (vers 1275)* (1950), 13r, 15r, 98v also showing windmills, and K von Amira (ed.), *Die Dresdener Bildhandschrift des Sachsenspiegels* (1902), I, plate 49 and II, plates 2, 110, 154 (from c.1220).

31 Takács, Füleky, 'Remains of the medieval canal system in Carpathian basin', 65–77.

FIG. 11 Gearing on a twelfth-century water mill, reconstructed at Guédelon, Yonne. (Photo: author)

centre of twelfth-century conflicts with the German emperors) spent resources to construct new canals and divert rivers. In 1162, the city constructed a moat 6 kilometres long and 18 metres wide; a canal (the *naviglio grande*) from the river Ticino, 25 kilometres long, was begun in 1177 and used to power watermills, irrigate fields and enable the transport of stone for building projects. Early examples of "crude locks" were being built by 1200. Similar, though smaller-scale, waterworks were under way at Padua and Cremona. Wet moats were already commonplace around the numerous *castelli* of the Po valley.[32]

Although these engineers remain largely anonymous, the skills required for canal building would be similar to those needed for constructing aqueducts, cisterns or wells and ensuring that wet moats remained filled. Thirteenth-century evidence from Perugia showed the city recruiting six, of whom four were monks or friars who had trained as metal or construction workers before turning their hands to hydraulics. In 1152, monks at the monastery of Maria Laach (near Andernach, Rheinland Pfalz) began work on a tunnel to run 880 metres through the hill to prevent the risk of flooding and to irrigate large areas of land around the monastery.

32 Simeon of Durham, *A history of the kings of England*, trans. J Stevenson (1855), 188; Magnusson, Squatriti, 'Technologies of water in medieval Italy', 224, 229–31, 248 (Perugia).

It remained serviceable until 1844. The construction ran at the lowest workable height and achieved the shortest route, an extraordinary engineering achievement that had, in fact, only one parallel in the region.[33]

The skills required to achieve this feat must have included surveying – errors in judging the levels or angles for such a work would have been fatal to the project. There is little explicit referencing of other situations involving this crucial skill. The use of a measuring cord for work to bring water to the lands of the abbey church (by a canal or channel) was reported in a charter for St Bertin in 1171. The chronicle of Lambert of Ardres showed particular interest in the subject. In 1139 it told that a castle at Ammerval was thrown up overnight by the castellan of Bourboug who had sent a surveyor to measure out the ditch and mound according to geometrical proportions. In 1200, the count of Guines built a new castle at Ardres, summoning workers to dig a ditch, and used Symeon, a "master of geometrical" work, who worked according to the "rules of geometry", to measure out the site with a rod (*virga*), destroying many properties in the process.[34]

What did these surveyors do when not plotting a castle ditch – and how was it a count could obtain Symeon's services with so little apparent difficulty? They were involved in establishing the plans of new urban centres or rebuilt ancient ones. Strict regularity of street and plot design has been identified in many new towns of this century. This could be achieved by ropes and stakes, but making the plan required prior knowledge of the desired shape and knowledge of how to turn that into reality. Lambert's allusion to masters of the "rules" of geometry does suggest at least a basic professional awareness of simple geometry. It has been pointed out that while many of the resultant towns were rectangular, they were not regular; that rectangularity was in any case the simplest approach to designing a town; and that the use of standard measurements based on familiar distances used in agriculture might render skilled surveyors unnecessary. Thus the debate remains open, but that surveyors with skill did exist is clear. New towns created in Poland as part of new settlement had regular plans and may well have been set up using surveyors brought from further west.[35]

33 K Grewe, 'Der Karlsgraben bei Weissenburg und der Fulbert-Stöllen von Maria Laach', in U Lindgren (ed.), *Europäische Technik im Mittelalter 800–1400* (1997), 111–15, here 114–15.

34 All three are cited by Mortet, Deschamps, *Recueil*, I, 121 (St Bertin); II, 51–2 (Ammerval); II, 189–91 (Ardres). Lambert of Ardres' chronicle is to be found in *MGH* SS XXIV.

35 K Lilley, 'Taking measures across the medieval landscape: aspects of urban design before the Renaissance', *Urban Morphology* 2/2, 1998, 82–92; P Hindle, *Medieval town plans* (2002), argued that the rectangular shape did not require surveyors (p. 53); B Zientara, 'Socio-economic and spatial transformation of Polish towns during the period of location', *Acta Poloniae Historica* 34, 1976, 57–83, at 73–4; A Czacharowski,

Finally, renewed bridge-building activity during this period allows an insight into the engineering skills available and the perception that the construction work carried out using them suggests an advance in technical abilities over those available to previous generations. The challenge of bridging major rivers was not beyond the ambition of wealthy rulers: in 1034, Bishop Reginhard of Liège organised to build a bridge across the river Meuse despite the difficulties and the great labour involved. More than a century later, the then archbishop of Cologne planned to bridge the Rhine from that city to Deutz, in stone, and only his death led to the abandonment of the project.[36] In Italy, Roger (king of Sicily – of whom more below) ordered a bridge built across the river Volturno at Capua. The new bridge at Avignon (built 1177–87) stretched 900 metres and involved formidable technical challenges, including the construction of stone piles, possibly through resurrecting the ancient Roman method of coffer dams, though renaissance of this technology is usually dated to the fifteenth century. The twelfth century appears in fact to have been a golden age of bridge construction, with more than one hundred identified in the territory of (modern) France.

"Old" London bridge was also begun at the same time, the piers here being constructed without coffer dams but instead by driving piles into the river bed at low tide, the pile drivers being mounted on barges. Replacing a decrepit wooden bridge that had nonetheless served for some two centuries, the new bridge stretched 276 metres and was supported on nineteen stone piers set on the 'starlings' driven into the riverbed. It was more than 9 metres wide and was protected to the south by a gate with a portcullis at the end, then by another tower with a drawbridge at the seventh arch. Begun in 1176, the work proceeded on average at the rate of one arch every twenty-one months, winter tides preventing work except between June and October. The technical challenges addressed by the organiser of the work, the chaplain of St Mary, Colechurch, named Peter (who died four years before it was completed in 1209), were formidable as the depth of water midstream was 4–5 metres even at low tide. The arches were all of different spans with a mean of 8 metres and each had to be supported somehow before the next one was constructed. The pile-driving was done by engines called gins, tripod frames erected on the barges driving heavy weights supported on ropes. This immense

'Sociotopography of medieval and late medieval towns in the north European zone as exemplified by Torún', same vol., 121–9; D Price, 'Medieval surveying and maps', *Geographical Journal* CXXI, 1955, 1–10. J Schulz, 'City plans', in J Friedman, K Figg (eds), *Trade, travel and exploration in the Middle Ages: an encyclopedia* (2000), 110–11 points out the rare examples where actual town plans have survived or been known to have existed, and makes the point that they would naturally have been discarded afterwards.

36 Lehmann-Brockhaus, *Schriftsquellen zur Kunstgeschichte*, nos 1835–6 (p. 369, from the Vita Reginhardi) (Liège); no. 730 (p. 151, from the Annales Egmundani) (Cologne).

construction would have called for a large team of skilled artisans, but such a pool evidently existed, and was very expensive: the bridge was financed by the monarch, the archbishop of Canterbury, the mayor of London and pious contributions. The tidal nature of the Thames ensured that continuous maintenance was essential, but it is significant testimony to the design that only two collapses took place in the medieval period, in 1281/2 and 1437, on both occasions following a period of neglect.[37] Meanwhile, engineers specifically called by the title of *pontanarius* were employed to build bridges at Arles (in 1178) and at Agen (named in a charter of Richard I in 1189).[38] These engineers were identified in the records by their full names. While experts point out that successful bridge construction did not require knowledge of stress levels, awareness of the hydraulic issues was critical and understanding of geometry was also necessary. The evidence across Europe therefore suggests that another area of engineering specialisation, alongside the building of the new Gothic architecture, was emerging during the twelfth century.

The best example of bridge building for military purposes was the construction of a bridge on the Seine, placed across ships, when the French king conquered the Anglo-Norman defences on Les Andelys prior to being able to besiege the fortress of Château Gaillard. Towers of oak were then erected above to provide a platform for archers or crossbowmen, enabling the French to capture these critical outer defences after a ferocious battle. The chronicler naturally ascribed credit to Philip Augustus; that the king's engineers were responsible seems more likely.[39]

As will be seen, there are accounts of the construction of siege engines, although it remains difficult to establish what technological progress also took place in the design and construction of the siege towers, battering rams and mantlets (variously "sows", "cats" and other anthropomorphic names), and the plentiful manuscript illustrations were not intended to be engineering drawings, but as will be seen in the following pages such devices became instruments of choice and were often decisive during the twelfth century. Intelligent human beings learn from experience and that experience became plentiful.

Manuscript illustrations also provide evidence for the shape and size of artillery (see Figs 3 and 4). These, and the records of expenditure on the parts of stone throwers and ballistas, suggest not much changed. However, some historians argue that contemporary sources identify an increasing power in stone-throwing

37 Brooks, 'Medieval European bridges', 28–31; Watson, *Old London Bridge lost and found*, 28–33; J Mesqui, *Le pont en France avant le temps des ingénieurs* (1986), 144–5; E Maschke, 'Die Brücke im Mittelalter', *Historische Zeitschrift* 224/2, 1974, 267–73; D Harrison, *The bridges of medieval England. Transport and society 400–1800* (2004), 87–91, 112–14. The evidence for London is in Watson et al., *London bridge. 2000 years of a river crossing*, here 83–91, 105–9, 122–38.

38 Mortet, Deschamps, *Recueil*, LXI (v. II, 134) (Arles), LXXVI (v. II, 159–60) (Agen).

39 Guillaume le Breton, *La Philippide*, trans. M Guizot (1825) 181.

artillery. This might have been expected and although there were practical limits to the development of the mangonel/petrary, it is probable that their builders would have learned how large they could actually be made and the optimum number of rope pullers. It is true that the number of chronicle references to bombardment by "huge" stones grows. One explanation offered was that a device called (by its inventor, Paul Chevedden) a "hybrid trebuchet", combining traditional human power with a counterweight, had evolved. I do not believe for many reasons that such an engine existed, but I do accept that artillery almost certainly did become more effective over this period of intensive innovation and progress, however, not to the point of being able to seriously damage a well-built wall.

Another weapon of importance that showed evidence of technological improvement over these years was the crossbow. If the weapon had always been available, it certainly became much more prominent during the twelfth century, leading to the well-known and widely ignored papal condemnations of its use in 1139, repeated by Innocent III later in the century. Muslim armies welcomed it for use in siege warfare and treatises on the weapon appeared under the 'Ayyubid sultans. By the end of the century, western rulers were making heavy use of crossbows and the manufacture of crossbow bolts was being organised on an industrial scale. The weapon required neither great skill nor great human strength to operate, but in contrast, a level of technological expertise was required for the manufacture of the various metal parts and for the assembly of the weapon itself.[40] Evidence of technical development, alongside the evolution of a more effective hand-held weapon to justify its expanded use, included the development of larger, two-footed versions, too large for manual use, and the development of mechanical devices to aid the spanning of the weapon, giving it much greater force than was possible with one spanned only by human strength. Cranks and (or) stirrups appeared for this function during the twelfth century, although their major evolution would take place in the following century. These weapons needed craftsmen to build them and more sophisticated government structures to oversee their manufacture and delivery.

40 DeVries, Brown, *Medieval military technology*, 2nd edn, 42; J Alm, *European crossbows: a survey*, trans. H Bartlett-Wells, ed. G Wilson (1994), 6–15; N Pétain, 'Philological notes on the crossbow and related missile weapons', *Greek, Roman and Byzantine Studies* 33/3, 1992, 265–91; R Loewe, 'Jewish influence for the history of the crossbow', in G Dahan (ed.), *Les Juifs au regard de l'histoire* (1985), 87–107, for the possible role of Jewish communities in the reappearance and use of the crossbow in western Europe; D S Bachrach, 'Crossbows for the king: the crossbow during the reigns of John and Henry III of England', *Technology and Culture* 45/1, 2004, 102–19, for the manufacture of crossbows under the Angevins; V Serdon, *Armes du diable. Arcs et arbalètes au moyen âge* (2005), 41–3.

MILITARY ENGINEERING IN PRACTICE

I have mentioned castles built according to geometric principles as a mark of the impact of new learning in Europe. There were also other changes in castle building that had a clear military function. It is important to stress this point because the traditional view of castles as being solely military buildings has been discarded, and understanding of the multiple roles of these constructions seen as so emblematic of medieval society has been and continues to be developed. Nonetheless, many castles still served military roles at times and the twelfth century witnessed a number of improvements in this aspect, which would continue to undergo development and would really see fundamental changes in the thirteenth. There is an enormous literature. Here I will mention a few significant changes, with examples from Britain: the building of defensive, projecting towers on curtain walls that offered flanking not available with plain walls (Framlingham, Orford, Windsor, Dover); the widespread adoption of arrow loops (a supreme example being Henry II's Avranches tower at Dover) [Fig. 12]; the employment of wooden hoarding to allow command of the foot of the wall; and the beginning of the use of more permanent protection in the form of machicolations. By the end of the century the first twin-towered gatehouses offered increasingly significant defences for the entrance (Chepstow in Wales, dated to 1189), while the addition of barbicans in front of the gate was just beginning (Dover, again). By the end of the twelfth century significant changes in overall castle shape and design were also apparent, with examples where the motivation was clearly military. Richard the Lionheart's unusual Château Gaillard in Normandy may be seen as an attempt to incorporate the military lessons this warrior had learnt from his campaigns both during the third crusade and against his rival in France. Sometimes the names of the builders of royal castles were recorded (see below), but for most castles they were not. How far the patron was responsible for the result, or how far the engineer in charge of the actual work based it on his own ideas, remains unknowable: where the former was an active warrior, an exchange of ideas may well have been productive. The innovations introduced in castles by great lords with international experience of warfare, like William Marshal who was responsible for the circular great tower and inner ward of Pembroke castle (Wales) and the gatehouse at Chepstow in the last years of the century, shows that it was not only monarchs who were leading castle design.[41]

The twelfth century in Europe was marked by regular conflict of which much was recorded in chronicles. While much of the evidence comes from the Middle East, however, it should be noted that the technology for constructing military

41 See Purton, *HEMS*, 280 (style), 12, 130, 232, 282 (machicolations), 376–9 (arrow loops); Hislop, *Castle builders*, presents a valuable summary of current interpretation and dating, esp. chapters seven and eight.

FIG. 12 Avranches Tower, Dover castle, Kent, provided with multiple arrow loops. (Photo: author)

equipment was available across the continent of Europe: the accounts of the campaigns of the early kings of Poland against their pagan Pomeranian neighbours make a number of references to machines and engines in use to capture enemy strongholds, including a siege tower that was raised by King Bolesław III (1102–38) so that it dominated the fortress at Czarnków and led to its surrender.[42] It was also an age during which crusades pitched western against Muslim technology from the Atlantic coast to Syria, with the Byzantine empire not always willingly engaged, and all of these events also attracted written records. In addition, for the first time, government records survive that are invaluable in understanding the real world.

The first crusade began a long period of intermittent conflict and occasional new crusades from the European heartlands that involved the transport of immense numbers of people across enormous distances to fight wars in an alien land. The logistical challenges posed are acknowledged only in passing in the many eye-witness accounts that have survived, written both by western clerics and,

42 *Gesta principum Polonorum. The deeds of the Princes of the Poles*, trans. P Knell, F Schauer , book II, c. 44, 200–1. The first time that "engines" are mentioned is under the year 1091 but thereafter they appear often.

less often, by those on the receiving end (the crusades themselves being of much greater significance to Christendom than to the majority of the Muslim world). Those challenges included being able to cross numerous rivers and being able to organise construction of great numbers of ships. During the second crusade, the army led by King Louis VII had to negotiate both the Rhine and the Danube: the former required the gathering of a vast fleet of ships, because there was no bridge, while the latter was crossed via bridges erected for Conrad III, whose armies had left earlier, and the fine bridge at Regensburg. Conrad III also employed a fleet to transport his baggage along the river down towards Bulgaria. In 1189, during the third crusade, the host assembled by Emperor Frederick Barbarossa made use of ships constructed "by carpenters with extraordinary skill" during their passage through Hungary. No names for these people, then, but evidence of wide-scale construction skills being available to rich monarchs.[43]

THE NORMANS IN THE MEDITERRANEAN – "DOCTISSIMI ARTIFICES ERANT"

A significant component of the successful host that eventually conquered Jerusalem in 1099 comprised Normans, including Bohemund who became prince of Antioch and whose father was Robert Guiscard, who acquired the duchy of Apulia (modern Puglia) by conquest. His brother Roger became count of Sicily. The military skill of these warriors was widely acknowledged. This necessarily included siege warfare where engineering capability was at a premium. There are plentiful contemporary accounts that appear to justify this reputation, including the claim that the princes themselves were skilled military engineers. This arises in the context of the overthrow of the existing regimes (Byzantine provincial governors and city states run by dukes of Lombard name and origins and the emirs of Muslim Sicily) and the construction of the Norman kingdom of Sicily during the eleventh and twelfth centuries. However, it has also been claimed that much of the fleet and the army, including the engineers, that served the first king, Roger (reigned 1130–54), was provided by the Muslim population whose rulers had run Sicily since they took it from the Byzantines in the ninth century. The final event in their conquest had been the capture of Syracuse in 878, achieved, according to Ibn ʿIdārī, by using (unidentified) "new" siege engines never seen before. They went on to introduce more sophisticated irrigation onto the island.[44] Whoever provided the engineering skill, the Norman-led armies were formidable, and the ability of

43 Odo of Deuil, *De profectione Ludovicii VII in orientem*, ed. V Berry, 22–3, 24, 33; Anon, *The history of the expedition of the Emperor Frederick*, ed. G Loud (2013), 59.

44 Quoted in L Chiarelli, *A History of Muslim Sicily* (2010), 43. Irrigation at 214–15.

the Normans to learn from the talents of such a diverse melting pot of civilisations explains much of their success.

The records of the Norman campaigns identify repeatedly their use of engineering skills to conquer cities, although it seems that they may have been quick learners from others rather than innovators. As early as 1041, following several victories over Byzantine armies, their aptitude in taking fortresses was praised – the words in the subheading above are from the Norman chronicler Malaterra who had moved to Italy in the late eleventh century. Richard at Capua in 1062 (who became prince of that city) used stone throwers, and Duke Robert did the same again at Bari in 1068. The capture of Palermo in 1072 was achieved using fourteen scaling ladders. The Normans' necessary access to naval power was discussed in chapter three. In other sieges the Normans resorted to the construction of siege forts to achieve an effective blockade. All these operations involved technical knowledge and skills.[45]

In 1131, the Normans constructed a wooden siege tower against the tower of Ravello, held by the men of Amalfi, and also nearly knocked it down with stones from artillery. In 1132, the king had first had the bridge over the river Sarno destroyed while he besieged Nocera, forcing his enemies to wait for it to be repaired – which it was, but they lost the ensuing battle. At the siege of Montepeloso in the same year, Roger had a large siege tower built that was moved slowly around to a weaker spot in the defences from where his soldiers shot down the defenders while his Saracens filled the town ditch and then set about destroying the outer defences using what was described as a long pole with a "huge" iron hook. This must have been suspended from a clever construction allowing it to be operated from a distance, although the garrison managed to cut it three times before it finally succeeded. The role of the Muslim soldiers in this account does not demonstrate that they were carrying out skilled tasks, but within the description it is evident that significant engineering skills were in use in Roger's army.

His opponents, including the rulers of Naples and Benevento, responded with their own attacks on defended sites with their own engines. In 1137, the emperor (Henry III), who had intervened to try to stop the advance of Norman power, and the people of Bari succeeded in capturing the Norman-held citadel by undermining it. Two years later, Roger returned to attack Bari and constructed no fewer than thirty "siege towers". This large number renders this translation

45 Amatus of Monte Cassino, *History of the Normans*, trans. Dunbar, ed. G Loud (2004), book IV, c. 28 (pp. 120–1), book VI, c. 19 (157–8). The translator confusingly uses the French translation which anachronistically describes the stone throwers as trebuchets. There is confirmatory evidence of the successful deployment of siege towers, battering rams and stone throwers (*not* trebuchets!) in other contemporary sources: G Malaterra, *De rebus gestis Rogerii Calabriae et Siciliae Comitis et Roberti Guiscardi Ducis fratris eius*, ed. E Pontieri, *RIS*, V pt. 1, book I, c. X, p. 13; book II, c. 45, pp. 52–3. See also J J Norwich, *The Normans in the south*, 132.

suspicious: the Latin simply translates as wooden engines and could also mean mantlets or protective shelters (such as those described by "Heron") under cover of which the attackers sapped the wall.[46]

It would be consistent with the Norman lords' devotion to bellicose pursuits that they had clear ideas of the capabilities of the various types of engine that their craftsmen might construct under their instructions, suited to the location in which they were operating. Reputations as engineers did not attach naturally to princes or magnates, so it may be wise to surmise that there was something material underpinning their fame.

Another to share this reputation was the Norman lord Robert of Bellême, who would play an important part in the crusader capture of Jerusalem. His origins in the tenth-century family of the king of France's *balistarius* were discussed in chapter three. Now, the twelfth-century chronicler Orderic Vitalis recounted his part in the siege of Bréval alongside the king of France and the duke of Normandy in 1092 as the man most "ingenious" in artifice who, specifically, constructed machines that were advanced on wheels against a wall or hurled enormous stones; he is also described as schooling his men in how to destroy walls and palisades.[47] The author apparently deliberately puns on the Latin words with the same root for ingeniousness and engines.

ENGINEERS IN THE CRUSADING ARMIES

The many accounts of the first crusade in particular, but also of other campaigns carried out under the banner of the cross, reveal much about the sophistication of siege engineering but also suggest that engineers from Italian cities played a large part.[48] The array of skills available to crusader commanders appears to have been substantial and one chronicler noted that Bohemund's army included many craftsmen.[49] At the siege of Nicaea (Iznik) in 1097, two men from Swabia named

46 E Curtis, *Roger of Sicily and the Normans in lower Italy 1016–1154* (1912), 308, repeats the claim originally made by M Amari, *Storia dei Musulmani di Sicilia*, in 1868, about the Muslim composition of Norman armies (v. III, 688–9). The source is A Telese, 'History of the Most Serene Roger, first King of Sicily', in G Loud (ed.), *Roger II and the making of the kingdom of Sicily* (2012), 81, 88–9, 92–3; 'Chronicle of Falco of Benevente', in the same, 205, 242; 'Annalista Saxona', in the same, 275.

47 Orderic Vitalis, *Ecclesiastical history*, ed. Chibnall, book VIII, vol. IV, 288–9.

48 R Rogers, *Latin siege warfare in the twelfth century* (1992), is an expert detailed study with much to inform about engineers and engineering skills. See e.g. 21–2, 131–2, and of particular interest, the connections between shipbuilding and siege engine construction skills underpinning the leading role of Italian maritime cities noted here (esp. at 201–3).

49 Ralph of Caen, *The Gesta Tancredi*, trans. B and D Bachrach (2005), c. 8, p. 30.

Henry of Esch and Hartmann paid for a mobile covered shed of oak beams which was completed to the last detail "with workmanship and bindings", but the price they paid was less than that of the twenty soldiers who were killed when it collapsed on top of them as they drove it up against the wall. Another effort involved constructing engines from carts "adapted by the skilful use of timbers and hempen ropes and straps of bulls' hide". Then a Lombard, described as a master and inventor of defences and siege works, offered (for pay) to construct an engine to bring down a tower and created a large machine to cover a successful sapping operation.[50]

The numerous sieges of the first crusade all involved the construction of siege equipment and in particular stone-throwing artillery, mobile siege towers, shelters (commonly called "sows") to provide close cover for undermining or sapping, and battering rams. At Antioch, the Norman Bohemund volunteered to construct the besiegers' fortified encampment because he had "the men capable of executing the work". At Jerusalem, the shortage of timber was a massive problem and soldiers were sent to scour the countryside for wood. The role of master craftsmen in constructing engines was reported. Ralph of Caen described how the collected timber was provided to the craftsmen who for five weeks were engaged in constructing the artillery and siege towers, "the posts, flooring and hurdles ready to be connected, bound and raised" before, famously, the whole massive erection was shifted to a weaker point in the defences overnight, the manoeuvre that made possible the bloody capture of the city. Two accounts described the success of a ram in breaking through the outer defences but that it was damaged by fire from the defenders and had to be destroyed by the attackers to enable the siege tower access to the inner wall. A thirteenth-century version of the lost *Chanson d'Antioche* named those responsible for engineering the wheeled rams as Nicholas de Duras and Gregory, the original failure as they were set alight by the defenders, then success at the second attempt. They also claimed the credit for the decisive siege tower: "Nicolas et Grigoire l'ont bien carpenté." These two heroes were apparently responsible not just for the tower, but also for a large stone thrower that battered the walls.[51] The account by Raymond of Aguilers described the appointment of Gaston of Béarn to supervise the construction of siege equipment (the reason given was his "ability and honesty"), while Count Raymond of Toulouse also put William Ricau in charge of

50 Albert of Aachen's *History of the Journey to Jerusalem*, trans. S Edgington, book II, c. 29 (vol. I, 65–6) and book II, cc. 30, 32 (vol. I, 66–7) and cc. 35, 36 (vol. I, 70–2).

51 Ralph of Caen, *Gesta Tancredi*, c. 121 (p. 134), c. 121 (p. 137), c. 123 (pp. 138–9) (Jerusalem); Albert of Aachen, *History*, book VI, cc. 2, 10, 11 (vol. I, 212, 216–17); *Mémoires de Pierre Tudebode (ou Tudeboeuf) sur son pèlerinage à Jérusalem*, ed. S de Goy (1877), V, I, 142 (Bohemund at Antioch), XV, I, 267 (Jerusalem); Richard le pèlerin, *La conquête de Jerusalem faisant suite à la chanson d'Antioche*, ed. C Hippeau (1868), ll. 3081–90, 3096–7, 4071–3, 4093–5, 4117–21.

similar operations on the side of Mount Sion. The artisans were paid from "public collections", while the Genoese sailors led by William of Embriaco salvaged ropes, hammers, nails and tools from the ships wrecked at Jaffa.[52] Together, these accounts confirm the role both of masters in charge of military engineering who were not always the noble leaders and of the (paid) artisans who executed their commands.

Individuals were named when the new king of Jerusalem besieged Tyre in 1111 and a household knight called Eustace Granarius constructed two siege towers (unsuccessfully, as both were set alight by the defenders).[53] At the next (this time successful) siege of Tyre (1124), the Venetians conveyed the materials for constructing siege engines and the men operating them were "instructed by experts in the art of throwing missiles". A later entry in this chronicle implied these experts could not have been present for long, because the besiegers had to send to Antioch for an Armenian named Havedic since no one in the camp "possessed the expert skill necessary for aiming and hurling".[54]

SHIPS AND THE CRUSADES

Insights into technical progress and engineering skills can be gained from studying the ships used by crusade leaders as an alternative to the arduous and dangerous overland route. By the middle of the twelfth century, and led understandably by the Italian maritime cities, technological breakthroughs had been achieved to enable the transport by sea of large numbers of horses. A new ship design was necessary and was used equally by Byzantine and Muslim rulers, and while there had been examples of horses conveyed by sea in the eleventh century, with the Normans particularly having mastered the art, by the twelfth this had become a more specialised art. The Venetians had overcome the problem of conveying sufficient water for both horses and crews for direct voyages (not involving frequent stops for refreshing supplies) by designing sufficient space for

52 Raymond of Aguilers, 'Historia Francorum', in C Tyerman (trans.), *Chronicles of the first crusade* (2004), 325–6.

53 Albert of Aachen, History, book VI, c. 2 (I, 212); book VI, c. 10 (I, 216) (Jerusalem); book 12, c. 6 (II, 184) (Tyre); H Gibb (ed.), *The Damascus Chronicle of the crusades extracted and translated from the chronicle of Ibn Al-Qalanisi*, 122–3, gave the dimensions of the besiegers' siege towers which were as high as the walls of the city, and also explained how the defenders managed to destroy them completely with incendiaries, and how a battering ram was thwarted by pulling its head away with grappling irons, more than once (p. 124).

54 William of Tyre, *A history of deeds done beyond the sea*, trans. E Babcock, A Krey (1943), vol. II, book 13, c. 6 (p. 10), c. 10 (p. 15); 'The first and second crusades from an Armenian Syriac chronicle', trans. A Tritton, *Journal of the Royal Asiatic Society*, 1933, 95–6.

containers by the 1120s. The names of those behind the technological progress are not known, but the ability of twelfth-century shipbuilders to rise to the new challenges posed by their rulers on all sides of the Mediterranean and to enable successful overseas expeditions seems to have been rapidly diffused, which explains the successful naval expeditions undertaken by Italian cities, crusader states in the Middle East, crusading kings like Richard I, and counter-attacks by a revived navy established by Saladin.[55] The particular skills involved in shipbuilding were also linked more directly to military engineering to the extent that they called upon similar expertise. Both ships and siege towers were large wooden structures involving the selection and effective working of very large timbers. There is direct evidence from Genoa of the connection between master shipbuilders and the master carpenters working on land.[56]

ITALIAN ENGINEERS AT HOME

The prosperous Italian cities played a major role in the success of the first crusade and they also provided some of the engineering expertise (though evidently not all). The level of technical ability available in Italian cities was demonstrated repeatedly in a contemporary chronicle's account of the war that led ultimately to the capture of Como by Milan in 1127 – from the first onslaught when undermining and escalade were attempted in 1120, to conflict in 1122 when "sufficiently ingenious artificers" from Genoa were engaged and stone throwers built to hurl great stones against the walls, to the defenders themselves acquiring the skills to construct ballistas which they used to defend against attacks by battering ram (1124), right through to the final siege where once again the "ingenious" Genoese were hired because of their renown in the skills of constructing siege towers and artillery ("artifices noti qui sunt ingeniosi"), while those of Pisa were of similar repute ("nec non satis ingeniosas"), along with miners skilled in undermining walls. They constructed a

55 J Pryor, 'Transportation of horses by sea during the era of the crusades', *Mariners' Mirror* 68, 1982, 9–28, 103–25; the same, *Geography, technology and war. Studies in the maritime history of the Mediterranean 649–1571* (1988), 63–6; M Mollat, 'Problèmes navals de l'histoire des croisades', *Cahiers de Civilisation Médiévale* 10(39), 1967, 345–59; A Ehrenkruetz, 'The place of Saladin in the naval history of the Mediterranean sea in the Middle Ages', in S Rose (ed.), *Medieval ships and warfare*, 235–51, recounted the re-establishment of an Egyptian fleet on the orders of the sultan after 1177 at Alexandria to challenge crusader control of the sea, the building of a fleet of galleys and transports called *ṭarrida* and successful raids on Acre in 1180 using ships carrying catapults to propel combustibles. However, their performance during the crusader siege of Acre was "dismal".

56 Rogers, *Latin siege warfare*, 203 and the Genoese sources listed there.

siege tower of immense timbers "sagaciter" (wisely) and two "cats", and used a ram with a strong roof to bring down the wall of the city.[57] Clearly, Genoa and Pisa were the cities that contemporaries knew could supply their needs for craftsmen capable of creating the most modern siege equipment.[58]

Italian master craftsmen were in regular employment at home and abroad and some names have been preserved. While the Normans were establishing their monarchy in the south, cities were growing as self-governing entities in the north. Early records from Florence confirm the use of siege artillery in their intermittent conflicts with Siena, as for example in 1147.[59] But the strongest evidence comes from the well-recorded efforts of Emperor Frederick Barbarossa to turn his title of "king of Italy" into reality, a task that entailed conflict with wealthy Milan, a city keen to establish its own local hegemony. Milan could certainly call upon engineering talent among its citizens, but the emperor found similar talent available among Milan's rivals, his allies, chief among them Cremona. Other cities, not wanting to be ruled by a German emperor, allied themselves with Milan, and Frederick became involved in two decades of conflict in Lombardy involving numerous sieges. In the end his enemies triumphed.

In 1156, Tortona was stormed by imperial forces, and the emperor's army used a battering ram and stone throwers: the chronicle accounts include the reference to the ram "being suspended to increase its impact". This simple statement suggests a technical advance because battering rams would often have been no more than a long pole tipped with iron and carried bodily by a team of men. The chronicler here confirmed that this must have been a mobile, roofed shed, in the form in which rams are usually understood. Furthermore, the words used refer not to protection for the team of operators but to the dynamic of the ram itself, an aspect of physics they could only recognise from observation. The Milanese rebuilt the city's defences and began to construct "the strongest bridge" nearby, confirming that skills in crossing rivers were essential on the Lombard plain. Two years later, the emperor himself took his army across the river Adda, having previously destroyed a bridge across the Ticino, in order to assail Crema.[60]

57 'Anonymi Novocomensis Cumanis sive prema de bello et excidio urbis Comensis ab anno MCXVIII usque ad MCXXVII', ed. J Stampa, *RIS* V (1724), ll 390–3, 436–8 (pp. 421–2), 637–41, 47 (p. 427), 900–4 (p. 433), 1822–32, 1877–88, 1903–4 (pp. 452–4).

58 Discussed in A Settia, 'Un "lombardo" alla prima crociata. Tecnologie militari fra occidente e oriente', *Società, Istituzioni, spiritualità. Studi in onore di Cinza Violante* (1994), II, 843–55.

59 'Sanzanomis Gesta Florentinorum', in O Hartwig (ed.), *Quellen und Forschungen zur ältesten Geschichte der Stadt Florenz* (1875), 7.

60 *Barbarossa in Italy*, ed. T Carson (1994), book I, ll. 366–77 (p. 13); G Fiamma, *Manipulus Florum. Cronaca milanese del trecento*, trans. R Frigeno (1993), cc. 175–82, pp. 22–4, 38–9; Otto Morena, 'De rebus Laudensibus', *MGH* SS XVIII, 594.

The siege of Crema (1160), initially carried out by the forces of its bitter rival Cremona and subsequently reinforced by imperial troops, also involved a substantial deployment of siege technology. This included a wooden siege tower of large proportions ("nothing like it had been seen this side of the sea" – Otto Morena. He described it as more than 40 feet (*brachiis*) high and wide enough for twenty men at a time to charge out from it). Two cats were constructed, one larger than the other, and these were moved against the walls once the ditch had been filled in. The siege tower was used to lay a bridge across to the wall, but the defenders held out and eventually succeeded in destroying the bridge. They had also attempted to mine under their own walls to reach the cats and were successfully repelled by the besiegers countering their mine. The technology on display was the best that could be brought to bear by Europe's most powerful ruler, but the builder of the mighty tower was described as from Crema, a master more "ingenious" than any other in the city, and capable of building mangonels and petraries and all other kinds of equipment used at the siege: he was brought over (bought over?) to the side of the emperor who welcomed him "with great joy" and, more materially, a fine warhorse and 12 pounds of silver. His name was Marchesius.[61] Later in the same year, it was recorded that the city of Lodi was provided with walls and a native of Cremona was in charge of the construction: his name was also recorded, and this engineer, Tinto Musa da Gata, went on to be ennobled by the emperor, who elevated him to the rank of count and endowed him and his heirs with feudal rights over a substantial estate including a former Milanese castle, in return for his excellent service to his cause.[62]

The Milanese themselves called upon native engineering skills when in 1160 a squadron of one hundred specially constructed armoured fighting carts equipped with scythes caused the retreat of the army of the emperor. They had been constructed by a Master Guintelmo, and the historian A A Settia has suggested the handing on from generation to generation of a technical notebook with the specifications (which derived from ancient practice) and that Guintelmo himself may have established a dynasty of military engineers who functioned in Lombardy. The suggestion certainly fits with a likely model of how skills were acquired and handed on, as does the possibility that these masters were also sufficiently literate to follow a text, although the illustrations of such books as the *De rebus bellicis* might have served just as well as the words. The lethal carts of 1160 were described rather differently by the Milanese chronicler Fiamma and in the pro-imperial *Gesta*

61 Otto Morena, 'Historia', ed. F Güterbock, *MGH* SS NS 7,611–19; *Barbarossa in Italy*, ll. 2990–3110.

62 Otto Morena, 'Historia', *MGH* SrG NS7, 117; 'Die Urkunden Friedrichs I 1158–67', *Die Urkunden der deutschen Könige und Kaiser*, *MGH* Diplomata regum et imperatorum Germaniae, 10, pt. 2 (1979), 79–80, 103–4 (dated 17 May and 30 December 1159, the second granted in the camp outside Crema).

Frederici but both accounts confirm the basic attributes of the device.[63] Fiamma's detailed description states that they were drawn by teams of six horses, themselves protected by armour. He described the design of the cart in terms of a ship with the men inside operating the scythes like oars. It is worth noting that Fiamma was writing many years after the event he described and one of his accounts reports 300 of these machines but the other reports 100, so caution is advisable in taking either figure at face value.[64]

Part of the second crusade decided to carry out its operations in al-Andalus rather than the holy land and did so with great success, and eye-witness accounts survive. A multinational force (chiefly men from England, Flanders and Cologne, under the command of relatively lowly knights) managed to capture Lisbon for King Afonso of Portugal and they did so by constructing effective artillery ("Balearic mangonels" operated by teams of fifty rope pullers at a time), a ram and a siege tower 95 feet (29 metres) tall. The last two weapons failed – the ram and the tower did not prove as mobile in the sandy soil as the builders must have expected and were burnt by the defenders. A further set of attempts involved constructing a second tower, this one 83 feet (25 metres) high, and new stone-throwing engines, one built by the seamen. A number of mines were dug but were either discovered and wrecked or otherwise failed in their objective. The siege tower was cut off by the tide but was heroically defended until it was possible to re-establish contact with it, although the engineer building it was wounded at this time, the chronicler recorded. The city surrendered when the siege tower had been pushed to within a metre of the wall top, from where a bridge could be lowered to allow the besiegers to cross. The engineer, it was specified, came from Pisa. The possession of these engineering skills was decisive, and had not previously been available in western Iberia, where evidence of siege warfare other than by blockade is weak. It is also of interest that Flemings and Rhineland Germans should be the ones credited with skills in mine warfare.[65]

By the time of the third crusade, two more generations had passed and the resources (including technological) of the wealthy and powerful kings who set off

63 A Settia, 'I carri falcati di maestro Guintelmo fra antichità e rinascimento', in the same, *Comuni in Guerra. Armi ed eserciti nell'Italia delle citta* (1993), 277–87, here 279, 81, 85.

64 Fiamma, *Manipulus Florum*, trans. R Frigeno (1993), 56–7; *Chronicon Maius*, ed. Ceruti, 236–8.

65 Anon, *De expugnatione Lyxbonensi*, ed. C David, 134–7, 142–7, 158–65. The Pisan engineer is referenced at 142 and (possibly, if it is the same man), at 162 ("artifex noster"). The king paid for the tower. The siege is discussed by M Bennett, 'Military aspects of the conquest of Lisbon, 1147', in J Phillips, M Hoch (eds), *The second crusade. Scope and consequences* (2001), 71–87 and Rogers, *Latin siege warfare*, 182–9. The rate of shooting claimed for the artillery – one shot every eight seconds over ten hours – surely raises questions about the reliability of the account.

to recover Jerusalem following Saladin's victory at the battle of Hattin (1187) and recapture of the capital of the crusader kingdom were of a different scale. Saladin had successfully captured city after city, using the full panoply of siege technology available, especially large stone-throwing artillery, although in many cases the decisive weapon proved to be the sap or the mine (Tiberias, Jerusalem itself, Lattakia, Darbsak, Kawkab). His own skills in siege warfare were to be matched by one of his most formidable opponents, Richard I (the Lionheart), who practised them en route to Palestine at the capture of Messina with his ram and wooden tower. One chronicler described him as a king than whom "no one was better at capturing cities and overthrowing castles".[66] The evidence is powerful that he was capable of giving close attention and detailed directions to the craftsmen employed to construct his siege engines.

ACRE 1190–91

There is plentiful evidence, if usually incidentally, of the resources commanded by the most powerful rulers of the late twelfth century. At the siege of Acre, from May 1190, three large wooden siege towers were constructed by the "artificers" so as to resist enemy attempts to burn them, while the archbishop of Besançon and Count Henry of Champagne each paid for the construction of a battering ram. In June 1191, while stone throwers of unprecedented size were weakening the walls, King Richard had made an engine "vulgo dictam berefridum" (i.e. a siege tower) – although it is mentioned in only one source – and all the while besiegers were digging under the walls and in some cases meeting the defenders as they countermined. In the same month, the crusaders captured a Muslim ship carrying many weapons including "Greek Fire" in phials. Most of the crew were killed but Richard spared thirty-five who were "skilled in the making of machines", a significant reference to the value attached to skilled engineers by this time. The references to Greek Fire are also interesting. In April, a present from the caliph arrived for the defenders in the form of "two loads of Greek Fire and a number of specialist artificers". Another source was Damascus, which was apparently "passionate" about incendiaries and the specialists needed for them. The governor of Acre having said that he had already tried the weapon without success, another attempt was ordered and the artificer, called 'Ali, son of a master coppersmith of Damascus, hurled pots of inflammables against the side of the crusader siege tower facing him, then ignited the mixture with a further pot that had been lit already. These details are sufficient to confirm that this was not the same Greek Fire that the Byzantines had developed in the seventh century, but petroleum-based incendiary grenades

66 *Chronicon Ricardi Divisiensis de rebus gestis Ricardi Primi Regis Angliae*, ed. J Stevenson (1838), c. 26, p.24.

that had to be set alight after striking the target. The evidence also confirms that a group of specialist operators from civilian backgrounds would be recruited for military operations.[67]

For the most part, these artificers remained nameless in the sources, but their presence – and their value as assets – is clear from both the Christian and Muslim sources.

ENGINEERS IN GOVERNMENT RECORDS

If chroniclers were not sufficiently interested in the commoners who built the rulers' engines, those who recorded their expenditure prove to be a more fruitful source of information. This section will focus on what can be retrieved from the Anglo-Norman kingdom under the Angevin dynasty that took the throne following the civil wars of the preceding reign of King Stephen (1135–54). His succession had been challenged by Henry I's surviving daughter, Matilda. It is now customary to dispute the portrayal of this period as truly, as previously dubbed, "the anarchy", but it is true that the Great Roll of the Pipe, as the exchequer records were known, existed in 1130 but does not reappear until the first year of Henry II (1154–89).

Prior to this, it is not possible to know how the government of the Norman conqueror and his sons dealt with their military engineering needs. The one man named as "ingeniator" in the 1086 Domesday Book was Waldin, a tenant in chief from Lincolnshire, and there is no information about what he had done to earn the title. It is an assumption, though possibly a justified one, that he must have been William I's chief engineer who served the king in his various campaigns, but there is nothing to prove it. Twelve people named in Domesday were identified as "balistarius" or "arbalistarius", some tenants in chief, others tenants of another, though all have Norman names. It is possible that the reason for the title was that the men in question held their land by dint of service due as crossbowmen, although it was long ago recognised that the actual service quickly fell into disuse. It is just possible that, in the absence of any other signifier of a role, the term had a wider meaning as someone responsible for the king's artillery, but the numbers

[67] *Itinerarium peregrinorum et gesta regis Ricardi*, ed. W Stubbs (1864), I, c. xxxvi, p. 84, cix, p. 111, II, c. xlii, pp. 207–9 (ship), III, c. vii, pp. 219–20 (berefridum); Ibn al-Athir, *Chronicle for the crusading period from al-kamil fi'l Ta'rikh*, ed. D S Richards, Pt 2 (2007), XI, XII, 320, 347, 372–3 (incendiaries); Bahā' al-Dīn Ibn Shaddād, *The rare and excellent history of Saladin*, trans. D S Richards (2002), c. 80 (p77) (Jerusalem), cc. 89, 95, 96 (pp. 85–9), 109 (p. 151) (ship). On the 'Greek Fire', see the analysis by J Haldon (bibliography).

suggest that the narrower interpretation is more likely.[68] There are also three tenants identified as carpenters, whose particular service, if any, in 1086 is also not evident.

The Pipe Rolls are the next source of information.[69] The only survivor of the pre-Anarchy rolls dates from 1130 and records payments to several masons for building work: the amount paid suggests these were master masons in charge of construction; and also £10 12s 11d was paid to "Gaufr. Ingeniatori" as wages and several payments for specific works.[70] When the series of rolls resumed in the first year of the reign of Henry II, there were numerous references to payments to masons and carpenters for much building work, often on royal palaces and castles, and to *balistarii*. This king was involved in warfare throughout his long reign and seems to have been entirely familiar with the requirements for a successful operation.[71] The chief interest of the financial records of his and his sons' reigns for this study lies in the payments for a number of men identified as *ingeniatori*. The range of functions for which they were paid confirms their role as multi-skilled experts in building as well as in the construction and use of military equipment. The earliest to be mentioned, and the longest serving, was called Ailnoth, who was paid between 1158 and 1190. His works ranged from construction and repairs of royal dwellings to the destruction of the castles of the rebellious earls of the rebellion of 1174–75 at Walton and Framlingham. Year after year, Ailnoth was paid wages at the rate of £7 12s 11d. Yvo the Engineer first appeared in the year 1172/3. In the following year he was specifically recompensed for hiring carpenters to construct siege engines for the king's army at Huntingdon, which was assembling during the rebellion. Another *ingeniator*, named Maurice, appeared for the first time in the accounts for 1181–82, alongside a fourth named Urric. A master mason named Elyas and another engineer named Reginald appeared in the first year of Richard I, alongside the final payment to Ailnoth of the same wages as he had received in 1158.[72] From the evidence of what is recorded about what they were being paid for doing, it seems there was some degree of specialisation: Ailnoth and Maurice, for example,

68 K Keats-Rohan, *Domesday people. A prosopography of persons occurring in English documents 1066–1166* (1999), I, 167, 200, 208, 216, 217, 248, 258, 307, 308, 327, 347, 370, 446 (Waldin), 460; J H Round, *The king's sergeants and officers of state* (1911), 14–16. W Porter, *History of the corps of Royal Engineers* (1889), understandably asserts that Waldin was the first to hold what must have been a continuous chain of royal engineering officers (I, 7–8).

69 See Appendix for a tabulation of the Pipe Roll references discussed in this paragraph.

70 *Pipe Roll of 31 Henry I, Michaelmas 1130*, ed. J Hunter (1833), 1, 22, 40 (masons), 143–5 (Geoffrey).

71 See now J Hosler, *Henry II. A medieval soldier at war, 1147–1189* (2007).

72 D Renn, 'Ailnoth 'the engineer' and Orford Castle', *Castle Studies Group Journal* 26, 2012/13, 201–1, discusses the career of this engineer.

seem mainly to have been employed on building work, the latter being prominent in the construction work at Dover castle where his service was specifically noticed, while Urric, who was being paid for his sustenance in 1184–85, although he was granted lands (presumably so he could support himself) in the following year, seems to have been a master of siege engines.

The advent of King Richard and his determination to go on crusade led to great military preparations, and the problems caused by the dissent led by his brother John during his long absence led to the first reference to a new engineer, Richard, and to mention of Reginald. They were particularly connected to the construction and operation of a petrary and a mangonel and the repair of artillery stored in the Tower of London (1193). Roger Enganet was paid for transporting arrows, bolts and incendiaries (*igne graeco*) from London to Nottingham in 1194, while Elyas the engineer was paid for taking a mangonel and a petrary from Windsor to the same destination. Similar service was called for on other occasions after the end of this rebellion, against external foes such as on the Welsh border, while Elyas was particularly engaged on repairs to castles, suggesting again a degree of specialisation among what now appears to have been a small corps of specialist engineers retained on the royal payroll, who were all paid a standard salary each year. A number of other recipients appeared from time to time, sometimes on only one occasion, when they were paid for specific work. During the early 1200s, under King John, there was William Báard (1205) and Albert (1206 and 1207). Fortinus made a single appearance in the accounts for 1204.

Urric's continued role as a royal military engineer was also picked up in the financial records of the Norman duchy. Working in the royal arsenal at Chinon, he was paid £25 7s 2d by the seneschal of Anjou for the construction of engines on 30 May 1200, alongside instructions to pay other named masters for the making of crossbows. Urric was sometimes termed a carpenter, sometimes a master and sometimes an engineer in these various accounts but it is likely they refer to the same man.[73]

Despite the number of "engineers" around the end of the twelfth century, for most of the time there appear to have been only one or at most two people holding the title in royal service – others who appeared in the accounts once or twice were presumably hired for specific services, but not retained in the same way that Ailnoth, Maurice, Richard, Reginald and Urric were. The retained engineers were paid at a healthy rate of 7d per day, even if it did not change from the first record in 1130. Although a long way down the social scale from the count created as a reward for loyal service by Frederick Barbarossa, these records confirm that during

[73] *Rotuli Normanniae in turri Londoniensis asservati Johanne et Henrico Quinti Angliae regibus*, ed. T Hardy (1835), I, 24, 27; and 'Rotuli Normanniae in turri Londoniensi asservati Johanne Angliae regis ab anno MCC ad annum MCCIV', *Société des Antiquaires de Normandie. Documents historiques*, ed. Léchaudé-Anisy (1845), I, 98.

the twelfth century engineers had become recognised specialists and that kings were happy to pay them a retainer to keep them in their service. As indicated, it also seems likely that their particular expertise was recognised in some division of labour between overseeing construction work on the one hand and organising the royal siege artillery on the other.

Another area where the Angevin monarchs' Pipe Rolls provide valuable evidence is the increasingly regular use of master miners for royal military operations.

MINERS AND MINE WARFARE IN THE TWELFTH CENTURY

The military role of miners has already featured in the accounts of the many siege operations of the first three crusades, on the side of the crusading armies and those of their opponents [Fig. 13]. The role of mining in the armies of the Byzantine empire appears to have been an unbroken tradition from the time of late antiquity, although even where much detail was provided (as in the tenth-century records) nothing specific was said of the miners, and it may be that they were drawn from the soldiers of the army rather than recruited from those who made their living from mining: indeed, it has been identified that in contrast to elsewhere, mining in the empire was carried out as a secondary source of income by peasants at a very low technological level by this time (the territory containing the main iron mines having been lost to the Seljuk Turks in the eleventh century).[74]

The Arabic sources are similarly unhelpful in explaining where the unquestionably proficient miners of their armies came from and on what terms they served. Ibn Shaddād made passing reference to the suborning of Aleppan sappers by Richard I; they were captured when Acre fell and were then employed by the king to bring about the capture of Dārūm (May 1191).[75] Aleppo was also identified (by Ibn al-Athir) as a major source of foot soldiers for the Muslim army, and other Arabic writers identified the source of the miners as north Syria (particularly Aleppo) but also Diyarbakir and Mosul, and the distant region of Khorasan. What is clear from the descriptions is the often critically important part played by both surface sapping and underground tunnelling in the capture of crusaders' castles and cities throughout the course of the twelfth century.[76]

74 K-P Matscke, 'Mining', in A E Laiou (ed.), *The economic history of Byzantium* (2002), I, 115–20, here 116–17.

75 Bahā' al Dīn Ibn Shaddād, *The rare and excellent history of Saladin*, c. 210, p. 203.

76 See the excellent discussion by N Prouteau, '"Beneath the battle?" Miners and engineers as mercenaries in the Holy Land (XII–XIII centuries)', in J France (ed.), *Mercenaries and paid men. The mercenary identity in the Middle Ages* (2008), 105–97, here 108.

FIG. 13 Soldiers sapping a tower: on a capital in the Castle Museum, Foix, Ariège. (Photo: author. With permission of the Château de Foix, département d'Ariège)

In Europe, the apparent long hiatus between the last definite examples of military mining in late antiquity and the next reference at Laon in 938 was discussed in chapter three. Clear evidence for what happened in the following period comes from the kingdom of England, where the survival of documentary evidence provides conclusive proof that by the twelfth century, if not by the late eleventh, the role envisaged by Vegetius for miners had been (re-)established but not in the form envisaged by the Roman. The miners serving the kings of England were not soldiers at all, but people who made their living from the excavation, treatment and sale of metals from the earth. Whether this model applied elsewhere is not always clear from the more limited evidence, but inferences can be drawn.

William the Conqueror shared the military knowledge and prowess of the other Normans then creating a new kingdom in the Mediterranean, but, perhaps because of the circumstances of the conquest of England, he does not seem to have used miners in his campaigns with just one exception: his eighteen-day midwinter siege of the city of Exeter, which rebelled against him in 1068. The story was told by the non-contemporary Orderic Vitalis (who was using a lost contemporary source); he mentioned the king's attempt to undermine (*suffodere*) the ancient Roman walls. Exeter surrendered before it could be captured and the chronicle does not indicate whether the mine was a surface sap or a tunnel.[77] Nor did Orderic have any reason to say who those doing the digging were. Apparently, this was the first time that William's new English subjects had been part of his army, and it is at least possible that the miners who attempted this operation were from the Forest of Dean.

Mining for coal and iron had taken place in this area since Roman times and although documentary evidence of their status only survives from the thirteenth century, there is substantial indirect evidence that these were "free miners" with the right to work wherever they chose within the forest, provided only that the king received a share of the proceeds. The area provided iron for the English navy under Edward the Confessor and it is speculated that the Conqueror may have reaffirmed their privileges on his progress in his first years as king. Later affirmations of the miners' rights and duties have been preserved, and evidence that the Forest of Dean was royal has survived from the eleventh century onwards. The castle of St Briavels located at its heart, built from 1131, became a major arms factory and arsenal for the royal armies. Much of the working was done on the surface but there were also deep mines, some of which were worked for centuries and were responsible for the immense networks of tunnels and caverns still preserved to the present day.[78]

The first explicit linking of miners in royal service to the Forest of Dean dates from the early thirteenth century but the Pipe Roll noted payments to teams of fourteen miners in the service of the archbishop of Canterbury at the castle of Welshpool in 1196 and an additional twenty-four in royal service.[79] The thirteenth-century records can be read back to conclude that there was a long history of royal military service for these miners, reinforced by the fact that their homes and workplaces were royal, and the constable of St Briavels would have been the official who would have mobilised them when so ordered. However, it is well attested that other mining communities were flourishing in the Weald of Sussex and in Derbyshire and Durham, as well as in the tin mines of Devon and Cornwall, so it is possible that there were other places from which military miners were recruited, but there are no records naming these as such and certainly none when the references to the Forest of Dean become commonplace in the thirteenth and fourteenth centuries [Fig. 14]. There is evidence that iron from the Weald was bought for military uses but not, it seems, the service of the miners themselves.[80]

77 Orderic Vitalis, book IV (ed. Chibnall, vol. II, 210–11). R Higham, 'William the Conqueror's siege of Exeter in 1068', *Transactions of the Devonshire Association for the Advancement of Science* 145, 2013, 93–132.

78 I Cohen, 'History of iron working in and near the Forest of Dean', *Transactions of the Woolhope Naturalists' Field Club* XXXIV pt. III, 1954, 161–77; C Hart, *The free miners of the Forest of Dean and hundred of St Briavels* (2002), 3–11, 22–29; 141. The iron mines of Clearwell are now a museum.

79 Pipe Roll 8 Richard I, 42.

80 H Cleere, D Crossley, *The iron industry of the Weald*, 87–8, 103–4.

Fig. 14 A Forest of Dean miner on a plaque in Newland Church, Gloucestershire. (Photo: Tim Sutton)

In England there is little evidence of mining being used for military purposes during the twelfth century until the brief civil war between Richard I and his brother John. The exception was when their father Henry II ordered the levelling of Hugh Bigod's donjon at Bungay castle after he defeated the rebellion of 1174–75. The evidence for this (preserved because the final destruction was not actually carried out) remains in the form of the tunnel under one corner of the great tower that can still be seen, and for which the only reasonable explanation is that it was excavated as the first stage in a plan to undermine the corner. Most of those who have studied the castle generally accept this explanation for the existence of this tunnel.[81]

The wars of Philip Augustus of France (1180–1223) and his successful campaigns to exploit the divisions among and between Henry II and his sons offer further evidence that mining had become an integral part of warfare in the late twelfth century. However, the sources do not say where the miners came from, although the terminology implies strongly that they were mineworkers as distinct from soldiers.

81 Radulf de Diceto, *Ymagines historiarum*, ed. Stubbs (1876), I, 385; Matthew Paris, *Chronica Majora*, ed. Luard (1874), II, 294; H Brown, 'Some notes on Bungay castle', *Proceedings of the Suffolk Institute of Archaeology and Natural History* XXII, 1936, 109–19.

Guillaume le Breton's eulogy of the victories of Philip contains a description of the capture of the castle of Boves in 1185. Protected by the shields held by the French knights, miners chipped away at the stonework of the wall at the foundation. They used tree trunks and shaped timbers to prop the wall as they proceeded to remove more than half the depth of the foundations; the fire was set when they judged they had dug away sufficiently and the result was a spectacular collapse, the smoke and dust blotting out the sun according to the poetically inclined chronicler. Soldiers stormed in and the garrison retreated to their citadel. At the famous siege that led eventually to the capture of the magnificent and brand new fortress of Château Gaillard in 1203–04, erected at the orders of Richard I in the 1190s, a "machine" was built to protect the miners as they dug away at the foundations of the inner wall. On this occasion, the Anglo-Norman garrison countermined and drove out the French, but the combination of undermining and bombardment by artillery proved sufficient to create a breach.[82] If it was unlikely that there were miners among the garrison, the countermine must have been dug by soldiers with some previous experience of doing so. There is frustratingly little evidence of the organisation of mining in the French kingdom at this time, except that much of it was in the hands of (especially Cistercian) monasteries. It is therefore not clear where Philip Augustus recruited his experts for his numerous expeditions.[83]

Ordinary miners found themselves in military service across Europe. In 1168, Widukind of Desenberg broke the peace made with his overlord and was besieged there by Henry the Lion, duke of Saxony. The castle having resisted artillery, a chronicler recorded that Henry summoned the *viros industrios* from Rammelsberg who carried out an "unheard of" feat by burrowing into the castle rock and finding its well, which they then intercepted and thus brought the now dry fortress to surrender. The Rammelsberg was an area of silver mines over which Henry exercised right as imperial advocate, so here is a further example of a ruler mobilising a mining community to meet a military need. The castle itself was actually a small stone-walled enclosure with a central tower, strong chiefly as the result of its location.[84]

A few years later, in 1176, in conflict between the king of Denmark and the town of Stettin, the Danes having brought various siege engines to bear also sought to

82 Guillaume le Breton, *La Philippide*, trans. Guizot, 51–3 (Boves), 209 (Château Gaillard).

83 P-J Hesse, 'Artistes, artisans ou prolétaires? Les hommes de la mine au moyen âge', in X Barral I Altet (ed.), *Artistes, artisans et production artistique au moyen âge* (1986), I, 431–73, here at 435–7.

84 'Helmoldi presbyterii Buzoviensis Cronica Slavorum', *MGH* SrG, ed. B Schneidler (1937), book II, pp. 210–11. Jordan, *Henry the Lion*, 43. F Krahe, *Burgen des Deutschen Mittelalters* (1994), 146, describing the site as rebuilt in the fourteenth century. Desenburg is in Warburg-Daseburg, Nordrhein-Westfalen.

undermine the walls, and the chronicler this time used the same term used by Vegetius to identify the miners as *cuniculor(ii)*. This time there was no indication of where they were brought from.[85]

The twelfth century was a period when rulers actively sought to consolidate their rights to mineral wealth (a process that had begun much earlier, for example, with the Saxon emperors and their oversight of the mining industry of the Harz mountains), and the first European examples of the codification of the arrangements date from the eastern Alps and northern Italy (both Lombardy and the Trentino) in the late twelfth century, whence date the Codex Wangianus of 1185 and charters of Frederick Barbarossa of 1182 and 1189. The mines referred to in these texts included deep mines with long underground galleries. There seems as well to have been considerable freedom of movement for those with mining skills: rulers encouraged immigration of those with the skills required, many of them of German origin.[86] Reference has been made already to mining as part of the multiple siege warfare skills available to the rulers of Italian cities during this century. This then confirms the continuation of ancient iron and silver mining in Lombardy and the Alpine valleys and, although without firm proof of the connection, suggests it may have been these mining communities that supplied the miners who carried out the digging for Milanese, Genoese or other paymasters when they were hired for warlike purposes.[87]

While the rulers' motivation for developing this industry was economic and fiscal, it is not surprising that from time to time the skills these workers had acquired in their quest for silver, lead, tin or iron would be called upon for attacking a hostile fortress. Some technical development had also taken place that improved productivity or the quality of the resultant metal emerging from the furnace, but there had been no change in the technique involved in mining itself.[88] As already

85 Saxo Grammaticus, *Gesta Danorum*, ed. J Olnik, H Raeder, book XIV, xliii (vol. I, 488–9).

86 P Braunstein, 'Les statuts miniers de l'Europe médiévale', reprinted in *Travail et enterprise au moyen âge* (2003), 141–63, here 146–8; R Kink (ed.), *Codex Wangianus. Urkundenbuch des Hochstiftes Trent. Fontes rerum Austriacarum*, pt. 2 v.5 (1852), 64–6; G Trener, 'Le antiche miniere di Trento', *Società degli Alpini Tridentini* XX, 1896–98, 27–89.

87 M Tizzoni, 'Mining and smelting in medieval Lombardy', in E Herring et al. (eds), *Papers of the fourth conference of Italian Archaeology*, 4 pt. 2 (1992), 231–4.

88 On the Harz mines, see J Nef, 'Mining and metallurgy in medieval civilisation', in M Postan, E Rich (eds) *Cambridge Economic History of Europe*, II (1952), 430–89, here 435–8; L Klapparf, 'Studies in the development and structure of early metal production', in C Segers-Glocke, H Witthöft (eds), *Aspects of mining and smelting in the Upper Harz mountains* (2000), 1–23; A Bingener, 'Medieval metal trade in and around the Harz mountains', in same vol., 119–40; C Bartels, 'Mining towns in the north-west Harz district', in P Anrieter et al. (eds), *Mining in European history and its impact on*

noted, there was also major mining activity in the Rhineland, from whence some of those who transferred their skills to military purpose at the siege of Lisbon hailed, while the first documentary evidence of coal mining in the bishopric of Liège, not now but in later centuries famous as a source for military mining experts, dates from 1228.[89]

Since many of these mines had existed already for centuries, the apparent failure of rulers to recognise the potential of the human resources they controlled may call into question the extent to which they knew of Vegetius – or alternatively that it was not recorded and that the useful dictum that absence of evidence is not evidence of absence should be applied here. It is also not impossible that this was one of the lessons brought back from the experience of crusading that so many European rulers, nobles and knights had undertaken. Certainly by the end of the twelfth century, mining was once again established as an important element of the military engineering resource available to commanders and it will be seen how much more frequently it would be called upon in the centuries to come.

The particular skills required of the underground miner were not written down comprehensively in any text that survives and the earliest that does dates from the sixteenth century (G Agricola's famous *De re metallica*), unfortunately for the medievalist. One can only assume that the techniques and knowledge presented here had continued little changed over many centuries, with the exception of learning how to make use of gunpowder. In contrast to other sectors, it does seem that there had been almost no technological advance in mining over many centuries, even if the status of the mine worker had improved since the days of the Roman empire.[90] The development of mining after the eleventh century therefore depended on the demand of rulers for silver (in particular) and iron and the availability of labour to explore potential sources for the ores.

CONCLUSIONS: THE KNOWLEDGE OF THE MILITARY ENGINEER

The miner called upon for military service may have possessed knowledge only of the same techniques as had been in use for centuries, but what did the engineer called upon to construct a siege tower or stone thrower, or to bridge a

environment and human societies (2010), 35–44. For a broader overview, J Gimpel, *The medieval machine. The industrial revolution of the middle ages*, 2nd edn (1988), chapter 3.

89 H Kranz, *Quellen zum Lütticher Steinkohlbergbau im Mittelalter* (2000), 13–14.

90 Gimpel, *The medieval machine*, 59–72; J Landers, *The field and the forge. Population, production and power in the pre-industrial west* (2003), 64–5. Agricola's *De re metallica* was published in 1556 and is available in an English translation by Herbert and Lou Hoover.

fast-flowing river, or to plan a defensible encampment have by way of knowledge not possessed by their predecessors? I have suggested that even though the knowledge possessed by such people was not derived from book-learning or teaching but from master-to-apprentice routes of transmission, nonetheless the practical knowledge itself must probably have absorbed new techniques through observation. This supposition would be no more than that if it were not the case that the historical evidence strongly suggests improvements in the technology. It is unlikely, anyway, that the novel building styles emerging during the twelfth century in particular, and continuing through further steps as time went by, would not have influenced all except the most conservative of craftsmen. The master carpenters and master masons who learnt these techniques on the sites of the new cathedrals (for example) were the same people who were called upon to build or repair royal castles, and to carry out military engineering tasks when ordered to help besiege other castles or cities.

There were educated princes and nobles – a few have been mentioned. There were also educated clerics known to have turned their hand to making or using engines – more will appear in the following century. One can imagine them instructing or discussing construction with the lay engineer or master craftsmen, thus establishing another route of transmission of knowledge. Some (clerical) chroniclers were sufficiently interested to offer detailed descriptions in their writings, as we have seen. Monastic libraries did contain relevant texts (Vegetius among the most often copied). But there is no evidence confirming the practical link between education and such texts, and little evidence of who else received an education at all.

The campaigns reviewed here have shown a more ambitious approach to the technology of siege weapons, in particular the large store of technical knowledge required to construct a siege tower that worked in practice: that is, under attack from an enemy. Such an apparently simple device as a battering ram in truth involved a number of skills just to construct the mobile shelter, to protect it from attack from above, while the question of how the ram itself was suspended was not as easy as might appear.[91] The accounts cited here confirm that many of these technical challenges were successfully met during the twelfth century, without it being possible to know more about the engineering work. Advances in timberwork have been identified by scientific study of preserved woodwork but no timber has survived from these engines of war, and this was a deliberate choice. There was no point in leaving a siege tower or battering ram sitting against the wall of a conquered city after its job was done, and the timbers may well have found use

91 On which see the fascinating study by R Aris and B Bachrach, '"De motu arietam" (on the motion of battering rams)', in K Elworth et al. (eds), 'Differential equations, dynamical systems and control science', *Lecture notes in applied mathematics* 152, 1994, 1–14.

somewhere else afterwards. But it is probable that the greater skills in carpentry demonstrated in other environments were also made use of in the construction of these large devices, even if it is now impossible to identify whether the siege towers and rams used in the first crusade were a technical improvement over their tenth-century predecessors described in Francia or in Byzantine use. The success rate of such forms of attack tells nothing about how well the engine had been built because such analysis is impossible to separate from the whole context of the siege operation. The siege tower used in the final capture of Lisbon in 1147, a device that was subject to prolonged attack and survived being cut off from its own people before being ultimately decisive in bringing about surrender, is testimony to a well-designed and well-built machine, and is only suggestive of superior construction skills.

If the skills of the engineers who built and commanded the artillery or the siege towers have been lost to us, and if the claim made here of improving status for engineers might be argued to be simply a consequence of the better survival of records, it is true that definitive proof does not exist. But the evidence of significant technological advances across many fields that has been summarised here is unarguable: therefore the people responsible had achieved higher levels of skill than previously. Combined with the increasing references in chronicles (including a small but growing number of named individuals), it looks as if growing complexity and the associated skills now commanded greater recognition by the employers of military engineering skills. The thirteenth century would see further development of these trends. Named engineers became more common and the progress in recognising their status accelerated. Mining continued to be a weapon of choice where it was appropriate and records of use demonstrate how important a part of the engineering resources available to rulers it had become. In terms of siege technology, the arrival of the counterweight trebuchet led rapidly to a transformation in the power of artillery that would last unless the final step of gunpowder weapons in the following century, and the construction and use of the trebuchet offer evidence for the progress of the skills of what became increasingly specialised engineering. For the first time, too, this weapon has been subjected to in-depth study by those able to explain its dynamics, and the consequences of their work include being able to explore the greater skills required of its builders and operators.

5

Engineers in Demand

Innovation and Development in the Thirteenth Century

THE TECHNOLOGICAL DEVELOPMENTS OF the twelfth century, alongside renewed scientific progress, continued throughout the following century. Greater survival of government records confirms that what was happening under the Plantagenets was also to be found under French kings and the Angevin rulers who seized the inheritance of the Hohenstaufen emperors in southern Italy.[1] At a geopolitical level, the entire century was marked by the devastating impact across much of eastern Europe, the Middle East and China of Mongol invasions, and how these nomadic warriors made use of the technology of the people who fell beneath their yoke offers insights into levels of knowledge, as well as the value attached to people who could use it. In military technology, the century witnessed the widespread adoption of the counterweight trebuchet, artillery with revolutionary impact that came to dominate sieges across Europe and the Muslim world. Gunpowder was also first identified in Europe later in this century, although it was already well known in China.

In this chapter I argue that alongside increasingly centralised government there were substantial technological advances (and I examine whether they related to scientific knowledge). Consequent upon both trends, there is evidence of rulers paying more attention to having a body of engineering expertise on which they could call. Greater recognition of the value of engineers and master craftsmen, and (arguably) a greater differentiation among engineers, accompanied these processes.

This was a period in which royal governments in Europe began to become more centralised and more reliant on professional bureaucracies to carry out their functions. This was a slow process and not without interruptions, while in the Holy Roman Empire local princes, bishops and increasingly independent cities themselves consolidated a long-existing de facto right to self-government with only token acknowledgement of the place of the emperor. Frederick II (1212–50), the ultimate failure of whose efforts to assert imperial rights in Italy led to the destruction of his dynasty and the long interregnum that significantly undermined

1 Thankfully published before their destruction in the twentieth century.

the power and prestige of German emperors, was a man whose wide range of interests included science and technology. One aspect of the process of developing a more bureaucratic and less personal government, the origins of which date back a long way, was the increasing use of paid soldiers to replace armed forces dependent only on feudal levies, with even the aristocracy now expecting reimbursement for service. Paid members of these armed forces always included the engineers, humble origins making them reliant on serving for pay and keep, and civilian miners.

ADVICE TO PRINCES AND CODIFICATIONS OF CUSTOMS

The reappearance of a genre of writings by intellectuals seeking to promote good governance by educating the ruler in ancient good practice took seriously the prince's duty to defend the realm, and the Christian examples used classical texts – Vegetius continued to be the key teacher – to assign important roles to fortifications and the means of defending and capturing them. Did engineering fit into these recommendations?

One of the earliest of this new generation of writings, offered to a prince of the 'Ayyubid house around the year 1200, called on the ruler to recognise the importance of paying wages to the masons, those who served the ballistas, crossbowmen, throwers of naphtha and the miners. Necessary equipment for the prince to maintain included sapping machines, timber for the artillery, plenty of ropes and cords, and "Greek Fire". The author had been present at the capture of Belvoir in 1189 and was therefore presumably familiar with good practice.[2] The "Saxon Mirror" written by Eike von Repgow around 1230 was an original text based on a similar model. In Spain, the *Poridat de las Poridades* dating from the middle of the century was the first translation into a vernacular language of the Arabic version of the popular text falsely ascribed to Aristotle and usually rendered as the "Secret of secrets". Its advice was amended to be up to date about technology. In this case it included how to use engines (with particular reference to the large "crossbows" newly popular with rulers) to attack and defend fortifications. "Secrets" had been introduced to western Europe in around 1130 but a complete text was only available from around 1220. It continued to be popular, and more than 600 manuscripts survive.[3]

2 J Sourdet-Thomine, 'Les conseils du Sayh al-Harawi à un prince Ayyubide', *Bulletin d'Études Orientales* XVII, 1962 – chapter XIII, 227–8.

3 *The Saxon Mirror*, trans. M Dobozy (1999); Seudo-Aristoteles, *Poridat de las Poridades*, ed. L Kasten (1957), here at 57. S Williams, 'The early circulation of the pseudo-Aristotelian secret of secrets in the west: the papal and imperial courts', *Micrologus* II, 1994, 127–44. A translation into English was made (in Ireland) by James Yonge around 1420: L Kerns (ed./trans), *The secret of secrets (secreta secretorum)*, (2008).

The writer of a similar text (in about 1240) for King Hakon IV of Norway (1217–63), who took part in the fifth crusade, was well informed about the capacity of the trebuchet and other siege artillery, the usefulness of the iron-headed battering ram and the value of defensive mines in the destruction of attacking engines. Engineers were not mentioned. Was their presence assumed?[4] From the second half of the century dated the *Siete Partitas* ascribed to King Alfonso the Wise of Castile (Alfonso X, 1252–84). Of interest here were the requirements on governors to retain not only knights but also crossbowmen and supplies of victuals including timber for the manufacture of hoarding (*propugnacula*) and wooden towers; they were also required to retain a supply of engines in fortified towns, especially those near a frontier, a list that specifically includes machines that hurled stones by counterweights.[5]

Ptolemy of Lucca's *De regimine principum* represented the completion of an unfinished work begun by St Thomas Aquinas and stated that the diggers, carpenters and stone-cutters should be under military discipline because their role in building defences was critical. In this he was reminding his readers of the teaching of Vegetius on the importance of the fortified camps built by Roman armies, but with the unspoken recognition that the artisans in his day would be civilians.[6] Of a similar character, not least in the reach of its influence, was the work of Aegidius (Giles) of Rome, addressed to King Philip IV of France (1270–85). It remained in use, including in translations into vernacular languages such as English, French and Swedish, well into the following century and survives in 350 manuscripts. Book III was devoted to military affairs, and was no less reliant than the others on the teachings of Vegetius. The latter's advocacy of defensible camps was repeated, as was advice on the best time to set a siege and the different ways it might progress. Many of the recommendations required engineering skills from the attacking army: diverting rivers, setting up timber defences for the besiegers' earthworks, setting up ballistas and stone throwers, constructing scaling ladders. Mines (using the term for rabbit holes picked by Vegetius, *cuniculos*) should be dug by qualified people and made to work by the traditional methods of installing wooden props then burning them. For the artillery, Giles described and named the trebuchet, but went on to describe variant designs, suggesting acquaintance with current military technology. In this area, he was required to venture beyond the advice of the Roman. Giles' text suggested that he was aware of everything of which his contemporaries were capable, although whether his offering of the simplest

4 *The King's Mirror – Speculum regale – Konungs Skuggsja*, trans. L Larson (1917), chapters 38–9 (220–6).

5 *Las siete partitas del rey Alfonso el sabio*, ed. G Lopez (1847), title XVIII, laws 9 (179), 10 (180), 12 (182), title XXIII, laws 23–4 (276–77).

6 Ptolemy of Lucca, *On the government of rulers. De regimine Principum*, trans. J Blythe (1997), 286 (book 4, c. 28).

advice for judging the height of a wall – to measure it by its shadow – demonstrated lack of knowledge of more advanced techniques or reflected common practice is unclear, but in this he was echoing his main source.[7]

Finally, the text written by Walter de Milemete in around 1326 for the young prince who became Edward III of England (1327–77), chiefly famous because a manuscript of the work dating to this time contained the first illustration of a cannon, was also presented as a version of the "Secret of secrets" and contained contemporary as well as classical information. This text (which did not mention the gun) described the use of a standard repertoire of siege equipment requiring engineering skills: mobile wooden towers, stone throwers including a trebuchet, several large crossbows or ballistas, mining operations and scaling ladders alongside more exotic weaponry including a flying machine that dropped bombs – perhaps one of those flights of fancy that can be found in not a few of these treatises, and that can sometimes cause difficulty in determining what really existed.[8]

What all this suggests is that respect for tradition meant that if teaching was to be seen as authoritative, it needed demonstrably to derive from the glorious past. More recent technology needed to be absorbed into the writing, and was. Few of the new writings made explicit reference to the importance of the people responsible for designing, building and operating the military equipment recommended, perhaps because these people would have been Roman soldiers in the original authorities, but recognition of their role was at least implicit. Similarly, in the laws written down in the name of these princes, instructions to those in command of fortresses required them to employ skilled men. Everything confirms that this was now commonplace

SCIENCE AND TECHNOLOGY

This period was marked at its beginning by a new technology that was revolutionary in its impact (the counterweight trebuchet) and at its end by the arrival of gunpowder artillery, which changed the character of warfare for ever. Some historians have acknowledged the impact of the trebuchet as well as gunpowder, but as "technological revolutions" in medieval terms cannot be described in the same way as equivalent changes in modern times, the case for this assertion remains to

7 Aegidii Romani, Bituriensis Archiepiscopi, *Libellus de re militari veterum auctores praesertim medii aevi*, ed. S Hahn (1724), I, 21–3, 44–6, 47–54. Aegidii Romani, *Opera omnia, I: Catalogo dei Manuscritti, de regimine principum*, ed. F del Punta, C Luna (1993), v–x. A version in Swedish was printed as late as 1634 (ed. L Moberg, 1964).

8 *The treatise of Walter de Milemete 'De nobilitatibus, sapientiis et prudentiis regum' and a selection from 'De secretis secretorum Aristotelis'*, ed. M R James (1913), 133, 136–7, 147–50, 154 (the flying machine) and 179. The first illustration of a cannon is on p. 181.

be made.[9] I will argue that the engineer(s) who invented the first counterweight artillery set in progress a significant change in military technology that was sustained by further improvements during the thirteenth century, to a point that the weapon transformed siege warfare and remained in use for more than a century after the emergence of its eventual replacement, the gun.[10]

Can technological progress be related to scientific progress? Some scientific developments of this era had a potential impact on engineering. Two men from the start of the thirteenth century had an important role, both linked by an association with the court of Frederick II. His reign was marred by continuing conflict with the Italian cities of the north, in a repeat of the struggles waged by his grandfather, but his reputation as a ruler interested in scholarly and scientific subjects was part of his claim to be *stupor mundi*. Michael Scotus (d. ?1235) was the emperor's court astrologer and was responsible for translating Aristotle on the natural sciences into intelligible Latin. He was an important link in the transfer of scientific knowledge between the Castilian court at Toledo, where translation of Arabic works continued, and the imperial court in Italy.[11] Michael was an associate of Leonardo of Pisa (c.1170–c.1240), who became famous under the name of Fibonacci for spreading the use of Indian numerals (from Arabic scientists). He wrote a new work on mathematical calculation, the *Liber abaci*, and also produced the most comprehensive encyclopaedia of mathematical knowledge, which included an explanation of the "golden mean"; it has long been suggested that this ratio is the basis for the design of the extraordinary eight-sided Castel del Monte built at the emperor's command in Puglia [Fig. 15]. Alongside his formula for calculating the numbers of rapidly breeding rabbits – Fibonacci numbers are still acknowledged in modern mathematics – and of immediately practical benefit were works on commercial arithmetic and practical geometry (1223), demonstrating how to measure land dimensions and area and the use of the quadrant for measuring heights.[12]

9 On the subject of 'medieval military revolutions', see C Rogers, 'The military revolution debate' (1995, republished 2010).

10 M Denny, *Ingenium. Five machines that changed the world* (2007), includes the counterweight trebuchet and discusses it at 80–7 and 166, and the early champion of studies of medieval technology and its wider social significance, Lynn White jnr, makes frequent reference to it in his many writings (see Bibliography).

11 C Burnett, 'Michael Scot and the transmission of scientific culture from Toledo to Bologna via the court of Frederick II Hohenstaufen', *Micrologus* II, 1994, 101–26.

12 Summary biographies by C Burnett (344–5) and M Mucillo (172–5) in T Glick et al., *Medieval science, technology and medicine: an encyclopedia* (2005). His main works have been translated into English by L E Sigler: Leonardo Pisano Fibonacci, *Book of squares* (1987) and *Fibonacci's Liber Abaci. A translation into modern English of Leonardo Pisano's Book of Calculation* (2002). The emperor's discussion of scientific subjects was described by C Haskins, 'Science at the court of Frederick II', republished in *Studies*

FIG. 15 Castel del Monte, Puglia, Italy. Frederick II's octagonal castle has been the subject of much speculation. It was built following complex geometric design throughout. (Photo: author)

Other scholars, such as the German Albertus Magnus (1200–80) and the Englishman Robert Kilwardby (d.1279), proposed classifications which recognised the legitimate place of the crafts in the range of human knowledge.[13] Kilwardby, a Dominican scholar (archbishop of Canterbury in 1272), covered the "mechanical arts" and specified that study with practical purposes, knowledge of arithmetic, astronomy and geometry were essential for human existence as they informed such tasks as building, navigation and armaments (as well as agriculture, music and medicine).[14] The overlap was spelled out in other scientific writings towards the end of this period. Raymond Lull (1232–1315), from the kingdom of Mallorca, unusually devoted space to the "art of carpentry" and its wide range of essential human applications, including military purposes, while separate chapters of his book *The tree of science* discussed warfare and the application of geometry for measurement,

---

in the history of medieval science, 242–71. On the influence of Fibonacci, D Struik, *A concise history of mathematics*, 4th edn (1987), 80–1.

13 Whitney, 'Paradise restored', 111–12.

14 R Kilwardby, *De ortu scientiarum*, ed. A Judy OP (1976), esp. cc. 372–9, 393–4, 400 (pp. 131–40).

including that of towers.[15] The Franciscan Roger Bacon (c.1215–c.1292) was the first European to describe the ingredients of gunpowder, and how they worked to produce fire and explosions, in his *Opus tertium*. Bacon's *Communica mathematica* spelled out the practical use of geometry in planning "cities, camps, houses, and towers", in making canals and conducting water, in designing "ingenious" bridges and lifting heavy weights:[16] this is "an organic vision of technology as applied science in the service of man's earthly life".[17] I cannot prove a direct link between this scholarly learning and practical applications to everyday technology, but I suggest that many educated rulers and clerics with secular responsibilities will have absorbed such teachings. It is a small jump to suggest it might have influenced their practical decisions, and it is possible to show that technology was using similar principles.

ENGINEERS, CRAFTSMEN AND THEIR SKILLS

Growing cities and expanding governments created a large demand for craft skills, so that the little documentary evidence that does survive confirms increasing numbers of skilled artisans, who would have been available for many tasks including military engineering. In London, the demand for craftsmen often exceeded supply, leading to large numbers of foreign artisans being recruited. The consequence of this mobility of labour was also inevitably a diffusion of technology, and this may have been more important than any other route in ensuring the spread of building and carpentry techniques of significance in military engineering. Evidence from Paris demonstrates the increasing numbers of artisans working in trades providing for military purposes, both directly (for example, crossbow-makers, armourers and archers), and for building trades (there were 95 carpenters registered in 1292 but 108 in 1300, and an increase in masons from 104 to 122 over the same period). The position of master carpenters and master masons was particularly recognised in the payment of higher wages, with additional emoluments such as horses being part of the package.[18] There is evidence that there were fraternities of carpenters

15 Raimundi Lulli, *Opera Latina. Arbor scientiae*, ed. V Varneda, Corpus Christianorum, CLXXXA (2000), I, 236 (book V, De arbore humanali), 239 (book V, De arte militare), 242 (book V, De arte geometriae).

16 Roger Bacon, 'Opus tertium', in A Little (ed.), 'Part of the *Opus tertium* of Roger Bacon', *British Association of Franciscan Studies* IV, 1912, 51: the original here reads: "Quod si fieret instrumentum de solidis corporibus, tunc longe major fieret violentia"; Bacon, 'Communica mathematica', in R Stede (ed.), *Opera hactenus inedita Rogerii Baconi* (1940), fasc. XVI, 42, 43–4.

17 Whitney, 'Paradise restored', 144.

18 N Fryde, 'Technicians, the crown and the city of London in the thirteenth century', in M Arnoux, P Monnet (eds), *Le technicien dans la cité en Europe occidentale 1250–1650*

in London in the thirteenth century, and although masters in this trade were few, they were particularly likely to be called for royal service.[19] There, increases in the wages paid in particular to master craftsmen followed the static levels maintained throughout most of the twelfth century.[20]

Particularly revealing of the importance attached by rulers to craft skills was the exchange value of such people in times of war. While knights and nobles could normally expect to be ransomed by their captors, this approach did not usually apply to people of lower rank. Richard I's attitude to captured Muslim artisans at Acre has been described. The same approach was shown by both Muslim rulers and the usually intransigently ideological Knights Templar in the thirteenth century, when the Mamluk sultan Baybars was denied the chance to exchange captured Muslim craftsmen for Christian captives because the craftsmen were too valuable to their captors. It is evident from other recorded incidents from the crusading period that the religious affiliation of master carpenters and master masons mattered less than their willingness to serve whoever held them. The principles of a Christian crusade, or of jihad, did not trump these practical concerns either for the ruler or the artisan.[21]

Skills in both carpentry and masonry were required for any major building project. Analysis of surviving timber structures dating from the thirteenth century has led to the conclusion that there were further technological advances in woodworking during the century, from the vital initial decisions on how to manage woodland such that suitable timber could be harvested without exhausting the supply to making use of smaller pieces (for the same reason), and also more ingenious design of complex timber structures to ensure stability.[22] Accumulated knowledge of the properties of different timber determined the choice of tree for a particular task, and this would be of growing significance as technical demands became more critical: it was one of the characteristics of successful construction

(2004), 15–23; G Fagniez, *Études sur l'industrie et la classe industrielle à Paris au XIIIe et XIVe siècle* (1877), 8–10, 191–203. There are some comparable examples of the organisation of trades in the twelfth century in the Muslim world in N Elisséeff, 'Corporations de Damas sous Nūr al-Dīn', *Arabica* III, 1956, 61–79.

19 B Alford, T Barker, *A history of the Carpenters' Company* (1968), 13–16.

20 P Latimer, 'Wages in late-twelfth-century and early-thirteenth-century England', *Haskins Society Journal* 9, 1997, 185–205.

21 Y Friedman, *Encounters between enemies. Captivity and ransom in the Latin kingdom of Jerusalem* (2002), 100, 114 (with reference to Abu Sharma's *Livre des deux jardins*) and A Forey, 'The military orders and the ransoming of captives from Islam (twelfth to early fourteenth centuries), *Studia monastica* XXXIII, 1991, 259–79, here 276 (reprinted in Forey, *Military orders and the crusades* (1994)).

22 Courtenay, 'Scale and scantling', 42–75, studies the roof structures of Cressing Temple barn (Essex), dating from 1257–90, the roof of the nave of Notre Dame of Paris (c. 1220) and the Byloke hospital in Ghent (1251–55).

of the new trebuchet. The quality of metalworking was also important. It was not only that metal parts were central to the effectiveness of crossbows and the larger engines that used the same principles of propulsion; they would also be of significance in the construction of the largest siege engines – although here, the masters generally preferred to use other timber to make the pegs that held them together rather than metal that was subject to corrosion. This accumulation of knowledge was presumably passed on in traditional style from one generation to another.

Carpenters were now the people most likely to double up as royal engineers, even to the extent of being in charge of work apparently more appropriate to masons. During the century, the technical progress connected with the change from the Romanesque to the Gothic architectural style, as discussed in chapter four, continued to evolve as the Gothic form itself developed, while the mobility of the builders continued to ensure that technical improvements could be diffused rapidly (however much the detail may have remained a treasured secret of particular builders). In this field, the greater heights sought for the towering new cathedrals after around the year 1200 posed challenges never previously encountered for masons and mason-architects, such as the stress caused by much higher winds at high level. The change in the design of roof vaults at this time has been ascribed to builders identifying a solution to cope with the higher stress, and modern engineering has confirmed that the change from sesquipartite to quadripartite vaults and the use of flying buttresses succeeded in reducing the longitudinal stress by half. Without the mathematics needed to work this out, the medieval builder still arrived at the right answer,[23] further confirmation of advances in skills and understanding. As for building skills applied to the military aspects of castles or urban defences, it is only possible here to summarise. Both in Europe and in the Middle East castles became more elaborate, offering not only potent signals of power and status and greater comfort for their occupants, but also frequently combining these with more sophisticated defensive features such as twin-towered gatehouses with drawbridges, portcullises and machicolations, "bent entrances" (a speciality of Muslim architecture), significantly better designed outer defences for entrances with separate barbicans, better designed arrow loops, and in the most important structures walls of (on average) twice the thickness of (most) twelfth-century examples (sometimes provided with internal galleries). Where terrain allowed, concentric design, which enabled defence to be conducted simultaneously from high inner and lower outer walls, reached perfection.

23 See N Coldstream, *Medieval architecture* (2002), 60–1, 65–70; also Mark, 'Technological innovation in High Gothic Architecture', 11–26.

THE DRAWINGS OF VILLARD DE HONNECOURT AND WHAT THEY REVEAL

The unique survival from the thirteenth century of the so-called "notebook" of Villard de Honnecourt has provided evidence for what common sense indicates must have been the case: that those in charge of erecting cathedrals and castles must have had an illustration of the end product, and that the specific features had a complexity that implied the same. These were not scaled plans, but they did show proportions and details. The master was contracted to build according to the patron's specifications, and amongst other considerations it was necessary to identify the quantity of materials required in order to organise the logistics and control the cost. The drawings will have been done in the masters' lodges and remained there but had no purpose after completion, while the actual construction of parts would have been traced out in real size on the ground (as had long been the practice).

Villard's work was done between 1225 and 1235 and identifies that he worked on the cathedral at Reims. Alongside drawings and instructions for many architectural features were tools such as lifting devices, a hydraulic-powered sawmill and piles for bridges constructed under water, as well as the base of a trebuchet. Also illustrated were many of the craftsmen's tools: compass, square, rule, plumb line and level, and sighting devices including the alidade. The drawings demonstrate that Villard was both an "architect" and an "engineer" in modern terms.

Villard was literate. The drawings are surrounded by his notes and explanations, in the vernacular, and suggest a familiarity with the basic geometrical knowledge of Euclid described in the school books often mentioned – the square and the right-angle triangle (that an illiterate craftsman could scale up to any size), and the proportional drawing known after its originator as "Vitruvian man". But it has also been argued that the drawings indicate an understanding of a more advanced geometry of pentagons and hexagons. Beyond these conclusions about Villard's level of education, all else is speculation. It might be rash to think that the masters and operatives working on these projects could read the notes – rather than using the design as a template – but the fact that there are written notes implies that Villard expected someone to read them: they contained information necessary for the correct interpretation of the drawings.[24]

24 On these facts and explanations, see F Bucher, *Architector. The lodge books and sketchbooks of medieval architects* (1979), I, 10–13; J Gimpel, 'Villard de Honnecourt, architecte-ingénieur', in *Villard de Honnecourt: carnet XIIIe siècle*, ed. A Erlande-Brandenbourg et al. (1994), 27–38; A Erlande-Brandenbourg, 'Villard de Honnecourt, l'architecture et la sculpture', same volume, 17–25; R Bechmann, 'Les dessins techniques du carnet de Villard de Honnecourt', same volume, 39–50. There is an English translation: *The medieval sketchbook of Villard de Honnecourt*, ed. T Bowrie, new edition (2006). For more recent analysis, N Hiscock, 'Architectural geometry and the portfolio of Villard

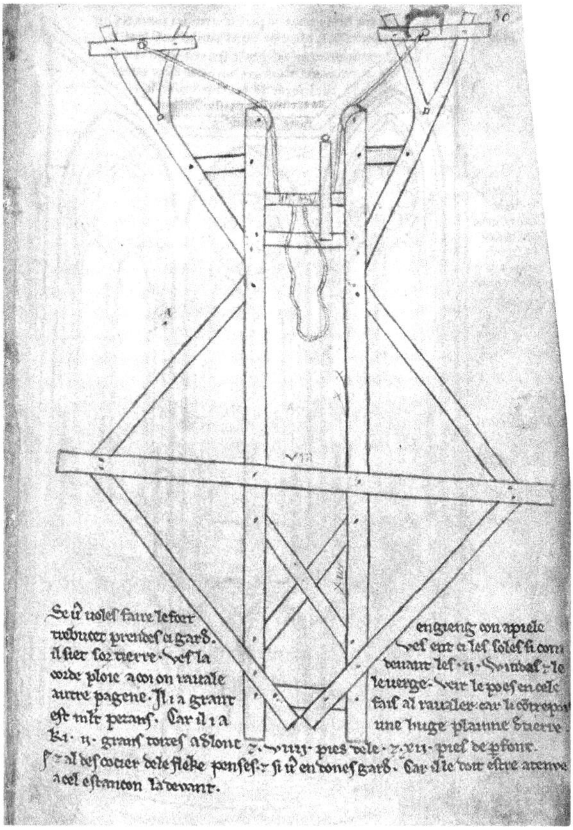

Fig. 16 Trebuchet base, with dimensions, in Villard de Honnecourt's notebook. The page showing the rest of the engine has been lost. (Bibliothèque Nationale de France, Paris, Français 19093, fol. 30, licence RDV-1702-000435)

If Villard's notebooks were not unique, the conclusion is that thirteenth-century engineers around Europe were capable of rather more sophisticated engineering (and architectural) challenges than their predecessors in previous centuries. If Villard was educated in a monastic school he could have had access to books containing the scientific learning some of his drawings suggest he had acquired. The alternative explanation, that he had learned these techniques through oral transmission, is countered by his evident ability to write, at least in French.

The transferability of knowledge and skills between ecclesiastical and secular tasks was nothing new and the inclusion in Villard's (evidently incomplete) sketchbook of a drawing of the base frame for a trebuchet confirms the continuation

de Honnecourt', in M-T Zenner (ed.), *Villard's legacy. Studies in medieval technology, science and art in memory of Jean Gimpel* (2004), 229–63; R Beffeyte, 'The oral tradition and Villard de Honnecourt', same volume, 93–119. On the role of metalworking in the notebooks, otherwise neglected by scholars, C F Barnes jr, 'A note on Villard de Honnecourt and metal', in R Bork (ed.), *De re metallica*. Avista studies in the history of medieval technology, science and art 4 (2005), 245–54.

of this interchangeability into the thirteenth century: no premature specialisation was taking place here. The accidental survival of this drawing (the text around it confirms that there had originally been another page at least, illustrating the rest of the engine as the wording indicates) has been of great benefit in understanding the reality of the new artillery and helps reconstruct its revolutionary impact [Fig. 16].

CRAFTSMEN AND THE ORGANISATION OF ARMS MANUFACTURE

Formal organisation of arms manufacture by princes was becoming established in a way comparable to the classical Roman *fabricae*. D S Bachrach's exhaustive studies of the accounts confirm that crossbows were made, and repaired, for English monarchs by teams of salaried craftsmen under a few masters throughout the thirteenth century, and they were based at a number of royal castles, starting at the Tower of London but also (under King John) at Nottingham. One of the well-paid crossbow-makers was Peter the Saracen and another was Benedict the Moor: neither ethnicity (nor religion) mattered as much as skill. Bachrach has identified references to twenty-eight royal crossbow-makers serving under John and Henry III. The iron mines of the Forest of Dean, whence came the men who served the crown in military mining, also provided the raw material for the establishment of a royal manufactory for crossbow bolts in the castle of St Briavels, the construction of which from the basic ore was not straightforward, and the records confirm (and name) one of the masters employed there under Edward I was the "king's quarreler".[25]

The once-Angevin royal castle at Chinon was also an arsenal in the thirteenth century. The French king made grants of land to a carpenter named Eudes in 1205; records show the work of a master carpenter named Martin de Bilenges in 1247–48, and another named Richard whose responsibilities included the rebuilding of the walls of Chinon (1298–1305). In Poitou, the castle of Niort was a centre of arms production. In 1244 and 1245, a master carpenter named Laurence, also described as *attiliator*, was paid to construct six ballistas and repair six others, and payment

25 D S Bachrach, 'The royal crossbow makers of England 1204–1272', *Nottingham Medieval Studies* 47, 2003, 168–97; 'Origins of the crossbow industry in England', *Journal of Medieval Military History* II, 2004, 73–88; 'The royal arms makers of England 1199–1214: a prosopographical survey', *Medieval Prosopography* 25, 2004, 48–74; 'Crossbows for the king: the crossbow during the reigns of John and Henry III of England', *Technology and Culture* 45/1, 2004, 102–19; and 'Crossbows for the king: the crossbow during the reign of Edward I of England (1272–1307)', *Technology and Culture* 47/1, 2006, 81–90. A Webb, 'John Malemort, king's quarreler. The king's great arsenal – St Briavels and the royal Forest of Dean', *Journal of the Society of Archer Antiquaries* 31, 1988, 40–6).

of a debt in wages of 2s 6d a day was recorded on his death. In 1246, the accounts made specific reference to "newly constructed engines … specifically ten mangonels made, and their ropes" in the forest, although here the master was not named. In 1248 records showed the building of a "wooden castle" transported to La Rochelle. In 1267, a master Walter (another *attiliator*) was in charge of a team manufacturing crossbows and quarrels at Saintes and the following year a master Garnier, "expert in the art of carpentry", was retained for service in the coming crusade of King Louis IX.[26]

Royal military industry in France continued under St Louis's successors. Frequent payments were made at the arms factory in the Louvre to masters identified as *attiliator*, as well as at other centres, John of Melun (in 1298) at 6d per day, Gilbert, "visiting" (that is, inspecting) the "works and the artillery", William at Montargis paid for timber, glue, ropes and sinews for a springald and horn for a giant crossbow, along with a substantial sum for the accompanying iron. In 1305, Master Peter Poignant, a carpenter, and Nicholas Voisin were well paid for their work on the king's artillery, while Gilbert made another appearance. Others were specified working in Poitou and at Toulouse in 1294–95 and at Carcassonne in 1302–03. Here, the leading engineer was Master Albert, who was entitled to his own valets and in charge of several *attiliators*; he was sent down to Marseilles to manufacture more engines. An account of 1310–11 described master John of Medun as the "master of the king's works" in Toulouse and Albi, his responsibilities including building (mills, the new hall at Toulouse and homes for the royal engines). A similar range of jobs was ascribed to the works at the new port-city at Aigues Mortes (1302–03), including construction in the town, the castle (the Tour de Constance), the lighthouse and lead bought for the work of the artillery.[27] The French crown, like the English, could call upon a corps of skilled operatives ranging from the servants of artisans to master carpenters and those with sufficient knowledge to be appointed to oversee the work of the others, while a few at a higher level were entrusted with oversight (and very likely also the design) of a wide range of varied royal engineering tasks.

Records therefore confirm that there was a large and growing number of expert craftsmen involved in military work during this century, among whom were not

26 H-F Delaborde, C Petit-Dutaillis (eds), *Recueil des Actes de Philippe Auguste, roi de France* (1943), II, 499; A Molinier (ed.), *Correspondence administrative d'Alfonse de Poitiers* (1894), I, 68 (no. 106), 554 (no. 864); A Bardonnet (ed.), *Comptes d'Alfonse de Poitiers 1243–47* (1875), IV, 24–5, 75 (Laurence), 157 (Walter), 203–4 (new engines); N Prouteau, 'La forteresse-arsenal de Chinon au XIIIe siècle', in *Chinon. Le destin d'une forteresse*. Société des amis du vieux Chinon, hors série, 2011, 15–28; M-P Baudry, *Le château de Niort* (2013), 30–1.

27 All from R Fawtier, F Maillard (eds), *Comptes royaux (1285–1314), Recueil des historiens de France. Documents financiers III* (1953–56), vol. I, 14, 115, 150–2, 175–6, 486, 630 (Carcassonne), 661, 666 (Aigues Mortes); vol. II, 204 (Master John).

a few who would also feature as engineers with a more enhanced role, although the distinction was not always clear. Some were important enough to be named. It is possible that one reason for increased status, recognition and possibly more specialisation was that they had to master new or more sophisticated technology than their predecessors, or that the technological changes called for better qualified people to start with. Therefore, the next step is to identify improvements in the technology they were responsible for directing when they followed instructions to construct arms and equipment or other machinery. This involves looking specifically at the "giant crossbow", the trebuchet and the springald. I will then consider the evidence from shipbuilding and surveying to identify whether there was technological progress in these areas that would have required new skills.

(a) Giant Crossbows

By mid-century, armies were making widespread use of large "two-footed" crossbows, the *ar[cu]ballista ad turnum*, which was really a piece of artillery requiring a tool to span it (probably using a winch), and a new bolt-throwing weapon in the form of the springald (see below). All these weapons, easy to use, required skilled labour for their manufacture and also continuous maintenance, and this was recognised in specific instructions from kings. English records show that, in 1253, one Raymond "the archer" was ordered to produce a "balista turonica" and send it to Bordeaux at a "just price", while a year later Master Henry, "atyllatori Regis balistarum", received payment for manufacturing twenty "balistas ad duas pedes" for which the king had provided sufficient timber, iron and strings, with a deadline of Easter.[28] The importance of this particular technology increased further under Edward I (1271–1307). The builder was now known as an *attiliator*, and a similar process can be identified in the French kingdom.

(b) The Trebuchet

Origin of the counterweight trebuchet

The counterweight trebuchet revolutionised siege warfare because for the first time it gave the besieger a realistic prospect of breaching a well-made stone wall by bombardment. Firstly, this could now be done because of the considerably greater size of the missile that could be projected by the new engine. Secondly, the trebuchet could provide an accuracy that guaranteed that the large stone balls would hit in and around the same spot time and again, rendering even the strongest masonry wall vulnerable. Thirdly, the largest engines could operate at a range of 300 metres (or

28 *Rôles Gascons*, ed. F Michel (1885), I, 357, 398, (nos 2688, 3131).

even more), putting them safely outside the range of most bows or crossbows.[29] In each of these considerations, it was significantly superior to the manually operated stone throwers that had offered the only artillery option for hundreds of years. The introduction of the counterweight weapon – it has already been shown that writers of military treatises quickly recognised its importance – did not end the use of the earlier forms of propulsion, nor did every trebuchet necessarily have to attain giant proportions: as with artillery in all ages, the instruments served a number of purposes including as anti-personnel devices, and the successful commander with the necessary resources continued to operate a range of differently sized artillery. The downside of the monster trebuchet was how long it took to load, and sometimes the rate of bombardment was more important than the size of the missile.

With medieval terminology words can be deceptive, and even the commonly accepted "trebuchet" has unclear origins. The situation has not been helped by the decision of historians to describe the manually operated weapon as a "traction trebuchet", even though this means using a term that was unknown to contemporaries and did not exist in this sense until the thirteenth century. It is probably too late to put this right. According to W Sayers,[30] the word *trebuchier* in Old French signified falling or tripping, and it was used from the late twelfth century to describe animal traps (tipping the victim into a pit) and ducking stools, the visual comparison to both the shape and the operation of the new weapon being evident. Alternatively, the term for the engine might have derived from its constituent parts, where the middle part derived from *buc* meaning trunk. Numerous variants are found in every romance language, such as *trebuca* (Occitan), *trabuquet* (French) and *trabuco* (Italian), and in German it was to be sometimes rendered as *tribok*, while the word was also quickly Latinised as *tre*[or *i* or *u*]*buchettum*. In Germanic languages the word *blide* quickly took over since it had served for a long time as a generic term for all catapults, while as the design of the trebuchet evolved across the thirteenth century, different versions themselves acquired new names, the most common being the *bricola*, *biffa* and *bride*. Seek in vain for any consistency. And finally, confirming the hopelessness of attempting to interpret the word used, writers in Arabic never used a term specifically to describe the new weapon, despite the first evidence for its existence deriving from an Arabic text.

Who first decided that attaching a counterweight to the front of the throwing arm of a stone-throwing catapult would achieve an improvement over the manually

29 It has been difficult to establish the range of medieval bows and crossbows but most experts would place even the formidable English longbow – let alone the much more common shorter bow in use elsewhere – at this figure, and this does not reflect that in combat conditions the maximum *effective* range was considerably less. On archery, see J Bradbury, *The medieval archer*.

30 W Sayers, 'The name of the siege engine *trebuchet*: etymology and history in medieval France and Britain', *Journal of Medieval Military History* VIII, 2010, 189–96.

operated lever artillery that had been in use for so long, and when did it happen? Historians have suggested various possibilities, including that common devices such as the *shāduf*, which raised water from wells or irrigation channels, relied on similar principles of balancing weights at the end of poles; or that some designs of crane used counterpoised weights. Perhaps an engineer, observing such a machine in operation, had the inspiration to apply the same idea to a stone thrower, and even if, as has been usefully explained by M Basista, different mechanics applied, the medieval engineer would not have understood that concept.[31] Before the Chevedden thesis that there had been hybrid weapons since the ninth century, there was some consensus dating its appearance to the late twelfth century and its widespread adoption to the first years of the thirteenth. The first explicit reference to the word trebuchet as a weapon was in an Italian text in 1189 (a source discovered by Chevedden) and it appears for the second time in 1199. The first identified in Britain was brought over by Prince Louis of France in 1217 during the civil war that followed Magna Carta. Here, its unusual character was confirmed by the near-contemporary French verse *History of William the Marshal*, the septuagenarian regent for the child king Henry III, which is full of the standard perriers and mangonels, but describes how one of the French prince's cogs at the sea battle of Sandwich came to grief because it was overloaded with horses and because *le trebuchez est dedanz*. This weapon was precious enough to have been dismantled for the return across the English Channel. In Germany, the *tribuk* made its first appearance in a description of Emperor Otto IV's siege of Weissensee in 1212.[32] Further early references in Genoese sources dated from 1224 and 1225.[33]

In addition to arguing the existence of a hybrid version, Chevedden also suggested that the time taken for the counterweight engine to reach the levels of

31 M Basista, 'Hybrid or counterpoise? A study of transitional trebuchets', *Journal of Medieval Military History* V, 2007, 33–55. The author also points out that the first examples of mechanical clocks where the escapement was driven by falling weights appear in the late twelfth century. However, while the source cited for this claim does demonstrate that clockmakers were also likely to be employed on a wide range of engineering tasks, including the military, it does not support this statement as this development did not occur before 1350: G Dohrn-van Rossum, *History of the hour. Clocks and modern temporal orders* (1996), trans. T Dunlop, 50–1, 105, 185–6.

32 *Codice diplomatico Eceliniano*, ed. G Verci (1779), 97 (1189: the reference was first discovered by Chevedden, 'Invention of the counterweight trebuchet', 101); 'Annales placentini guelphi', *MGH* SS XVIII (1863), 420 (1199); *Histoire des ducs de Normandie et des rois d'Angleterre*, ed. F Michel (1840), 188, 192, 195 and *Annales prioratus de Dunstaplia*, ed. H Luard (1866), 49–50 (England); *History of William the Marshal*, ed. A J Holden et al., vol. 3 (2006), 372–3 (line 17,391); 'Annales Marbacenses', *MGH* SS XVII (1861), 172 (Weissensee).

33 'Marchesi scribae annales', and 'Bartholomei scribae annales', *MGH* SS XVIII (1863), 155, 158.

destructiveness achieved in the thirteenth century argues for a longer period of development than is suggested by this dating.[34] However, if it is recognised that the decision to replace human pulling power with the counterweight represented not a process but a revolution in engineering, the question of the time gap becomes less relevant: instead, it is appropriate to substitute the time necessary to arrive at the most effective specification for the new engine. Nonetheless, it is important to consider the strongest candidates found by Chevedden for the earlier arrival of the new engine: the siege of Hungarian Zeugminion by a Byzantine imperial army in 1165 and the (Sicilian) Norman conquest of Thessaloniki twenty years later. The town now called Zemun and part of Belgrade was strongly sited and its citadel and walls would feature in the sixteenth century, but little is known of its defences in 1165. Emperor Manuel Comnenos (1143–80) led the siege along with his cousin Andronikos, renowned for his expertise in this kind of warfare, and was described as taking charge of a stone thrower in person, using the sling, winch and screw press to "shake" the wall opposite. Another reference is to stone shot weighing one talent (implying around 30 kilograms) that "weakened the wall's joints".[35] One clue is not mentioned in Chevedden's claim that these big stone throwers were counterweight trebuchets: the chronicler Choniates made clear that the collapse of the wall followed undermining or sapping. Neither the weight of the missile nor the description of the equipment used to load it (a large stone thrower might well require a winch to wind down, whatever its means of propulsion) rules out that this was a very large petrary.

In contrast, Eustathios, bishop of Thessaloniki, wrote to describe the shameful failure of the governor of the empire's second city to hold it against a Sicilian Norman army that had marched across the Balkans almost without opposition in August 1185; they were joined there by their fleet. The writer described the professional skill of the enemy and that they prepared siege engines (*helepolei*, the generic term for the most powerful weapons), including one of particular size (*megales mechanes*). The city fell when the wall on the eastern side collapsed. According to the account, this was only partly as a result of sustained bombardment: the key factor was sapping. The bombardment had been important – but as the bishop explained more than once in his text, it was causing the destruction of the battlements, while other smaller engines caused heavy casualties among the defenders and damage inside the city. This is in keeping with the idea of larger engines using the same technology rather than counterweight engines able to demolish well-built walls (such as the ancient walls of Thessaloniki). The confirmation is in the detail: the giant engine's ammunition comprised stones "fully as much as a man could lift".[36]

34 Chevedden, 'Invention', 72.

35 Niketas Choniates, *O city of Byzantium*, trans. H Magoulias (1984), 76.

36 Eustathios of Thessaloniki, *The capture of Thessaloniki*, trans. J R Melville-Jones (1988), 59, 68, 97–9; the Greek original is accompanied by the translation, making it

Such a description – how much could medieval man hoist from the ground? – is not a basis for concluding that the missile was so big as to prove the presence of a new weapon. In using medieval chroniclers' accounts, it is as unwise to accept what may have been hyperbole as scientific evidence as it would be to take at face value many descriptions of modern war by journalists. The most that can be said of Eustathios' account is that it is capable of bearing other interpretations. With that said, it would be possible in the time frame I have proposed for the Normans to have used the weapon at Thessaloniki in 1185.

The first *explicit* description of a counterweight weapon which survived remains that of Mur'a b.'Alí al Tarsūsī, whose treatise must have been composed before the death of its dedicatee, Saladin, in 1193.[37] In his account of what he called a "Persian (or Turkish) mangonel", he stated that this device was better than the traditional model "because it pulls with a constant force, men differ in their pulling force". This writer therefore grasped the significance of the new technology. However, his diagram of this new weapon, unlike his sketches of the shapes of the traditional stone thrower, was hopelessly unrealistic, and his account of how a crossbow mounted on the frame was to be used to assist in its operation makes no sense [Fig. 17]. Throughout the treatise, al Tarsūsī referred to the machines he described, including this one, as having been made by an Alexandrian named as Abu'l Ḥasan b. al-Abraqī al-Iskanderuni, and it may be that he had not actually seen this engine when he wrote about it (the others, he several times explained, were too common to require detailed descriptions). The implication is that this engine was also new to this writer.

The regular appearance of the name itself comes from the first three decades of the thirteenth century, and this if nothing else implies a new weapon at this time. If the issue of the date of invention cannot be resolved conclusively, a defensible case can be made for not earlier than the last decade or two of the twelfth century, with diffusion taking place over the following twenty years. Not even this level of (un) certainty can be suggested to answer the question of who was the first engineer to build one. The first appearance of the name as a weapon in Italian sources suggests a Mediterranean origin but whether this was Christian or Muslim, and if Muslim, in which state, cannot be answered. It is also possible that the invention took place in the Byzantine empire, where emperors such as Manuel I Comnenos continued to wage vigorous military campaigns, creating a climate that might have been conducive to innovation, although if it was not used at Zeugminion and was used

straightforward to check the words being used, none of which contain any surprises.

37 The first translation was by C Cahen, 'Un traité d'armurerie composé pour Saladin', *Bulletin d'études orientales* XII, 1947–48, 103–63; parts including the description of the siege artillery were put into English by B Lewis (ed./trans.), *Islam from the prophet Muhammad to the capture of Constantinople, I: Politics and war* (1987), 218–23; Chevedden, 'Invention', provided a better translation at 115–16.

only by the enemy at Thessaloniki, this claim is shaky. There is no decisive evidence to attribute first use to any one ruler.

Whoever used it first, they were responsible for a step change in the application of technology to warfare, and one that represented the application of new engineering skills to construct the weapon. While the human-powered stone throwers did require knowledge of the right proportions to be used in the construction of their frames (something spelled out in al Tarsūsī's descriptions of the other types of engines), the counterweight machine added new layers of complexity. In processes now lost, engineers over these few years successfully developed an engine where the proportions of the various elements to each other were not optional but essential: the proportion of the lengths of the throwing arm either side of the pivot or axle; the length of the sling as a proportion of the throwing arm; the size of the counterweight relative to the weight of the missile. The size of the frame was also not a matter of chance. It had to be large enough to contain the counterweight box as well as sustain the shock of its weight as it was first winched up above the level of the frame and then fell down on release. Alongside these calculations, there were other obvious challenges to meet: central to the entire operation was the axle or pivot on which the throwing arm turned. In one of the experimental reconstructions, the first effort to shoot failed when friction prevented the pivot from turning. The choice of timber for the various component parts might be critical – if the pivot failed as a result of the stress caused by the weight imposed on it, the effect on the engine and its crew might be catastrophic. There were further decisions needed: as well as the length of the sling, the angle at which it was secured to the arm and the angle at which it released its shot were both critical in determining the range. This was a detail that ideally needed to be decided before the machine was erected, too, as it significantly impacted on the range and therefore where to set up the engine.

It could be that all this was achieved by trial and error, but the sheer quantity of data needed to build an engine which would succeed does suggest something a little more sophisticated. Once established, the dimensions and the proportions for engines could be memorised and handed down in the traditional way, but it is surely telling that Villard de Honnecourt's surviving page of instructions for the construction of a trebuchet does contain specific dimensions and proportions for the frame and counterweight box of the engine, the latter needing to be 2 "toises" in length (translating into approximately 12.8 feet or 3.9 metres), 9 feet wide and 12 feet deep (2.76 and 3.7 metres respectively) [Fig. 16]. Villard's counterweight was to be made up simply of earth. Renaud Beffeyte made one of his test reconstructions of a working trebuchet according to Villard's formula and plan, including using a pentagram to provide the correct dimensions for the cross braces of the triangular frame.[38] Villard's dimensions accorded with geometric ratios. For an arm 10 to

38 Beffeyte, 'The oral traditions and Villard de Honnecourt', 93–119; the same, *Les machines de guerre au moyen âge* (2000), 13–18. Beffeyte's reconstruction at the Loch Ness

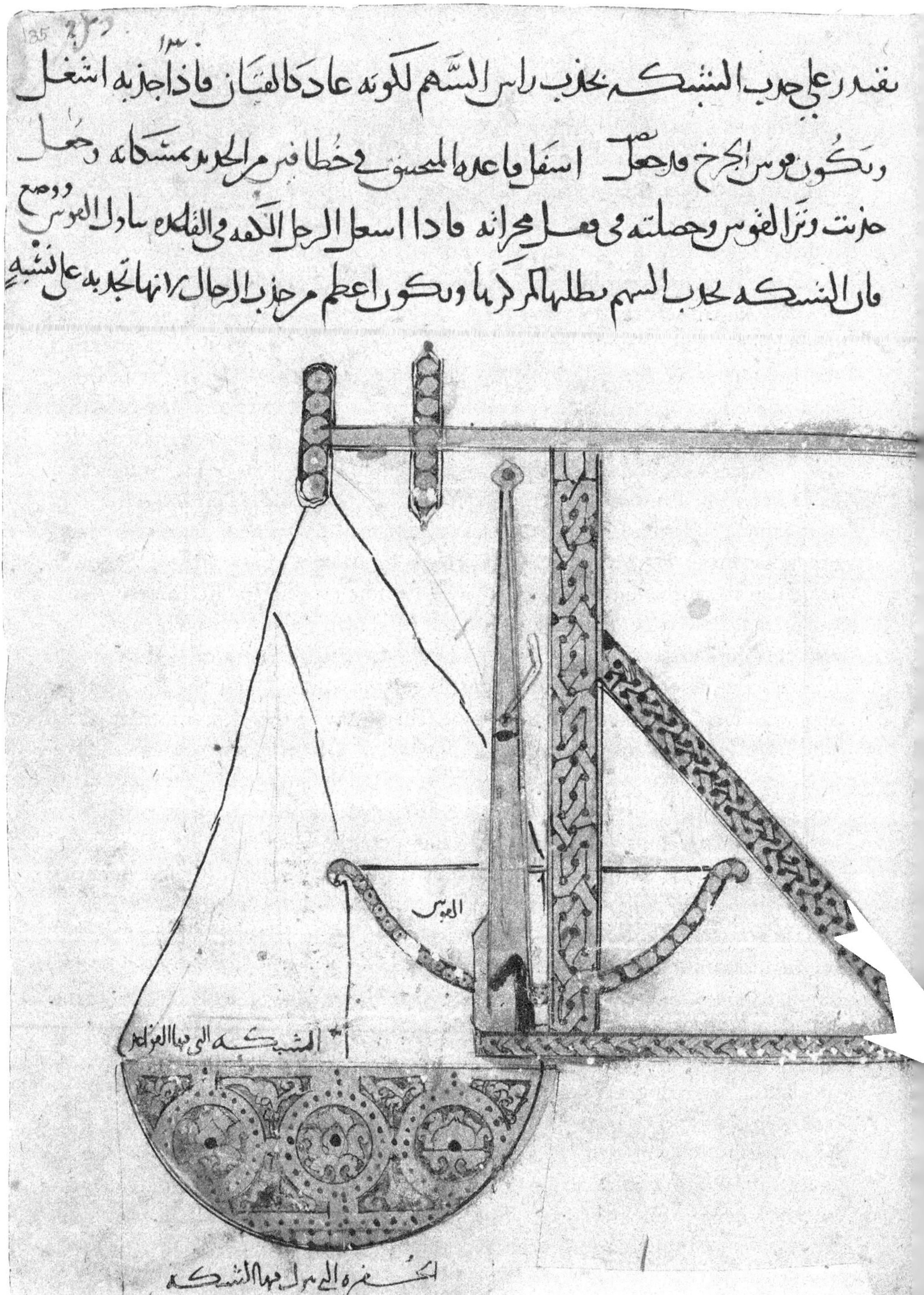

Fig. 17 The first surviving illustration of a counterweight engine, in the manuscript of al-Tarsūsī, Bodley MS Hunt 264 fol. 134v–135r. (The Bodleian Libraries, Oxford, with permission)

بعد ان يضع الحجر في الكفة ويعلق حلقها في الخطاف الموضوع في راس السهم فان الرجل
الكفة بخطاف من الحديد جعل في اسفلها في حلقة ثبتت في قاعده يمسك فوه الشبكه
وتره مع الحبال التي ترفع الشبكه في خطاف مثبت في الحبال فاذا اطلق الحبال بالشبكه
السله في مجراها ورمى عنها ثم تعود في ساعته الى الكفه فعلقها على ما تقوم في دفعه
واحده والرجال تخلف حبلها وهذه صورة ذلك ❁

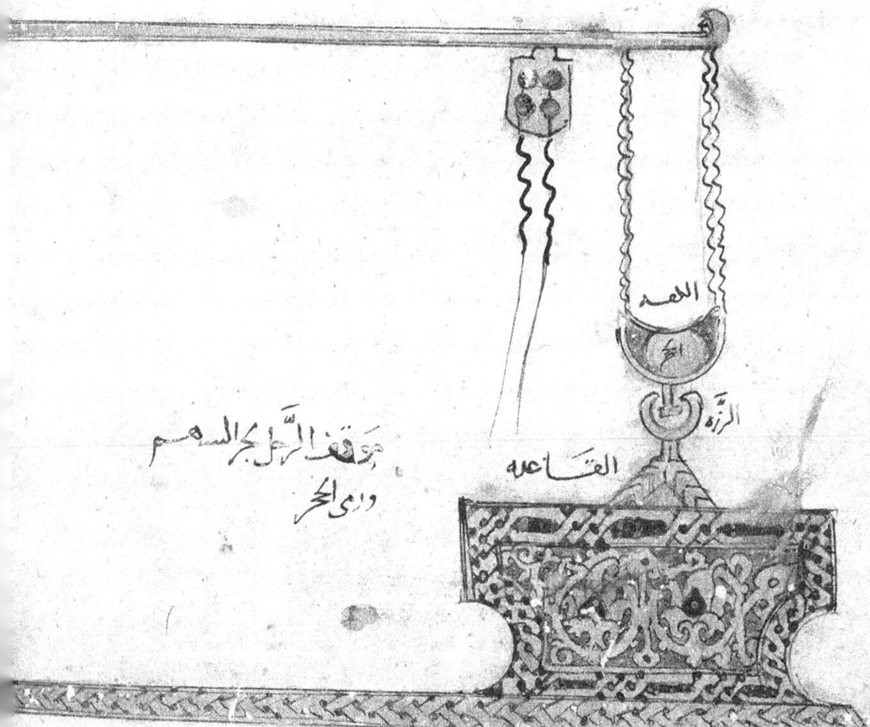

12 metres long used with a counterweight of 22 tons it would be necessary if the proportion of the two sides of the arm was 1:5 to exert a force in the proportion of 22/5 to lift the counterweight, which could be more easily done by dividing the individual effort by four through the use of four pulleys – just as Villard recommended.

Other decisions also influenced performance, and developments in the design of the engine following its introduction seem to suggest that engineers had thought through the consequences of changing the original model. One of the more significant was the step of having the counterweight housed in boxes either side of the arm, as opposed to having it in a single box suspended beneath. This shape evolved later in the century and worked best with a smaller engine using a single post to hold the beam. It was commonly called a *bricola*. The name first appears in written sources at Frederick II's siege of Brescia in 1238 but this does not mean that it had not been around previously, although the fact that it was distinguished from the trebuchet may confirm it was seen as a different device.[39] Having the counterweight swing loose was an early step that imparted greater force. Some modern reconstructions mounted their engines on wheels, which improved the operation by adding additional force to the projection of the missile; it is not clear that this reflected any medieval design. What did was the introduction of treadwheels either side of the frame to help with the single most effortful part of the loading operation, the hauling up of the counterweight after each shot: this design was used in the reproduction weapons on display at the Middelaldercentret in Denmark, subsequently copied at Warwick castle in England [also shown on the front cover and Fig. 18]. The use of treadwheels as a development of the windlass was identified in numerous manuscript illustrations of medieval building sites from around 1240 and fits well with the idea that the same technology may have been adopted for the purpose of simplifying the effort of preparing the heaviest trebuchets for loading. It is another example of technical progress with multiple applications across the thirteenth century.[40]

---

experiment in 1998 was a more reliable reflection of the design parameters than its US counterpart, designed by Colonel W Neal using a late-fifteenth-century Arabic text but adding wheels, something not seen in medieval sources. The ranges, size of missile and accuracy of both modern versions were, however, in keeping with expectations based on medieval evidence. See E Hadingham, '"Ready, aim, fire". A risky experiment reveals how medieval engines of war brought down castle walls', *Smithsonian* 30/10, 2000, 78–87.

39 In the 'Annales Placentini Guelphi', 479 and in Caffaro's Genoese chronicle. The process has been studied by P Chevedden in 'The artillery of King James I the Conqueror', in Chevedden et al. (eds), *Iberia and the Mediterranean World of the Middle Ages* (1996), II, 48–80.

40 A L Mathies, 'Medieval treadwheels: artists' views of building construction', *Technology and Culture* 33/3, 1992, 510–47, explored many aspects of the appearance and design of windlasses and treadwheels, their use as lifting devices, and the surviving examples

Medieval military engineers did not know mechanics, but each of the steps taken – starting with the invention of the trebuchet itself – suggest that they understood what would make the engine more devastating in its impact, or increase its range. We do not know if medieval military engineers studied geometry in monastic schools – unless perhaps they were the educated clerics recorded as operating artillery – but they knew how to design and construct immense engines using geometric principles and capable of sustaining great pressures; they used different types of timber so as to maximise effectiveness while minimising expenditure of time and money on the construction, while simultaneously sustaining the enormous force created by the weapon's operation. Unlike other medieval siege weapons, the trebuchet has attracted the interest of modern scientists as well as those who build reconstructions, and their studies all testify to the remarkable achievement of the trebuchet engineers.

MODERN TESTING OF THE TREBUCHET

Modern studies of the trebuchet confirm that its original builders knew what they were doing. The earliest recorded studies were carried out at the orders of Louis Napoleon Bonaparte (subsequently Napoleon III, 1852-70). Working with serving artillery officers, he undertook a thorough review of the medieval accounts and then tested them with models. One finding confirmed that the sling and the angle of its hook were critical elements, and changes in its length could double the range. Experiments carried out with the model made by a Captain Favé noted the enormous force brought to bear on the frame by the large counterweight, but confirmed the remarkable accuracy achieved, something noticed again in more recent experiments.[41] Computers have now secured similar results. The device has been chosen by science and engineering teachers to challenge their students, and the results published. The internet is now full of descriptions of experiments in the dynamics and mechanics of the trebuchet, alongside instructions for building

(mainly within the towers of European cathedrals). The earliest dated illustration of a treadwheel in Binding, *Medieval building techniques*, is from the middle of the century (no. 355). An illustration from a manuscript of Matthew Paris of c.1250 shows clearly a windlass-operated crane (number 168) while Villard's notebook shows something similar (number 469). There is an excellent illustration of a wheel operated by two workers side by side from a little later in a manuscript from Dijon (number 159). Remarkably, a late-thirteenth-century treadwheel with a diameter of around 12 feet (3.7 metres) remains in place at the foot of the immense spire of Salisbury cathedral, set up for its construction: brief description and illustration in R Spring, *Up the spire* (1982), 5–6.

41 Louis Napoleon Bonaparte, *Études sur le passé et l'avenir de l'artillerie* (1851), 26–30, 38–41. The model weighed 17,500 kilograms.

one at home. The scientific experiments are important because they confirm the characteristics that made the trebuchet such a significant improvement in technology and demonstrate the complex range of calculations needed to achieve optimum performance – no doubt the reason the device has become popular with modern teachers. A most comprehensive analysis of the criteria for optimum operation was done by Donald Siano in 1998; his paper has provided the basis for many subsequent analyses.[42] Taking this as a starting point, another set of calculations was done to compare the accuracy of the trebuchet with other weapons: it came out as hitting the same spot appreciably more consistently than the longbow and the musket. Alongside calculations of the variables – concerning torques and forces, mass and inertia – and analysis of both angular and rectilinear acceleration, the original calculations had shown the most effective ratios, including that the counterweight needed to be 100 times the weight of the shot and that the angle of release should be before reaching 45 degrees.[43] Comparisons of the theoretical results that can be calculated using the equations proposed by Siano confirmed a fairly close correlation with practical outcomes using models, the differences largely accounted for by factors such as friction and various angles of the sling hook. The significant improvement in range "efficiency" (that is, the actual compared with the theoretically calculated range using the energy input from the falling counterweight and its mass and that of the projectile) has also been calculated and showed that the counterweight engine was four times as efficient as its predecessors.[44]

The massively complex mathematics underlying these (computer-aided) calculations would have been incomprehensible to medieval engineers, but the evidence of trial and error must surely have been rooted in instinctive estimations: it is plausible to imagine the engineer wondering whether the way to increase the range was to lengthen the sling or change the angle of release, for example, and then testing the hypothesis. In this respect it is important to bear in mind that some computer simulations based on modern calculations have been shown not to reflect very accurately the reality when an actual (replica) engine has been tested. The medieval

42 See, for example, http://www.real-world-physic-problems.com/trebuchet-physics.html which is based on Siano's original calculations that relied on the use of Lagrangian equations: http://asme.usu.edu/wp-content/uploads/2013/09/trebuchet35.pdf (both accessed 12.4.2015).

43 R P Rhoten, 'The trebuchet: accuracy analysis of a medieval siege engine', *1999 ASME Design Engineering Technical Conferences*, vol. 2, 69–73. See also R Goulet, 'Medieval engines of siege warfare and modern engineering tools', *ASEE Annual Conference 2003*, paper 3566 (1–9).

44 J O'Connell, 'Dynamics of a medieval missile launcher: the trebuchet', *The Physics Teacher* 39/8, 2001, 471–3; R Rhoten, 'Trebuchet energy efficiency – experimental results', 44th *American Institute of Aeronautics and Astronautics aerospace sciences meeting. Technical papers vol. 13*, 2006, paper 775, 9355–9.

engineer relying on an appreciation of some instinctive "mechanics" was capable of creating an engine of unprecedented destructiveness and accuracy. As one of those who reconstructs modern replicas has identified, the real-life tests differ considerably from the predictions of the computer simulations or of small-scale models, as Captain Favé had already identified in the mid nineteenth century![45]

This applies to another important improvement of counterweight artillery over its manually operated predecessor and those with an alternative means of propulsion such as torsion or tension: its range. For attacker or defender, being able to hit the target at a distance far greater than could be reached by bow or crossbow brought obvious advantages. Some modern replicas have failed to achieve the ranges suggested from medieval accounts. This has been a consequence of failures of modern engineering rather than deficiencies in the originals, where the engineers were not concerned overmuch with health and safety considerations or the risk of harming paying visitors. Some, however, *have* reached remarkable ranges, with the longest achieved using quite small shot (445 metres with a 10-kilogram shot, for example, in a Czech reconstruction).[46] Surviving trebuchet shot are usually much bigger: from 50 kilograms up to 100 kilograms appear to have been "standard" ammunition sizes, and for these missiles a still substantial range of 200 or more metres may have been normal.

The greater ranges were not the result of later evolution: one of the rare incontrovertible accounts to confirm an enormous range described the attack by the archbishop of Cologne on that city in 1252 during one of their feuds. The churchman gathered a noble army and fleet, sailed down the Rhine, and set up a *blide* on the opposite bank at Deutz. From here, five shots were hurled at the city that caused a little damage to the house called Rodenburg. However, nothing more was achieved and the attack was called off. There are remarkable conclusions to be drawn from this account: the trebuchet had hit the same target five times; and shooting across the Rhine at this point meant it had achieved a *minimum* range of 450 metres.[47] The small damage caused also confirmed that at this range, only lighter stones could be propelled. The formidable range is verified by studies of a number of "siege forts" located as the base for either the

45 Peter Vemming, responsible for the magnificent trebuchet (and other engines) at the Middelaldercentret in Nyköbing Falster (Denmark) and its copy at Warwick Castle (England), personal communication.

46 I tabulated the sizes of counterweight, shot and range in Purton, *HEMS*, 385 (with references).

47 G Hagen, 'Dit is dat boich van der stede Colne', *Die Chroniken der deutschen Städte 12: Niederrheinische Städte*, I (1968), 43 (lines 740–53); also repeated with additional information in the fifteenth century: 'Koelhoffshe Chronik', same series, 13, p. 553. Discussed in M Feuerle, 'Das Hebelwurfgeschütz. Eine technische Innovation des Mittelalters', *Technikgeschichte* 69 (1), 2002, 1–40, here at 20–1.

Fig. 18 A springald (left) and a trebuchet, *The Romance of Alexander*, Bodley MS 264 fol. 201r. (The Bodleian Libraries, Oxford, with permission)

attack or blockade of a castle where it is known that artillery bombardment took place. One such example was Thurant, besieged by the archbishops of Trier and Cologne in 1247, and although the evidence for where the attack was based is not definitive, the siege fortifications were at a minimum more than 200 metres distant and one site is 500 metres away.[48]

(c) The Springald

The thirteenth century also witnessed the appearance of the springald, another piece of new artillery to stand alongside the two-footed giant crossbow and the counterweight trebuchet. This engine represented a reversion to an ancient form of propulsion: torsion, as used by the armies of imperial Rome and probably abandoned by the end of the sixth century (see chapter two[49]), but in a new shape. Liebel, a

48 Discussed by Feuerle, 'Das Hebelwurfgeschütz', 23–33; but now see O Wagener, 'Sieges, siege castles and the question of visibility – new research with the help of LiDAR scans', in R Atzbach et al. (eds), *Castles at war* (2015), 217–30.

49 D S Bachrach has argued that the consistent use of the two terms "mangonel" and "petrary" in English royal accounts over the twelfth century must signify two different

leading expert on the subject, identified a number of important characteristics: in place of the early Roman ballista that resembled a medieval crossbow in shape or the late Roman *onager*, both of which propelled stones, the springald was a new design. The coils that provided the torsion were held in a box-like wooden frame, and the propelling string was wound back using a metal screw, shown in surviving illustrations [Fig. 18]. Liebel suggested that the reappearance of this design might be related to the invention in Europe of the spinning wheel around 1250, when the first explicit references to the springald are found. This would have made much easier the winding of the skeins, but the dating evidence does not support the argument.[50] It

---

types of weapon, one of which, he claims, was torsion powered, but can adduce just one piece of evidence to support that claim, which is the requisition for a winch ("in duobus circulis ferri ad turnos mangonellos", from the Close Rolls for 1225). I do not see this as sufficient evidence on its own, and while it may be true that English royal clerks did distinguish between two engines, this does not prove they were of different types; while throughout Europe all other references indicate a much more promiscuous approach to using these two terms (D S Bachrach, 'English artillery 1189–1307. The implications of terminology', *EHR* CXXI, no. 494, 2006, 1408–29.)

50 P Baines, *Spinning wheels. Spinners and spinning* (1977), argues that even the simpler spindle wheel, although already well known in China and India, did not reach Europe until the late thirteenth century; the mechanisation of silk production may have been achieved at Lucca in the 1270s but the spinning wheel did not become widely used in Europe until the fourteenth century: 44–6, 82–5.

was a fearsome weapon, with great powers of penetration, and a computer model suggested it could shoot a 1.4-kilogram bolt a distance of 180 metres if mounted in a tower at an elevation of 15 degrees. The earliest reference is dated 1249, Henry III had them at his siege of Kenilworth in 1266 and there are frequent references to them in use or in storage in arsenals for the next century or so in England, France and Germany until they were rendered obsolete by progress in the quality of gunpowder weapons in the later fourteenth century.[51]

What the springald also suggests is a case of new technology applied to ancient knowledge. Contemporaries would have known Vegetius' description of ancient artillery, and medieval manuscripts not infrequently reproduced drawings of classical devices that no longer existed, so awareness of torsion power should not be a surprise. What the springald confirmed was that medieval engineers improved the design, recognising the potential of a different frame, a more effective means of stringing and possibly also superior metalwork (as Liebel suggested) to create a more efficient and reliable engine than that of the ancients. The disadvantages of torsion were also recognised in some of the evidence for the deployment of the springald. It may only usually have been deployed under shelter to protect the skeins from rain, with the evident limitation this imposed on the field over which it could be shot, as well as the common problem with this kind of weapon that it could not be depressed very far without the risk of the bolt sliding out.

The significant fact is that all of these weapons – giant crossbows, the trebuchet and the springald – represented not the continuation of ancient engineering, but improvements upon inherited knowledge.

(d) Shipbuilding in the Thirteenth Century

There is plentiful evidence that skilled men were employed on the wide range of tasks called for in constructing seaworthy craft and, in the military sphere, making them into effective war machines. The Pipe Roll for King John's reign (1199–1216) contains evidence for the building and fitting out of the royal fleet. The significant total of more than £6,000 was spent on building or refitting royal galleys and other ships between 1209 and 1211 at Portsmouth, Romney, Shoreham and other

51 J Liebel, *Springalds and great crossbows*, trans. J Vale (1998); E Gessler, 'Der Springulf, ein mittelalterliches Torsionsgschütz', *Basler Zeitschrift für Geschichte und Altertumkunde* 20, 1992, 89–203. This and other evidence for the use of the springald summarised in Purton, 'The myth of the mangonel', 79–90. Interestingly, the only specific evidence of the power of springald bolt contradicts the general understanding: in 1304, Thomas Gray's father survived being hit in the face by a springald bolt at the siege of Stirling (Gray, *Scalachronica*, trans. H Maxwell (1907, republished 2000), 25–6.)

ports, including constructing a protected dockyard at Portsmouth. Analysis of the numbers suggests that the galleys were only one third the size of those built under his grandson Edward I, indicating that the skills were later applied to much larger ships than the early-thirteenth-century navy. If this was carried out by shipwrights' rule of thumb, their abilities included being able to multiply dimensions proportionately.[52]

Edward I's military campaigns frequently called upon ships and many of them could justifiably be described in more modern parlance as genuine "combined operations". Accounts for 1294–95 show the construction of twenty new war galleys at Newcastle upon Tyne and other ships were built at Lyme and Southampton. The accounts name numbers of masters and confirm the range of skills required: carpenters, clenchers and caulkers under their masters received the best rate of pay but there were sixteen distinct functions altogether. The Lyme galley used fifty-four oars and was 22 metres long. It required 20,000 iron or wood fasteners. No treatise has survived explaining the design or how the many different parts fitted together. The masters carried the knowledge in their heads, including the best uses for different kinds of wood and the proportions between the dimensions for the hull, masts, rudders and many other timber constructions; this amounted to a formidable reservoir of facts. The end result was a fleet that was more than fit for purpose, using skills refined through experience.[53] Scholars studying the evidence suggested that the masters were highly skilled and may have been literate and numerate.

At the same time, the French king was establishing a navy of his own with an arsenal constructed as a base for his galleys at Rouen. However, the major development of this site would happen later and the main interest in French naval progress at this time relates to the Mediterranean, where the crusades called for by St Louis required the creation of a navy suitable to transport large forces across the sea to Egypt and North Africa. Evidence survives in the contracts made with shipwrights in Genoa and Marseilles. In assessing what skills were required to create the vessels, their size and their ability to load, convey and unload horses were critical. The contracts specified different types of ship, including galleys and the larger *tarides*, and they spelled out exact dimensions of many of the parts of each ship. The horse transport would be 35.7 metres long and 3.4 metres in the beam,

52 L Carr Laughton translated and analysed the texts in 'Naval accounts for 1209–11', *Mariner's Mirror* 28, 1942, 74–7; the originals were subsequently published by the Pipe Roll Society, *The Great Roll of the Pipe for 14 John*, ed. P M Barnes, 47–8.

53 I Friel, 'The documentary evidence for ship building in England 1294–c.1500', in C Villain-Gandossi et al. (eds), *Medieval ships and the birth of technological societies* I (1989), 139–49; S Rose, *England's medieval navy 1066–1509*, 80, 88–9; R Anderson, 'English galleys in 1295', *Mariner's Mirror* XIV, 1928, 220–41; J Tinniswood, 'English galleys 1272–1377', *Mariner's Mirror* XXXV, 1949, 276–315.

with two loading ports and twenty horse stalls. Those involved in the negotiations knew exactly what was needed and how to provide it, using a store of existing awareness and experience that leaves little room for doubt that these masters were literate, numerate and highly skilled at their trades.[54]

Many of these ships were adapted to carry artillery. The instructions for St Louis' fleet included some requirements to mount two mangonels on the deck, but this just continued a tradition that was known under King John (1214). However, it seems improbable that stone throwers would have been of any use in naval combat as opposed to shooting at a land target, as the Venetians may have done when they captured Constantinople in 1204. Ships designed specifically for fighting had long had timber "castles" erected bow and stern from where archers, crossbowmen and soldiers could engage an enemy ship. The springald also featured sometimes in these waterborne conflicts.[55]

Frederick II had (re)built a navy in the 1220s, with ships constructed and maintained at Messina, Naples and Brindisi, where strong arsenals were ordered to be built in 1239. The same types of ships were used as elsewhere in the Mediterranean, and when Charles I seized the kingdom, he maintained the naval tradition, including the use of horse transports. Detailed specifications were preserved in the Naples archives for the dimensions of galleys 40 metres long. Armament included what was described as "Greek fire" to be propelled from the deck by rocket (?) or by pots and bottles projected from catapult, though there is no evidence of such use.[56] In these cases, it is known who the builders or designers were. The evidence of these three navies – it could be replicated for other maritime powers (Venice or Genoa, for example) – confirms the presence of a large group of expert shipwrights, highly skilled and able to deliver large-scale shipbuilding. Their

54 Original publication of the contracts by M Champollion Figeac (ed.), *Documents historiques inédits tirés des collections manuscrits de la Bibliothèque royale*, I (1841), 507–615; L Belgrano (ed.), 'Une charte de nolis de S. Louis', *Archives de l'orient Latin*, II, Documents (1884), 231–6 (from Genoese archives); M Mollat, 'Le passage de Saint Louis à Tunis', *Revue d'histoire économique et sociale* 50/3, 1972, 289–303; R Bastard de Péré, 'Navires méditerranéans du temps de Saint Louis', same volume, 327–56; Pryor, 'Transportation of horses by sea during the era of the crusades', 102–25.

55 F Brooks, 'Naval armament in the thirteenth century', *Mariner's Mirror* XIV, 1928, 115–31; for a naval battle in 1293, J Hattendorf et al. (eds), *British naval documents 1204–1960* (1993), 20–1; for springalds in a sea battle between the French and the Flemings, G Guiart, 'Branche des royaux lignages Chronique métrique', in *Collection des chroniques nationales Françaises*, ed. J Buchon, VII (1828), lines 9068–585.

56 J Pryor, 'The galleys of Charles I of Anjou, king of Sicily 1269–84', *Studies in Medieval and Renaissance History* XIV, 1993, 33–102; W Cohn, *Die Geschichte der sizilischen Flotte unter der Regierung Friedrichs II* (1926), 127–36; J Göbbels, 'Die Militarorganisation im staufischen Königreich Sizilien', in A Esch et al. (eds), *Friedrich II. Tagung des Deutschen Historischen Institut in Rom 1994* (1996), 486–500, here at 495–9.

skills in carpentry in particular were of the same standards as their equivalents working on land.

Meanwhile, in northern Europe progress in the design of the commonly used clinker-built cog type of ship has been uncovered by maritime archaeology and images, demonstrating movement from simple craft to a "most sophisticated" ship, with shipwrights in the towns of the Hanseatic League using "quite a lot of skill and imagination in developing their big seagoing vessels".[57]

(e) Surveyors

References to the military employment of surveyors – such as their ancient role in laying out encampments and the ground plans for fortifications – become increasingly rare in the records. Perhaps others had taken over this role along with other responsibilities by the thirteenth century. Surveying itself continued to matter in other spheres: statutes laying down the rules for the measurement of estates were enacted. Study of the contents suggests no progress in the limited technology and even more limited mathematical knowledge of the people called upon to carry out these jobs, although it appears that more sophisticated tools were actually available: more than a hundred manuscripts survive of a treatise written in the second half of the thirteenth century that described the use of the quadrant to measure heights and depths along with nautical functions such as latitudes. But it has been argued that no practitioner would have known how to use it and only one thirteenth-century example of a quadrant is known (from Florence). There is no evidence to explain this disconnect between theory and practice.[58]

For new towns, however, skilled surveyors were necessary. Foundations continued unabated in the thirteenth century, around a thousand being established across Europe over this period. A significant number planted in south-west France were a feature of the plans of both the English and French rulers in the region, and

57 D Ellmers, 'The cog of Bremen and related boats', in S McGrail (ed.), *Medieval ships and harbours in northern Europe*, BAR International Series 66 (1979), 1–16, quotation at 14; the finding was confirmed by Danish wrecks: O Crumlin-Pedersen, 'Danish cog finds', same volume, 17–34.

58 See E Taylor, 'The surveyor', *Economic History Review* XVII/2, 1947, 121–33, for a study of the significance of the statute Extenta Manerii of 1276 and later developments. He argues that there was no progress in the accuracy of measuring techniques before the late fifteenth century and that the pole and perch maintained a long tradition of crude measurement restricted to straight lines. N Hahn, 'Medieval mensuration: Quadrans vetus and Geometrie due sunt partes principales', *Transactions of the American Philosophical Society*72/8, 1972, provided this analysis (pp. ix–xxix) and pointed out that the second document he was publishing was largely a copy of Hugh of St Victor's twelfth-century work (see p. 110).

the bastides that remain remind visitors of their regular geometry – parallel streets running in a grid pattern with a market square as the focal point, some of them fortified with walls and gates on a similar rectangular pattern, many others not.

There are a remarkable number of common features which are surely not accidental. The bastides of south-west France comply with regular proportions using the formula of $1:\sqrt{2}$. At Sainte-Foy-la-Grande (Gironde) (established 1255) and at Grenade-sur-Garonne (Haute Garonne) (founded 1290), the diagonal of the market square is the same length as the side of each of the spaces allocated to housing. The result is consistent with the advice to surveyors contained in the *corpus agrimensorum* and transmitted in the writings of Gerbert of Aurillac, Hugh of St Victor and the *Practica geometriae* of Fibonacci. Documents authorised the construction of the disappeared town of Baa near Bordeaux (1286) under the oversight of Richard of Escham and Master Gérard de la Tour, who was in charge of laying out the site. Pierre Guarrin, described as royal surveyor for Toulouse, was requisitioned to lay out Beaumont de Lemagne (Tarn-et-Garonne), a town founded in 1278–79, which was fortified. The geometry of these medieval towns is more complex than Roman grid towns, so they were not just copies of ancient models.[59] Another example studied and measured – Acaya, in Italy, founded on the orders of Charles of Anjou in 1268 – has a structure based on the squared *canna* (64 "palms" or 4.45m) and the whole "shows a strong arithmetic-geometrical coherence".[60] Here, as elsewhere, medieval techniques had progressed and there was a link between mathematical knowledge and the practice on the ground (literally).

Although the methods of laying out and measurement remained primitive (ropes and pegs and, as has been seen, perhaps quadrants), the skills of the surveyors themselves had developed.Even if it was the patron who determined what the town should look like, the evidence for the use of complex proportions (just as in architecture) is hard to explain unless someone involved knew what was required to execute them in practice. In support of this thesis, and slightly later in date, are several new towns founded by the city of Florence from the late thirteenth century to consolidate the power of the city. At the start, these towns

59 A Randolph, 'The bastides of south-west France', *The Art Bulletin* 77/2, 1995, 290–307; Coldstream, *Medieval architecture*, 124–7; F Divorne et al., *Les bastides d'Aquitaine, du bas Languedoc et du Béarn*, here at 44–5; C-L Salch, *Dictionnaire des châteaux forts et fortifications* (1976), 125 (Beaumont de Lemagne). Divorne established that 80 per cent of the bastides used a square plan and that a variety of people with different backgrounds (sometimes they were just described as clerks) were appointed to do the designing. In Poitou, the count appointed a knight, Arnold of St Leodegar, with a master, Odo, to measure the lands involved in a dispute over boundaries in 1267 (Molinier, *Correspondence administrative d'Alphonse de Poitiers*, I, 180, 182 (nos 291, 294).

60 G Mele, 'A geometrical analysis of the layout of Acaya, Italy', *Nexus Network Journal* 14/2, 2012, 373–89, here at 378.

had a clear military and political purpose. D Friedman's analysis confirmed that their plans, uniquely, were based on trigonometry rather than geometry. Control was in the hands of committees of citizens who contracted the actual work to experts including surveyors and building masters, some of whom also worked on the city's new cathedral. In this respect they resembled the people known from other Italian cities, such as the engineers Albert at Bologna, working on the cathedral, the cisterns and the city walls in 1245–67, and Antonio di Vicenza who in the fourteenth century worked on the fortress, the walls and the city's mills.[61] The new towns saw active military service, such as when Scarperia successfully held off an attack by a large Milanese army in 1351. This new town was the base for the carpenters and smiths responsible for Florence's "engines".[62] Evidently an overlap in engineering roles was present in this republic.

The impressive new stone castles erected on the orders of powerful monarchs such as Edward I in Wales required both a ground plan and people to lay out the plan on the ground. Those placed in charge must have themselves employed surveyors to mark out the ground, but these experts seem to have been anonymous, unlike Master Gérard in Aquitaine, or of junior status. However, the visual impact of these great constructions – deliberate and part of the intention – relied on harmony and proportion, as in all large-scale medieval buildings, and such qualities did not come about by accident. It is worth repeating that the fact that no scale drawings survive does not mean that the masters were not aware of the dimensions needed to achieve their design, and that to achieve these dimensions required drawings to enable the masons to create the necessary templates for the stonework. If these were drawings laid out at full size on the floor of the mason's lodge, someone must still have had the bigger picture in mind.

"World maps" (*mappae mundi*) were another new feature of the thirteenth century. Although their purpose was symbolic and religious rather than practical, the well-known Gough map provided distances between towns and the charts of sea journeys across the Mediterranean, the earliest of which also date from around 1300, have been described as being accurate and of actual use to seafarers. The many surviving itineraries generally gave the distances between cities or daily staging points to assist the traveller, but not maps. Medieval European topographical maps that have survived were not scaled, but rather represented visual impressions or pictures, in the same way that Villard de Honnecourt's sketches were pictorial even if proportional, but not scaled. The situation was a little different in the Muslim world where maps based on Ptolemy had been described in the eighth and ninth centuries, drawn for the caliphs, and the traveller Ibn Hawqal used them (in a

61 D Friedman, *Florentine new towns. Urban design in the late Middle Ages* (1988), 121–42, 153–7, 164–6, 256–8.

62 Friedman, *Florentine new towns*, 42–3. The siege of 1351 was vividly described in M Villani, *Cronica*, ed. A Racheli, book II, 58–77.

manuscript dated 1086). A tradition of geographical treatises flourished until the fourteenth century (Ibn Jubayr, Yāqut, Ibn Saīd and Abu'l Fida, for example).[63] It has been concluded that their purpose was administrative. In all, however, it appears that the history of cartography does not show a progression from the skills exhibited in the plan of St Gall during the following centuries to match development in other areas, and that if there was a renewed interest in providing maps from the late thirteenth century, it was a long time before a new, more scientific method was applied to this work.[64]

THE MILITARY ENGINEER IN THE THIRTEENTH CENTURY: RECOGNITION, STATUS, SKILLS AND SPECIALISMS

Thirteenth-century engineers were therefore faced with greater challenges to every area of their knowledge, skills and specialisms than ever, but they might now have had higher expectation of recognition and reward if they met them. It is these changes that make it reasonable to conclude that it is not just the chance survival of more records than in the twelfth century that accounts for this. Here I will look at evidence in particular from England, France, Sicily, Aragon and a number of smaller entities (in Italy and Savoy) where records permit analysis.

A striking feature was the length of service of a number of engineers. A few lost their lives in royal service but many were retained year after year. In France, the many successes of King Philip Augustus (1180–1223) necessarily involved reform of his army, and its engineering capacity was included within this. His work has been subjected to scholarly study for many years and the conclusions of this research agree on the main elements of his work in this sphere. The names of the engineers charged with the construction of his new castles, each dominated by their circular donjon, each of similar design and dimensions and now identified as *tours philippiennes*, were recorded, eleven of them identified as masters.[65] The work extended over twenty years. The names include a few particularly associated with military engineering, for example Master Guérin who was responsible for the royal *cabulus*, a new term for a stone thrower that appeared at the end of the

63 Nasr, *Science and civilisation in Islam*, 102–3.

64 Chapters by O Dilke, 'Cartography in the Byzantine empire' (I, 258–75), P Harvey 'Introduction' (I, 283–5), the same, 'Cartography in medieval Europe' (I, 464–501) and G R Tibbetts (II, 90–130) in J Harley, D Woodward (eds), *The history of cartography*, I (1987) and II (1992); Price, 'Medieval surveying and maps', 1–10.

65 A Erlande-Brandenburg, 'L'architecture militaire au temps de Philippe Auguste: une nouvelle conception de la défense', in R-H Bautier (ed.), *La France de Philippe Auguste: le temps des mutations* (1982), 596–7.

twelfth century. James the Conqueror, king of Aragon, relied on a corps of people to operate his siege equipment, but prominent amongst them was Master Nicoloso of Albenga who was involved in the capture of Mallorca in 1229 and was still serving at the siege of Burriana in 1238. He must have been of an advanced age when he was conveying the king's engines and manufacturing 800 of the shot for the engines from ship to shore as late as 1275.[66]

English kings seem also to have been loyal to a number of their engineers, retaining them in service for many years. Prominent amongst these were Bertram, Gerard, Jordan and Richard. Jordan was the earliest, the man who "made the king's trebuchet", the first for an English king, who continued to receive wages for several years thereafter.[67] Master Bertram, *ingeniator regis*, had a career lasting from at least 1253 to 1276. Sometimes he was recorded on his own, at other times with other engineers who appeared only once.[68] Edward I regularly paid Richard of Chester for a wide range of engineering services, including organising the bridge of boats across to Anglesey in 1282 and, twenty years later, bridges across the river Forth in Scotland. But his service started with the building work at Rhuddlan and Flint castles in 1277 (presumably under the supervision of James of St George) and continued with his role in commanding royal trebuchets, referenced at the siege of Drsylwyn in 1287 and in Scotland in 1304.[69] The bridges in particular should be recognised as an extraordinary achievement of design, planning and effective implementation.[70] Equally impressive was the career of Gerard (of Winchester), first recorded working on Corfe castle in 1242, who served in Gascony as a "master of engines", made engines at Newcastle in 1244 and was undertaking similar tasks up to 1256.[71]

66 B Desclot, *Chronicle of the reign of King Pedro III of Aragon*, trans. F Critchlow (1928–34), I, 87, 103–20, 132–6; *The book of deeds of James I of Aragon. A translation of the medieval Catalan Llibre dels Fets*, trans. D Smith, H Buffery (2003), 93–9, 158, 158–61; J Miret i Sans, *Itinerari de Jaume I "el Conqueridor"* (1918), 524 (1275 payment). For an analysis of the king's artillery, Chevedden, 'Artillery of James I'.

67 *Calendar of Liberate Rolls*, Henry III, II, 8, 26, 71, 81, 94, 121.

68 *Rôles Gascons*, I, 357 (no. 2689: Richard), 372 (no. 2828: the list), 447 (no. 3644: in charge of engines). Bertram's long career (he died in 1284) was at the centre of a survey of engineers serving King Edward by A Taylor, 'Master Bertram, ingeniator regis', in *Studies in medieval history*, 289–303.

69 C Caple et al. in *Excavations at Dryslwyn Castle 1980–85* (2007), 195; R Turner, 'The life and career of Richard the Engineer', in D Williams, J Kenyon (eds), *The impact of the Edwardian castle in Wales* (2010), 46–58.

70 R Allen Brown, H Colvin, A Taylor, *The history of the king's works, I: The Middle Ages* (1963), 354–7, 416–17. See also A Z Freeman, 'Wall-breakers and river-bridgers', *Journal of British Studies* X/2, 1971, 1–6.

71 *Rôles Gascons*, I, 126, 2051; *Calendar of Liberate Rolls*, Henry III, II (1240–45), 245, 255, 263, 323; IV (1251–60), 143, 248.

The role of Master James of St George, who was brought over from Savoy, as overseer of the works at the ring of majestic castles erected around north Wales from 1277 is well rehearsed (and has more recently been subject to re-examination),[72] but the other engineering achievements of the campaigns of conquest attract less attention. Among the first was the large-scale project to divert the river Clwyd and to canalise it to enable ships to travel the 2 miles (3.2 kilometres) upriver from the coast to the site of Edward's new castle at Rhuddlan. The accounts refer to the cost of hiring labourers to dig the new route, but someone else in the king's retinue must have planned the work, as well as the construction of a dock to receive the ships safely within the outer defences of the castle itself and the construction or repair of a bridge.[73] Master James was well rewarded and remained in service for nearly two decades.

Other wealthy rulers called upon the regular service of professional engineers to oversee and execute their siege operations, among them Charles I, king of Sicily (1266–85): in March 1279, Master Giovanni de Tullo (Jean de Toul from Lorraine), "engineer", was ordered to select his two best *machinisti* to construct war engines. He was also employed in castle building, along with another man from France, Pierre d'Angincourt, presumably named after his home town (in Oise), who was responsible for building castles and churches for the king between 1269 and 1284.[74]

The accounts of the county of Savoy, which was only just emerging as a regional power in the Alps, show that it was not only rulers of large realms who could sustain an engineering facility. Masters were employed by the counts for multiple tasks including changes in the design and construction of new castles (and repair of existing sites) and the building and maintenance of the siege artillery which Peter II (1263–68), Philip I (1268–85) and Amadeus V (1285–1323) were to put to effective use. Master Radulph was a master carpenter who was in charge of the count's artillery between 1257 and 1267, alongside a substantial list of other master carpenters whose services were recorded and preserved in the archives. Radulph operated from a major arsenal and factory for crossbows of all types in the (eastern) frontier castle of Chillon, where craft workers carried out their functions under

72 See most recently N Coldstream, 'James of St George', in D Williams, J Kenyon (eds), *The impact of the Edwardian castle in Wales* (2010), 37–45. Coldstream argues strongly that James was not the architect but rather "a consummate organiser with practical building knowledge" (44).

73 Allen Brown, Colvin, Taylor, *History of the king's works, I*, 318–22; a slightly different calculation of the number of diggers (worked out from their daily rate) is in A Taylor, *Rhuddlan castle* (guidebook), 11.

74 For short biographies of both men, U Thieme, F Becker (eds), *Allgemeines Lexikons der Bildenden Künstler von der Antike bis zum Gegenwart* (1907), I, 516; XVIII, 471.

his direction. This land-locked territory also deployed ships including a *galea* for military purposes on Lake Geneva.[75]

It seems that the chief determinant of whether rulers could maintain a permanent engineering capacity was wealth. The examples given already all relate to rich employers – kingdoms or wealthy cities. In these cases, engineers generally performed multiple functions required by the city, working on cathedrals as well as city walls, organising drainage as well as constructing artillery. Sometimes the same people were recruited by others – for example, Frederick II in his many Italian wars. An unnamed lay brother who was also a "master engineer" was responsible for constructing "engines and trebuchets and cats and rams" at Padua in 1250.[76] Frederick's failed siege of Brescia in 1238 attracted international attention. Imperial defeat was blamed on a Spanish artillery expert who, having been captured by the Brescians, served them instead. Calamandrino may have been invented for the purpose of explaining the setback, but if he was, records show the commune of that city recognising the services as a military engineer during the siege of a citizen named Master Alberto Scaiola, *inzinierius*, who was chiefly a surveyor.[77] Bologna was the home of a talented man frequently used by the city commune for advice and the oversight of building operations, including the new cathedral (from 1245), the bell tower, construction and maintenance of bridges, roads and diverting rivers. He was known as "Albertus inzignerius". Statutes of Parma in 1266 awarded higher status to *inzignerii murorum et manarie* (engineers of walls and mines) than to other masters.[78] The engineers of Italian cities were also called on to construct a battery of siege weapons for the many occasions on which they waged war upon one another and sent expeditions to besiege an enemy. Of course, in Italian cities, this practice had become established during the previous century, but it was a

75 L Blondel, 'L'architecture militaire au temps de Pierre II de Savoie: les donjons circulaires', *Genava* XIII, 1935, 271–321, at 289–91. Blondel's views on the circular towers of Savoyard castles influenced thought for decades and were popularised in England by A J Taylor. A more nuanced understanding now prevails although the precise form of the link between Savoyard and English castle design continues to be debated. J-P Chapuisat, 'Chillon base militaire savoyarde au XIII siècle', in A Pathuel-Guillard, L Chavoutier (eds), *Soldats et armées en Savoie* (1981), 79–84.

76 'Annales Placentini Guelphi', 479–80; Salimbene de Adam, *Cronica*, ed. G Scalia (1966), I, 135, 572. The chronicles had by then started to use a new coinage to describe artillery bombardment: attackers "mangonelled" the city (*manganaverunt*).

77 A Settia, 'L'ingegneria militare', in P Toubert, A Bagliani (eds), *Federico II e le scienze* (1994), vol. 2, 272–89; for Alberto, P Grillo, 'Velut leena rugiens. Brescia assediata da Federico II Iuglio-ottubre 1238', *Reti Medievali Rivista* VIII, 2007, accessed online at http://www.rmojs.unina.it/index.php/rm/article/viewFile/127/108 (accessed 25.11.2016) .

78 S Neri, 'Albertus inzignerius, un grande costruttore del duecento', in F Bocchi (ed.), *I portici di Bologna e l'edilizia civile medieval* (1990), 125–33, here 128.

case of continuity and further development associated with mastering the new technologies of the thirteenth century.

Two other examples give evidence that cities also employed (local) engineers for military purposes. During the Albigensian crusade, the Château Narbonnais in Toulouse (1217) was besieged by the citizens, and the city's forces included two named engineers, Garnier and Bernard Parayre. The sources also say that everywhere the master carpenters were preparing the engines of war, in particular these two who were "experts in trebuchets".[79] When Charles of Sicily attacked the town of Messina following the "Sicilian Vespers" of 1282, the defenders used *pix et oleum* to burn his scaling ladders and a Master Bonaccursus operated a "terrible" ballista, a bolt from which transfixed two French nobles standing near King Charles himself.[80]

Not only, then, were there skilled engineers in growing numbers able to meet the demand from rulers (including city governments), but there were also hints of greater recognition of their skills. The Parma statute was mentioned above. In Aragon, Master Nicoloso of Albenga was able to make a presentation to King James's council with his proposals for a new type of siege tower. His reputation or previous service enabled him to talk directly to the monarch. Engineers were being given extra rewards for their service – Bertram (and another engineer) received gifts of robes for what he achieved at the siege of Benauges (1253). Whether or not Calamandrino was real, it was seen as important that he had an international reputation as a trebuchet engineer. Even lesser mortals than the famous engineers were having their service recognised: John of St Emilion, "our carpenter", who had lost a foot in royal service, was granted a pension with his wife by Henry III.[81] Whereas even master carpenters might be paid at 7d per day, higher level engineers were getting paid a shilling.

The evidence that supports the case for there being a larger pool of engineers – already demonstrated by the increase in numbers of craftsmen – also suggests a clearer differentiation taking place. This was not so much about specialist skills: although some of those identified here did specialise (as Gerard or the two masters of Toulouse appear to have done with trebuchets), others of similar status have been identified carrying out a wide range of quite distinct functions. It was rather

79 P Contamine, 'L'armée de Philippe Auguste', in R Bautier (ed.), *La France de Philippe Auguste* (1982), 579–83; E Audouin, *Essai sur l'armée royale au temps de Philippe Auguste* (1913), 100–2, 182–3 (cabulus). Audouin (in 1913) identified the cabulus as a torsion weapon, though later historians have not followed this line; Guillaume de Tudèle et l'anonyme, *La chanson de la Croisade Albigeoise*, trans. H Gougaud (1984), ll. 4363ff, 4375, 4480, 4489 (ram), 4672ff (incendiaries), 6828ff (trebuchet), 7555f, 9415f (the carpenters).

80 Bartholomoei de Neocastro, 'Historia Sicula', ed. G Paladino, *RIS* XIII/3, 28.

81 *Rôles Gascons*, I, 89, no. 659: John of St Emilion.

that a gap had opened up between "ordinary" and higher-ranked military engineers, with the former often mentioned as serving the latter. It is not now possible to identify what skill distinguished one from another or whether men could progress to higher status. It is not possible to make such distinctions because engineers such as Gerard, described in one place as *artifex*, even late in their careers were paid for such apparently mundane (albeit important) tasks as selecting pieces of timber. In Edward I's futile Flanders expedition of 1297, the *magister ingeniatori regis* in charge of nine carpenters servicing the springalds was named as Stephen of Northampton. He does not appear anywhere else.[82]

Nor is it just about distinguishing rank because certain individuals were named in the records. In 1304, Edward I famously assembled an enormous artillery battery to bombard Stirling castle. Many engineers were associated with this artillery including Thomas of Houghton, Adam of Glisson, Robert of Bedford and Masters Reginald and Nicholas. Another master (Walter of Hereford) was in charge of quarrying the stone for the missiles. Edward insisted on the garrison not yielding until he was able to shoot the last of the engines constructed, War Wolf. The story suggests that this was a larger type of trebuchet which the king was keen to see in operation; it had occupied five master carpenters, fifty ordinary carpenters and five pages, and £40 was paid to Thomas Greenfield for the work. Afterwards, it took a week to dismantle the engines.[83] All those named, too, seem to have been from England, suggesting there was sufficient local knowledge to make it unnecessary to import experts anymore. The detail is wonderful. But none of the named engineers appeared before or after this campaign in royal service, while those that were named evidently were in charge of teams of craftsmen who were anonymous. With the exception of a reference to a father-and-son team of masters of engines (Madio and Nicholas of Brindisi) from Puglia (in 1248–49), none of the numerous engineers who served in Frederick II's armies has a name either.[84]

82 B and M Lyon, *The wardrobe book of 1296–97: a financial and logistical record of Edward I's 1297 autumn campaign in Flanders against Philip IV of France*, Commission royale d'histoire (2004), 100–1.

83 Allen Brown, Colvin, Taylor, *History of the king's works*, I, 417–18; there are frequent references to payments to the engineers listed in G Simpson, J Galbraith (eds), *Calendar of Documents relating to Scotland, V: 1108–1516* (1986), 169–70, 259, 266, 272, 457, 465, 472 as their service continued. See also the documents in J Bain (ed.), *Calendar of documents relating to Scotland in the Public Record Office* (1884–87), vol. II, 389–405 ("Warwolf"), 406 (Greek Fire at Stirling). For other sources see Purton, *HLMS*, 87–8.

84 See Settia, 'L'ingegneria militare', 272–89; G Amatuccio, *Mirabiliter pugnaverunt. L'esercito del Regno di Sicilia al tempo di Federico II* (2003), 121–2, 164–8, 172–6; *Quaternus de excadenciis et revocatio Capitinitate de mandato imperialis maiestatis Frederici Secundi*, 7. This edition of a manuscript from Monte Cassino contains other references to *balistarii* and (unnamed) *magistri*.

It is also clear that all of the people discussed were commoners – with the one exception of Sir Anceau de Brie, who built a trebuchet in Frederick's service at Kantara castle (Cyprus) in 1229.[85] That also makes it impossible to identify any family background. Many were described as carpenters, confirming the leading role of that trade in producing masters who went on to military roles. With the one exception (from Puglia), none of the people named here inherited or passed on the skill (or at least the recognition of it) to descendants.

MINING IN ENGLAND, FRANCE AND ALSACE

The history of warfare under the Plantagenet kings of thirteenth-century England also provides a more substantial record than before of the role and status of mining, and miners, as a vital component of siege operations, and the experience in England is only more noteworthy because of the preservation of the records that make its study possible.

Unlike the progress in scientific understanding and advances in timber-working techniques, reflected in new weaponry, there was no recognisable change in the technology of mining, except – importantly – that digging underground, as compared with digging holes in the ground at surface level, necessarily continued to expand as surface deposits of the ores became exhausted. There is evidence for this in government records of the silver mine at Bere-Ferrers in Devon between 1292 and 1348, revealing the sinking of new shafts and construction of side galleries of considerable length, intermittently calling for new drainage tunnels to prevent flooding. The galleries were timbered, just like mines for war.[86]

Commanders' awareness of Vegetian advice continued from the operations of the later twelfth century. It must be imagined that the commander, relying on knowledge or experience, would have decided where the miners should start digging and how to protect the diggers or conceal the mine from enemy sight because no account tells of this. Knowledge of the geology of the site was needed and here it may have been the master miner who could advise on the practicability of digging, or the likely rate of progress.

A mining operation known to a wider public took place at Rochester castle in 1215. The occasion was the civil war that followed King John's rejection of the Magna Carta. The barons (based in London) occcupied the castle at Rochester, controlling a major bridge across the river Medway. With a garrison of at least eighty knights, plus crossbowmen and no doubt other lesser-ranked fighting

85 Philip de Novare, *The wars of Frederick II against the Ibelins in Syria and Cyprus*, trans. J La Monte (1936), ch. LXIV, p. 105.

86 From the exhaustive account by I Blanchard, *Mining, metallurgy and minting in the middle ages* (2005), v. 3, pp. 1617–23.

men, the castle was strongly held. Its main feature was an immense donjon standing in one corner of a large walled bailey, protected by a ditch deeper than it is today and probably with an outer bailey beyond. A bombardment by stone throwers having failed to make any dent in the strong stone defences, John turned to the miners. The accounts suggest that first they mined into the bailey, although there is now no evidence of where, and then, spectacularly, brought down the corner of the donjon. The defenders held out briefly behind the screen wall before surrendering. The siege is famous because of the king's order to find forty pigs to slaughter so their fat could be used to accelerate the burning of the props, and for the evidence of the effect, as the area of the tower that collapsed – a large section across the corner – can still clearly be seen in the masonry when it was rebuilt a decade later [Fig. 19]. It seems certain that this was in fact a surface sap, because the enormous plinth on which the donjon stands does not appear to have been disturbed. Viewing the site today, the challenge of excavating such a sap directly under the hoarding which almost certainly projected from the wall high above is apparent. Nonetheless, mining succeeded twice in quick succession, a testament both to the courage and the skill of King John's miners, who were ordinary miners from the Forest of Dean. The Pipe Roll records payment to masters Ernulf and William, with eleven miners, for service of 159 days at 4s 9d per day for the team.[87]

A larger team of thirty miners was paid for in 1219, and miners' service was required again to recapture the castle of Bytham in 1221. A large group of forty was sent from the Forest of Dean to London in 1222, but the next significant call for active use of their skills occurred in 1224, when one of King John's former mercenaries, Fawkes de Bréauté, rebelled against his son, Henry III (1216–72), leaving a garrison in his castle at Bedford. The rich contemporary accounts and the government records enable a detailed reconstruction of the events of the siege. Bedford was an old motte-and-bailey design but with substantial stone walls and a large tower on the motte. Despite deployment of artillery, it was the miners who brought down the bailey wall, and they followed up by breaching the defences on

87 Contemporary sources for the much-discussed siege included Coggeshall, Roger of Wendover, Walter of Coventry (copying the Barnwell Chronicle) as well as later writers. The order for the pigs was dated 25 November: T Duffus Hardy (ed.), *Rotuli litterarum clausarum in turri Londoniensi asservati* (1833), I, 238. D Renn, 'Refortification at Rochester in the 1220s: a public/private partnership?', *Archaeologia Cantiana* CXXIV, 2004, 343–64, presents a clear-headed modern view of what happened (and after) and I am grateful to have had discussion with him, and with Richard Dunn, director of the Royal Engineers Museum at Chatham, who has a deep knowledge of the place and who kindly shared his thoughts with me in 2014. See also Purton, 'Suffossores immiserunt: mines and mining in medieval siege warfare', in R Atzbach et al. (eds), *Castles at war* (2015), 187–201.

FIG. 19 Rochester castle, Kent. The corner of the great tower rebuilt after the undermining in 1215. The new masonry is visible. (Photo: author)

the motte to bring the siege to a bloody end. The Pipe Rolls recorded payments to Ernulf and William ("son of Lambert"), along with Roger and Ade Bruno, described as the "king's miners", who were paid at 5d per day, along with three cheaper assistants, and there were separate payments to John de Standon, a miner, for thirty-five days. John de Standon was later rewarded with 12 acres of land, and there is a single further reference to Ernulf, who was paid the large sum of 40s 10d in 1230: perhaps arrears of wages, or some kind of pension after long, loyal and dangerous service to the crown.[88]

Thus it is clear that mining was seen by the royal government as one of its most useful military options, that its success was reliant on having access to men who dug mines routinely, and that the master miners were well recognised and rewarded. Later references confirm the continued use of the Forest of Dean's miners for royal ends well into the fourteenth and early fifteenth centuries.[89]

But military mining was a dangerous occupation and Ernulf, Roger and their companions lived charmed lives in the service of John and Henry III. It was not always so. In 1287, when an English army was besieging the Welsh castle of Dryslwyn during a revolt against Edward I's domination, a tunnel driven under one of the curtain walls collapsed and among the dead were twelve miners, as well as several knights, who were named, that had gone in to inspect it.[90]

The English example is atypical only in naming several masters and identifying where they came from. French kings – certainly from the time of Philip Augustus – made regular use of miners for siege operations. Louis IX intended to take them on crusade to Egypt. Miners were listed in the host gathered for Philip IV's incursion into Aragon in 1285, in a combined wage bill also including carpenters, carters and masons, so their rate of pay cannot be discerned.[91]

The oldest references to mining in war in Alsace date from the late thirteenth century and some remarkable survivals have been explored by H Schoen. The castles of the region tend to be located on high peaks with steep sides. At Girsberg-Schänzel (Haut Rhin) the castle was besieged by Rudolf of Habsburg (elected king of Germany following the interregnum after the death of Frederick II) in 1289. The site still has artillery shot weighing up to 30 kilograms on the ground, but it also has tunnels and trenches carved into the rock. A similar mode of attack was identified at two other sites, Berwartstein (Rheinland-Pfalz), besieged in 1314, and Vieux Windstein (Langensultzbach), besieged in 1332: at both a base camp some 300 metres distant and a gallery dug from there used natural ravines towards the wall's foundations. Sometimes this involved excavating on very steep slopes, while there is evidence for niches for oil lamps and vents for the smoke. The Colmar Chronicle confirmed that silver miners were enlisted to undermine Girsberg. Eighty miners were employed by the bishop of Strasbourg's coalition army against Windstein, where the hole dug under the cliff was 2.4 metres wide

88 Pipe Rolls 17 John (1215), 13–14; 3 Henry III (1219), 8; 5 Henry III (1221), 74; 6 Henry III (1222), 40; 8 Henry III (1224), 91; 14 Henry III (1230), 97.

89 References to the Close Rolls given in Hart, *The free miners of the royal Forest of Dean*, 20–1.

90 Caple, *Excavations at Dryslwyn castle 1980–85*, 188–9, 197.

91 'L'abregement des despens faiz en la voye d'Aragon', Guignant and Wailly, *Recueil* XXI, 515–17.

and 1.8 high. The silver mines near Freiburg-im-Breisgau, where excavation has recovered thirteenth-century mine workings, offer a plausible location for the miners recruited for these operations.[92]

The discovery of these workings provides rare evidence of the actual operation of mining a fortification: normally, where the mine succeeded, it is either buried or else was subsequently destroyed by the victor. Here, the castles seem to have been abandoned. Between them, these examples from England, Alsace and the Rhineland demonstrate the technical skill of master miners, possibly expanded by the growth of underground mining during this century.

MILITARY ENGINEERING IN THE MONGOL AND MAMLUK EMPIRES

During this time, developments far to the east of Europe began that would have a devastating impact on both Christian and Muslim societies. Early in the thirteenth century Genghis Khan, having united the tribes of Mongolia under his sole rule, began the conquest of the northern portion of the Chinese empire before turning his attention westwards. He and his successors overran central Asia and Iran, in the process destroying the Abbasid caliphate, and penetrated deep into Iraq and Syria. They conquered several city-based principalities of the Rus' before devastating Poland and Hungary. During this procession of relentless and usually extremely bloody victories, they mastered countless strongly defended cities.[93] There were no cities in the Mongol homelands, so how did they achieve this?

The Mongol commanders demonstrated great capacity for organising campaigns and winning battles; they also seem to have appreciated their own weaknesses, and rapidly devised ways of overcoming them. The critical step was to take advantage of the populations they conquered. At the crudest level, they turned their captives

92 H Schoen, 'Quelques sites de guerre de sape médiévale des Vosges et du Wasgau', *Revue d'Alsace* 122, 1996, 127–36; C Gérard, J Liblin (eds), *Les annales et la chronique des Dominicains de Colmar* (1854), 130–1, 142–3; Matthiae de Nuwenburg, 'Gesta Bertholdi episcopi Argentinensis', *MGH SS NS* 4, II, 518; F Batt, *Das Eigentum zu Hagenau im Elsass* (1876–81), II, 63 (archive of agreement to supply diggers, masons and carpenters to bishop); A Brunn et al., 'Weitere montanarchäologische Untersuchungen im mittelalterlicher Bergbaurenier am Birkenberg bei St Ulrich', *Archäologische Ausgrabungen in Baden-Württemberg*, 1992, 380–4. The evidence is also helpfully assembled in T Biller, *Die Burgengruppe Windstein und der Burgenbau in den nördlichen Vogesen* (1985), including drawings of the underground works.

93 Alongside many histories of the Mongol conquests, I summarised their achievements in siege warfare in my *HLMS*, 1–54.

into human shields. A slightly more sophisticated version of this approach was not to massacre the artisans among the occupants of captured cities when everyone else was slaughtered. These captives were then put to work. It is evident that Mongol siege technology was largely dependent on such captives, something which also confirms that these cities had unwilling recruits in sufficient numbers to meet the demand. Early chronicle references to Mongol victories include descriptions of places taken by mining, and it seems unlikely that the miners were Mongols. Among many other examples, accounts of the Mongol invasion of Korea in 1231 demonstrate that they had access to siege towers and miners at the siege of Kuju in that year, which is also interesting in demonstrating the high level of skills possessed by the defenders on this occasion, one of the rare times in the early years of the conquests that the Mongols were forced to retreat.[94]

After Genghis, his vast empire was subdivided and subsequent conflicts contributed to a greater prospect of being able to resist. At the battle of 'Ain Jālūt in 1260, the victorious Muslim army was led by the Mamluk general Baybars, who went on to create a new dynasty in Egypt and Syria. Baybars and his successors then turned their attention to eliminating the fragments of the crusader holdings in the region, which culminated in the conquest of Acre in 1291, an event marked by the mobilisation of what was probably the largest array of stone-throwing artillery of all sizes (seventy-two are cited in one account) ever recorded outside China, including fifteen of the largest trebuchets. Large numbers of engineers were used and mining also played an important part in the Mamluk victory. The local Lebanese forests provided an ideal type of timber for the construction of the engines, one of which, nicknamed "Victorious", was conveyed in its dismantled state in (it was said) one hundred carts. The chronicler Abu 'l Fida, who wrote one of the many accounts of the event, was in charge of one of the great engines.[95]

The Mamluk state possessed arsenals like those of western Europe at Damascus and Cairo, and major fortresses, such as Krak des Chevaliers (following its capture from the Hospitallers in 1271), were used as forward bases for the storage of siege equipment such as stone-throwing artillery. The great size and immense strength of the crusader castles and city defences called for far greater efforts to capture them than were required in many contemporary military operations in Europe, and while the fact that the crusader states were unable to call upon external sources of relief was critical, the ability of the Mamluk regime to mobilise large and well-organised siege trains and to carry out effective mining operations tells of an efficient engineering capability at its disposal. The size and strength of the target

94 Described according to original sources in W Bartlett, *The Mongols from Genghis Khan to Tamerlane* (2009), 99–101.

95 For a summary see Purton, *HLMS*, 60–3.

fortresses had by now rendered ancient technology such as siege towers redundant in this region, requiring an even larger role to be played by artillery and mining.[96]

THE MONGOLS AND CHINA

Meanwhile, another of Genghis' grandsons, Kublai, completed the destruction of the Chinese Song empire and established in its place the Mongol Yüan dynasty. The new regime acquired the knowledge and technology of China which had for centuries far exceeded that achieved anywhere else. Information about the military engineering operations of the Mongols largely derives from the official history of this dynasty, compiled after the dynasty had been replaced (by the Ming in 1368). Nonetheless, this source confirms that the Mongols rapidly established an artillery corps and the timescale alone suggests that at the outset it was manned by Chinese, although the commanders were presumably Mongols. In 1214, for example, Ambughai was ordered to recruit 500 selected men who may have been Chinese or Mongol to be trained to operate the stone throwers. Xue Taraqui became "marshal of catapult operators" and on his death (in 1233) Zhang Batur took over. This suggests that the Mongols continued a tradition of professional commanders of the artillery who served until death and were then replaced, ensuring continuity. A total of 10,000 men was given for the size of this artillery corps. If this sounds excessive, the scale of military operations in the Chinese empire, the number of cities and garrisons, and the size of the population suggest it may indeed be accurate. The number of Mongol military operations recorded in primary sources that called for the whole range of engineering, such as bridge building, diversion of rivers, mining and the use of artillery, between them testify to the availability of large numbers of skilled operatives, and their commanders must have been appropriately expert themselves.[97]

96 Archaeological investigation of Krak des Chevaliers confirmed much of what is known about the technology of the late-thirteenth-century siege. A store of trebuchet shot weighing 100 kilograms was found at the place where it is thought that Baybars' battery was emplaced, at a distance from the castle of 300–350 metres. It was also suggested that the immensely strong walls of the castle suffered only minimal damage from these missiles, and that mining played a greater role here: J Zimmer et al., *Krak des Chevaliers en Syrie. Archéologie du sol et du bâti 2003–2007* (2013), 284–5, 305. This magnificent study sadly makes many errors in attempting to describe siege technology. The castle fell in the end because the garrison, isolated and without hope of relief, was persuaded by Baybars to surrender.

97 Chi-Ch'ing Hsaio, *The military establishment of the Yüan dynasty* (1978), 134; T Allsen, 'The circulation of military technology in the Mongolian empire', in N di Cosmo (ed.), *Warfare in inner Asian history* (2002), 265ff, 277–81; W Swietoslawski, *Arms and*

A great deal is known of the Chinese civilisation that the Khans inherited. Knowledge of mathematics, including use of the zero, had come from India and was far in advance of Europe. In terms of technology, cast iron had been in use since 400 BCE and cranks were invented by 1200. Treadmills were used to propel boats by the fifth century and skilled carpenters were constructing wheel-driven warships for the imperial navy by 1130. Large-scale irrigation had long been a fundamental underpinning of Chinese civilisation. State-owned mines for iron and coal were large employers by the year 1000. Despite the different structure of Chinese society compared with European, the status of craftsmen was comparable to their western counterparts in that mechanical skills were seen as vulgar, while the immense civil service established and consolidated (and famously open to anyone with talent) under the Tang dynasty (618–907) had been re-established by the Song dynasty from around 960.

The state's Confucian ideology prized scholars but not manual labour. The status of artisans (those paid and employed in government workshops and others working for themselves who could be pressed into service) was low. However, inventors and higher-ranking engineers were numbered among the bureaucracy, which explains the continuation of inventions. Lower-ranked but also educated engineers were occasionally recorded in history, such as Tang Dao who developed special weapons like fire-lances to defend Hanyang against the Jurchen invaders from the north during the epic siege of that city in 1127–32. However, the majority of such artisans, as in the west, were illiterate and learnt their skills by transmission or by rote.[98] Engineering skills could be found in a number of callings, such as that of the Buddhist monk Fu-jian. He had been in charge of constructing the siege engines for a rebel army besieging Tang-controlled Fengtian in 783; his device was an enormous mobile ladder, protected against fire, apparently 30 to 40 feet (9.2 to 12.3 metres) high and wide, and mounted on wheels. This siege also

---

armour of the Nomads of the Great Steppe in the time of the Mongol expansion (1999), 67–9. These works all use the Yüan Shih (Yüan dynastic history).

98 J Gernet, *A history of Chinese civilisation*, trans. J Forster, C Hartmann, 2nd edn (1996), 231–2, 320–1, 331; J Needham, 'Iron and steel production in ancient and medieval China', in Needham et al. (eds), *Clerks and craftsmen in China and the west. Lectures and addresses on the history of science and technology* (1970), 107–12); J Needham, 'Classical Chinese contribution to mechanical engineering', in same volume, 113–35; J Needham, 'Thoughts on the social relations of science and technology in China', *Centaurus* 3, 1953, 40–8); C Ronan, J Needham, *The shorter science and civilisation in China*, 4 (1995), 1–38 (Tang Dao at 14). V Conrui Xiong, 'Sui Yangdi and the building of Sui-Tang Luoyang', *Journal of Asian Studies* 52/1, 1993, 66–89, analysed the construction of a new capital starting in 605–06 with many sidelights on planning and construction skills. R Sawyer, *Fire and water. The art of incendiary and aquatic warfare in China* (2004), discusses Chinese texts and histories covering incendiary weapons and examples of campaigns where such weapons and also water engineering were used.

involved mining and counter-mining. It was eventually abandoned but its course confirmed the wide range of technology available in China and the presence of engineers to design and oversee its construction.[99] Records of the defence of the city of Hsiang-Yang and the defence of Te-an by Song generals against sieges by the Jurchen (Chin, before they became the first victims of Genghis' conquests) in 1206–07 demonstrated large quantities of standardised artillery (bamboo-poled stone throwers of different sizes, including, at Te-an, ninety-eight of the largest, the nine- to ten-poled "whirlwind", called *hsuan-feng p'ao*), fire lances, incendiary devices and so on. Among those named in these formal records were Chang Pin, who was directed by the general (Wang) to deploy a battery of stone throwers to demolish some of the siege works (he was successful) – it sounds as if he was a general officer rather than an engineer himself. The records also named Ch'en Yua, described as "a catapultist", who advised the commanders that he could destroy the besiegers' battery of five large engines with his own nine-poled catapult; his "team is outstanding, no stone misses its target", and he was as good as his word. Here, then, was a local commander of artillery cited in the record.[100]

The diffusion of knowledge and technology between China, India and places further west is hard to prove but is quite likely. The possibility that the Chinese invention of the manually operated stone thrower had been transmitted (possibly via the nomadic Avars) to Europe, the Middle East and North Africa, leading to the adoption of this novel form across these continents, has been mentioned. Kublai Khan's campaign provides one case of such transmission in reverse. Marco Polo's book of travels included an account of Kublai's siege of the Song city of Saian Fu in 1268—73, in which he claimed that he himself had contributed the design of a large stone thrower that smashed the defences. The dynastic history ascribed this to two Muslim engineers, named I-ssu-ma-ying (Ishmael) and A-lao-wao-ting (Alau'd-Din), who constructed trebuchets. The Persian chronicler Rashid al-Din stated that "before this time there was no big Frankish mangonel in Cathay" and that Talib "the mangonel (*mandjanik*) maker from Baalbek and Damascus with his sons" and assistants built seven great engines so effectively destructive that they achieved the surrender of the city.[101] The unprecedented size and impact of the newly introduced weapon was remarked upon here and in another siege in 1274. The official version is broadly consistent with the chronicle one. Regardless

99 B Wallacker, 'Studies in medieval Chinese siegecraft: the siege of Fengtian, AD 783', *Journal of Asian History* 32/4, 1998, 185–93.

100 C Hana, *Bericht über die Verteidigung der Stadt Tê-an während der periode k'ai-His 1205–1208 (K'ai His Te-an Shou-Ch'eng Lu,* by Wang Chih Yuan), esp. 153–6, 159–60; H Franke, 'Die Belagerung von Hsiang-Yang', in G Ulmen (ed.), *Society and history: essays in honour of K A Wittvogel*, 351–6.

101 *The book of Ser Marco Polo*, trans./ed. H Yule (1903), book II, c. LXX, 158–60; A Moule, *Quinsai with other notes on Marco Polo*, 70–6, for the other sources cited.

of the truth of who was responsible – Marco Polo's account is probably best seen as crude self-promoting propaganda – it is clear that despite Mongol encounters with the trebuchet in the Middle East and in Europe, they had not transmitted the knowledge eastwards and it was previously unknown in China.

Another of the products of inherited military technology was the immense fleet that Kublai had built in his catastrophically unsuccessful attempts to extend his rule to Japan. Failure was not the result of the naval technology taken over by the new ruler, because Chinese shipbuilding was then the most advanced and sophisticated in the world, rather it stemmed from the combination of poor decisions and typhoons.[102]

CHINESE GUNPOWDER

Chinese technical knowledge had long since included how to make gunpowder. It seems that the purposes to which the discovery was put for the most part comprised what was a form of incendiary device not so different from those regularly deployed by Muslim armies in the Middle East rather than a propellant, although Needham's analysis is now being challenged.

Those who have studied the original contemporary Chinese sources show how in the later years of the Tang dynasty, experiments with already familiar mixtures of what would become the ingredients of gunpowder (charcoal, saltpetre and sulphur) had produced what Joseph Needham described as "proto-gunpowder", an explosive that was then adapted to military use in bombs thrown by artillery or in bamboo fire lances, following presentations to the emperor around the year 1000. The first printed description appeared in a text of 1044 (the *Wu Jing Zung Yao* of Zeng Gongliang) which explained a number of different powder-using weapons. As a result of official endorsement, an entire government department was established overseeing thousands of workers in eleven large workshops responsible for manufacturing immense quantities of arrows and bombs adapted for gunpowder. However, these remained incendiaries or explosives, not the means of propelling other missiles. They were recorded at the siege of Kaifeng (1126) and many other sieges in the twelfth and thirteenth centuries. Fire lances appeared in 1132 and these were projectile weapons, in that they were bamboo poles stuffed with powder that were used to project bullets. This leaves unexplained the depiction of what appear to be similar devices on silk banners found in caves from the province of Kansu and reliably dated to the ninth century. Research suggests that the first bronze or

[102] For an account of these attempts based on recent marine archaeology, J Delgado, *Khubilai Khan's lost fleet* (2006) and the useful bibliography it contains. The khan was also responsible for failed naval expeditions against Java and Vietnam, suggesting an inability to exploit the skill of his largely anonymous shipbuilders to extend his empire.

iron barrels appeared under the Yüan in the late thirteenth century,[103] not so long before the first illustrated cannon in western Europe. The weapon's subsequent development under the Ming dynasty (from 1368) paralleled that taking place in Europe.

It has not been possible to identify the route of transmission although many were available, and it is not possible to rule out that Europeans discovered the formula for gunpowder independently (as nationalist-driven scholars of the nineteenth century sought to prove): Roger Bacon's words were noted above. What was different in Europe was the rapid application of this chemical reaction to a practical and destructive function, in contrast to China which took centuries to reach the same point.

The technology adopted in the early fourteenth century eventually ushered in a new age and the slow but irrevocable obsolescence of most previous military technology. It saw the appearance of new breeds of craftsmen to manufacture the powder and the guns created to make use of it, and of artillerymen to operate these weapons, achieving another specialism that was firmly rooted before the advent of the age generally called the European renaissance. This specialism would be taken up by generations of manufacturers, artisans, masters and experts who gradually developed into the professional artillery corps of early modern rulers. But for most of a century this process took place alongside the continuation of older technologies, and the engineers who had responsibility for executing them continued to provide services to rulers willing to hire the expertise expected of any modern army throughout the fourteenth and fifteenth centuries.

CONCLUSIONS

Even taking account of the increased survival of records, there is sufficient evidence that there were, in fact, more engineers in service from the thirteenth century than previously. Many of those named here were still responsible for a wide range of functions including building (ecclesiastical and secular), but others were described as having responsibility for the royal engines. Developments in technology, including the introduction of different types of large weapons (giant crossbows, the counterweight trebuchet and its subsequent variants, and the springald), may have led to some greater specialisation, a logical step even if it can only be a hypothesis. A king faced with the challenge of conquering cities and castles with now stronger

103 J Needham, *The guns of Kaifêng-Fu. China's development of man's first chemical explosive* (1979), 1–16; Zhou Jiahua, 'Gunpowder and firearms', in Institute of the History of Natural Sciences, *Ancient China's technology and science* (1983), 184–91; J Needham, Ho Ping-Yu et al., *Science and civilisation in China*, V part 7, 'Military technology: the gunpowder epic', 40–1, 161–71, 220–3, 261, 276ff (analysis of first Chinese cannon).

defences would naturally seek out experts who could accelerate such expensive and time-consuming operations, and such people existed. At the other end of this spectrum, men have been identified who did not promote themselves to the highest bidder, but who served long years as the "king's engineer", particularly in England and France where stronger, more centralised administrations were developing, although the more widespread use of the title in the English (and Gascon) records argues against the idea that a single individual held this as a formal position. The hands-on royal engineer, such as Andronikos Comnenos, was increasingly rare. That such titles appear in the records with increasing frequency is testimony to the emergence of this position as a specific role within some royal administrations. It is also true that in many cases they remain absent from the accounts even if they must have been present. The trend established during the thirteenth century would become more widespread as gunpowder artillery added new skills and new specialisms.

6

Old and New Technology and its Operators in the Fourteenth and Early Fifteenth Centuries

IT TOOK SEVERAL DECADES for gunpowder weapons to come into use everywhere in Christian Europe and the Muslim world, but once they were acquired there was never any chance of going back. Despite the considerable limitations of early guns, no ruler (king, sultan, prince or prince bishop, doge or city corporation) failed to take up this new technology. For most of the fourteenth century, major constraints were imposed by limitations in metalworking skills and unsolved problems in the combining of the ingredients and the process of making the powder itself. By the end of the century, both issues were largely resolved, allowing the creation of new generations of weapons during the following century that changed the face of warfare. Even before this point the manufacture of gunpowder and of guns had become a major industry, but the role of master gunner had yet to achieve the importance that would be attached to it as the weapons themselves began the journey towards becoming decisive from the middle of the fifteenth century.[1]

For most of the fourteenth century more traditional technology continued to be dominant and the engineering skills called upon by commanders during the previous century continued to be as valued as those of people making and operating the new weapons. In western Europe, the decision of the English king Edward III (1327–77) to reassert his claim to rule France led to the intermittent but seemingly interminable conflict now known as the Hundred Years War, generating regular bloody warfare (albeit usually with impact limited in time and place) from Iberia to Scotland. In the Baltic, the theocracy established by the Teutonic Knights, ejected like their fellow crusading orders from the Holy Land in 1291, ignited conflict with the pagan inhabitants of Prussia and Lithuania and, eventually, with neighbouring Christian states. In Asia Minor the dynasty founded by Osman in the early fourteenth century ultimately came to dominate

1 The developing impact of gunpowder on war has been well explored by, for example, Kelly DeVries. See DeVries, Smith, *Medieval military technology* (2nd edn, 2012) which has a list of his relevant writings in its bibliography. See also Purton, *HLMS*, 116–17, 173–5, 191–2, 214–15, 270–80 and 401–4.

the Muslim world as the Ottoman empire, on the way completing the destruction of Byzantium, briefly interrupted by the invasions of Timur, latest (and last) of the great Mongol conquerors, who marched undefeated across central Asia, India and Asia Minor at the end of the fourteenth century. The "great schism" within the Latin church that began in 1378 accelerated a crisis within that institution. Meanwhile, governance of China returned to a native dynasty when the Ming overthrew the Mongol Yüan dynasty (1368). There is evidence that there were no great differences between the technology of the states involved and of the people employed as engineers.

Famine and plague wreaked terrible slaughter and economic as well as human consequences, because much of the workforce disappeared, but noticeably had very little impact on the warmongering of the ruling classes.

Alongside the large powers, smaller political units had acquired engineering and military mining skills that in previous centuries might only have been available to the wealthier. The self-governing city states of northern and central Italy (now usually aristocratic lordships that had replaced the rule of city communes) had been joined by more (in practice) independently governed cities further north. Successful smaller regional powers like the (vis)counts of Béarn and Foix and the counts of Savoy managed to take advantage of the wars between their powerful neighbours and also extended their control by astute use of military force, including military engineering. The number of employers for people with the necessary skills was therefore expanding, but records demonstrate that demand was met with sufficient supply, although without revealing the background or origins of the new generations of engineers, except for the hints that traditional routes such as master to apprentice must still have been the most significant.

ECONOMIC AND TECHNOLOGICAL DEVELOPMENTS

The long period of population growth of previous centuries came to a violent halt in the fourteenth, with recurrent famines in the early decades and the "black death" in the late 1340s, a plague that was to return many times. Economically, it led to reduced demand and a reduced labour force, the latter resulting in an increase in wage rates. Naturally, this included skilled labour, with consequences for the diffusion of technology and the availability for hire of the craftsman. Studies suggest that "tramping" from one job to another independently of the masters became a regular way of life in France in the fourteenth century for journeymen as they found ways to meet local demand for their skills post-apprenticeship, and that this process undermined attempts by, for example, Italian cities to protect their technological advantages as diffusion of knowledge inevitably accelerated. Not a universal phenomenon and one that took different forms across the continent of Europe, this mobility of skilled labour must have had consequences for military engineers,[2] a group primarily drawn from carpenters, as the records have shown, but not exclusively, as with the example of Villard de Honnecourt. It has been argued that literacy among the masters was continuing to expand, citing Villard's notebooks as well as random examples of masters signing their names as witnesses on court documents. How far this reached, and what it signified in terms of how masters serving a military role executed their roles, is impossible to know, except to be fairly certain that they had not graduated from the universities which were now becoming established in Europe. There continued to be educated administrators with engineering skills in the Muslim world, where records show the relatively high status assigned to engineer-architects throughout these centuries, but it is hard to find any equivalents in the Christian world.[3] Nor is it known how far down the scale from literate masters the knowledge reached, although where family dynasties developed it was likely that there was transmission of education too. Literacy theoretically gave the master access to written knowledge, although it remains impossible until the fifteenth century to identify any specific example apart from the very local materials produced in a workshop such as that represented by Villard's "notebook". It is perhaps more likely that skills in writing were used to complete accounts or correspond with patrons or officers of the crown (or other ruler) rather than in the process of design and construction, since it is known that the former definitely happened while evidence for the latter is completely lacking.[4]

2 S Epstein, 'Labour mobility, journeyman organisations and markets in skilled labour in Europe', in M Arnoux, P Monnet (eds), *Le technicien dans la cité en Europe occidentale 1250–1650* (2004), 251–69; the diffusion of technology in Sweden had long been the result of the immigration of itinerant, mainly German, craftsmen to Swedish towns: A Götland, *Technology and religion in Medieval Sweden* (1993), 246. Supply of course relied on demand and the enormous number of projects for ecclesiastical building in the late twelfth and thirteenth centuries had generated the need for large teams of masons and carpenters, often recruited from a distance; this process continued with large-scale building work in the fourteenth century: M Davis, 'Masons and architects as travelers', in Friedman, Figg, *Trade, travel and exploration in the Middle Ages: an encyclopedia*, 380–2.

3 Examples from the Mamluk state of Egypt and Syria are given in D Nicolle, *Saracen strongholds 1100–1500. The central and eastern Islamic lands* (2009), 27–9.

4 J Harvey, *Gothic England* (1948), 20–2, 155–6; Britnell, 'Pragmatic literacy', in the same (ed.), *Pragmatic literacy east and west, 1200–1330* (1997), 3–24; J Harvey, *Medieval craftsmen* (1975), 43, 118–23; A Adamska, M Mostert (eds), *The development of literate mentalities in east central Europe* (2008), esp. 531–4, covering widespread lay literacy in Bohemia and Poland cultivated by rulers. The ability to read was suggested by the emergence during the fourteenth century of written treatises on, for example, estate management and testified by signatures of witnesses on surviving documents: see J Adamson, 'The extent of literacy in England in the fifteenth and sixteenth centuries', *The Library* ser. IV, X, 1930, 163–93.

FOURTEENTH- AND EARLY-FIFTEENTH-CENTURY WRITERS, AND THE PRACTICALITIES

The fourteenth century witnessed continued progress in scientific understanding, including in areas where there were practical implications. The Norman Jean de Meurs wrote many works and completed, in around 1344, one begun by another author which was a text on practical geometry. It included the statement that he had witnessed a stone-cutter making models (in stone) of both regular and semiregular polyhedra, an indication of the geometrical awareness of some craftsmen in the early fourteenth century.[5] Stonemasons undoubtedly had not read his works but already knew how to apply the same knowledge at a practical level, and probably possessed "pattern books" containing the templates for different geometric shapes to be used in design, as de Meurs's account seems to suggest.[6] Further hints about the spread of scientific ideas beyond the schools were provided in vernacular writing such as Geoffrey Chaucer's from the second half of the fourteenth century. Nicholas, the clerk in the Miller's Tale, was a "poure scoler/ had lerned art, but on his fantasye/ was turned for to lerne astrologye" (lines 3190–2) and in his room were "His almageste, bokes grete and smale/ His astrolaybe ..." (lines 3208–9). In the Summoner's Tale, the subject described himself as "No maister, sire, quod he, but servitor;/ Thogh I have had in scole that honour" (lines 2185–6). The author of the *Canterbury Tales* was evidently interested in such subjects. If literature did reflect reality, then perhaps it would not have been hard for masters of relevant trades to talk to those who had benefitted from school learning or knew of the astrolabe.[7]

Improvements in technology during this period were slow and incremental, and it is hard to point to any single breakthrough except in the area of gunpowder weapons. Here the technology needed to create a significant change was also slow, although there was continued development of the design of the trebuchet. In other

5 Glick et al., *Encyclopedia*, 284; *Johannes de Muris, De arte mensurandi*, ed. H Busard, chapter XI, proposition 5 ("Irregulare corporis tetracedron nullum esse"), p. 338. Another group of eminent scientists had taken a major step towards a better understanding of the science of motion by breaking with Aristotle's explanations and advancing propositions to explain how projectiles moved, and why they eventually stopped moving. Scholars such as Thomas Bradwardine (d.1344), John Buridan (c.1306–61) and William of Heytesbury (c.1313–72/3) were responsible for this progress, and de Meurs also contributed, but immediate practical consequences cannot be identified.

6 Coldstream, *Medieval architecture*, 42, 50.

7 G Chaucer, *The Canterbury Tales*, ed. J Mann; discussed in Glick et al., *Encyclopedia*, 99, 105, 221 (the named scientists) and 125 (Chaucer); *Johannes de Muris*, ed. Busard 13. His works *De moventibus* and *De ponderibus* were part of a major mathematical text completed in 1343.

FIG. 20 The Pont Valentré at Cahors (Lot) is 170 metres long. It was completed over the first fifty years of the fourteenth century. Built on piles in the river Lot, it was defended by three large towers that survive and a barbican, now gone. (Photo: author)

areas, progress previously identified in masonry and carpentry skills continued, and the regular demand from rulers for their military or other purposes ensured that a skilled craftsman could rely on regular employment that would keep his skills sharp and create opportunities for sharing advanced knowledge of techniques. Shipbuilding was an area where in both northern Europe and in the Mediterranean there was high demand, and the carpentry and additional skills of those paid to construct or maintain the ships requisitioned when they were needed to convey armies and their supplies across the seas ensured a large pool of skilled workers.

A similar sharing of military and commercial functions was reflected in the repair and construction of bridges [fig. 20]. The English crown was routinely required to carry out both types of operation. In 1313–14, for example, Edward II (1307–27) ordered John of Havering to repair the bridge over the Garonne at Agen, paid for by a levy on the town itself. Meanwhile in Paris in 1326, a team of carpenters, masons and government officials was assembled to inspect the Grand Pont.[8] It has been pointed out that the use of starlings (that is, creating a rubble-filled pier inside piles driven into the bed) rather than coffer dams was regularly preferred across the fourteenth century for building bridges because they were

8 Y Renouard (ed.), *Gascon Rolls preserved in the Public Record Office 1307–17* (1962), 339 (no. 1207); in 1308–09 the "master of the king's ships", Peter Langer, was provided with timber and workers to construct ships at Bayonne: p. 77–8 (no. 238); M Boyer, *Medieval French bridges*(1976), 150, 153–6.

cheaper and easier to build and required less technical skill, though maintenance was an issue. Easier to construct they may have been, but particular and specific skills were still necessary to create a bridge in the first place. Routine military use of bridges continued almost unnoticed in the sources, although it was recorded that Edward III had to repair bridges over the Seine which were broken down by the French during the campaign of 1346, while at the French siege of Aiguillon in the same year the duke of Normandy employed carpenters to attempt the challenging task of bridging the river Garonne and only failed because of sorties by the English garrison.[9] English armies had often to find ways to cross large rivers during these early stages of the Hundred Years War, as in the Black Prince's devastating *chevauchée* of 1355. Since it would have been self-defeating if the French had left boats lying around after they had demolished the bridges, it must be assumed that the English had sufficient skilled people either to execute repairs or to find other ways to convey a large army of men, horses and wagons across the big rivers of southern France.[10]

New treatises continued to be written taking the usual form of advice to rulers, and if the precepts of Vegetius continued to underpin almost everything written on the subject of warfare, awareness of contemporary practice ensured the relevance of these writings – in 1347, copies of Vegetius and Giles of Rome were ordered from Paris for the education of the future count Amadeus VI of Savoy.[11] For the modern student they can suggest developments in thinking and consciousness of what technology was now available. The importance of engineers in the service of the ruler continued to feature. Some texts focussed on advice about how to carry out a new crusade to recover the Holy Land for Christianity. That written by Marino Sanudo (also known as Torsellus), probably during the 1330s, was full of recommendations on the construction of weapons for siege warfare, with particular reference to the detail of stone throwers, and included the point that wise and experienced "masters and engineers" were essential to create the necessary equipment. Making use of this text, as well as the much-read thirteenth-century advice of Giles of Rome, Guido da Vigevano's *Texaurus* of 1335 was also meant to advise the French king Philip VI on how to carry out a new crusade. Although he was actually a medical practitioner (living in Paris), Guido de Vigevano's text (which does not survive in an original version) contained illustrations of boats, wagons and floating bridges for military purposes. Among the devices shown were

9 N Gribit, *Henry of Lancaster's expedition to Aquitaine 1345–46. Military service and professionalism in the Hundred Years War* (2016), 136–7. Some sources say the bridge was completed.

10 H Hewitt, *The organisation of war under Edward III* (republished 2004), 104, 112 (and the sources cited there).

11 Hayward, *Histoire de la maison de Savoye*, quoted in E Cox, *The Green Count of Savoy*, 51.

paddle wheels and a mechanical lift that was not just fanciful. It advised on how to assess the height of a tower using a different approach to the normal but still based on knowing the properties of a triangle to make the calculation using the height of the observer and sticks.[12]

Study of the tradition of medieval illustration suggests that to apply a modern appreciation grounded in criticising these survivals because they have hopeless perspective and no scale is ahistorical and misses the point. The illustrators of books like the *Texaurus* came from a tradition of drawing in which what is intended to be represented now by perspective was instead communicated using a style that sought to convey more than one angle of view in a single drawing. Some of the drawings in texts from the fourteenth and fifteenth centuries, analysed this way, can be shown to have illustrated every design point that a contemporary builder would need to know to set about their task. Sometimes it is possible to distinguish between these drawings and those designed to please a patron, where technical detail was sacrificed (the patron did not need it) and instead the machine was presented in an imaginary, picturesque scenario. The text accompanying Guido's work stated "a skilled worker will easily understand this because I cannot write it more clearly". The idea of a picture being clearer than many words was known in the fourteenth century, and these writers expected their designs to be turned into reality by engineers who could grasp the design principle they saw on the page, and applied their knowledge of materials to make it real. The same idea applies to illustrations in the texts of Kyeser and Taccola discussed below.[13]

Although some drawings show fantasies, these writings do help provide a good understanding of the technologies available at the time, while recognising that the dedicatee of the book was well advised to employ professional engineers to turn words into machines. Even more evidently of practical purpose were the architectural drawings that survive in ever-increasing quantity after the thirteenth century. The same points arise and understanding them relies on grasping each illustration's purpose as some were designed to impress putative patrons. But like Villard's drawings they often had a practical function. The implication was clear: there were a growing number of people able to envisage complex realities (the

12 Marinus Sanutus dictus Torsellus, 'Liber secretorum Fidelium Crucis super Terrae Sanctae recuperatione et conservatione', reprinted in R Schneider, *Artillerie des Mittalters* (1910), 94–6 (itself reprinted from Bongars' edition of *Gesta Dei per Francos* (1611), book II, pt IV, chapter 22); A R Hall, 'Guido's Texaurus, 1335', in B S Hall, D C West (eds), *On pre-modern technology and science: a volume of studies in honor of Lynn White jnr* (1976), 11–52 (the original text is at 16–30).

13 R Leng, 'Social character, pictorial style and the grammar of technical illustration in craftsmen's manuscripts in the late middle ages', in W Lefèvre (ed.), *Picturing machines* (2004), 85–111; D McGee, 'The origins of early modern machine design', in same volume, 53–84, here 56–73.

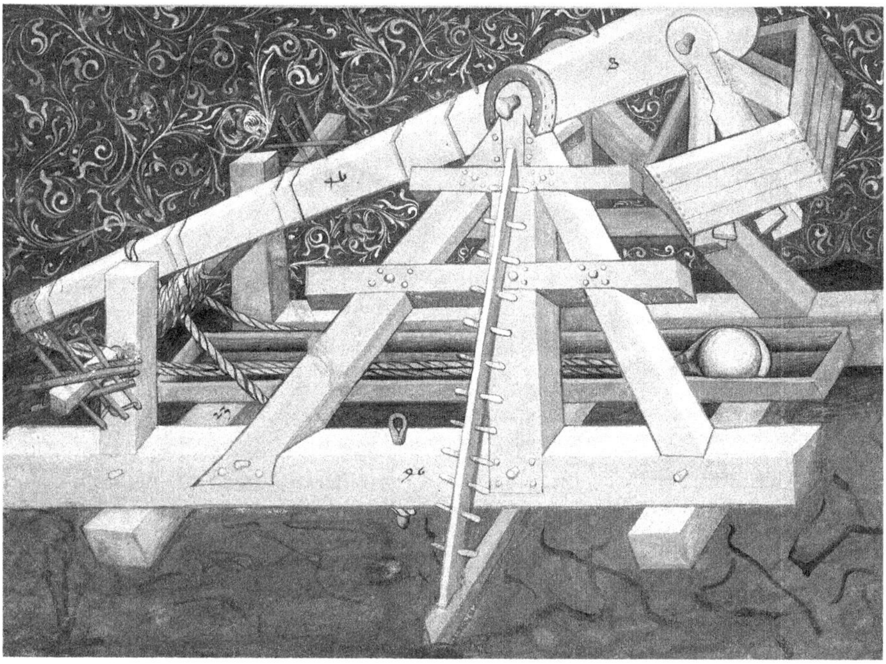

FIG. 21 Fifteenth-century drawing of a trebuchet with dimensions on each part, in Konrad Kyeser, *Bellifortis*. (Niedersächsische SUB Göttingen Cod. MS PHIL 63 (fol.30r), with permission)

components of a cathedral, or on a more modest scale a trebuchet), and there were plenty of engineers able to understand and convert the drawings into reality, most evident today in the survival of immense buildings from this date.[14]

Christine de Pisan, the unique example of a woman writer on these matters, further updated the advice of Vegetius in accordance with modern practice, in particular including gunpowder artillery, in her *Fais d'armes et de chevalerie*, written around the year 1410, while her near contemporary Konrad Kyeser of Eichstadt wrote *Bellifortis* around 1400, a richly illustrated work which was copied numerous times in the following century. The variety of counterweight stone throwers it described helps us to understand how this weapon had evolved since its first appearance [Fig. 21]. Both works were obviously continuators of a historic type of writing chiefly aimed at (securing the patronage of) wealthy rulers (in Christine's case the king of France but a copy was made for that monarch's soon to be arch-enemy the duke of Burgundy too) and mixed classical knowledge with modern awareness. Both authors demonstrated a good grasp of the developing role of

14 E Knobloch, 'Technische Zeichnungen', in U Lindgren (ed.), *Europäische Technik im Mittelalter 800 bis zum 1400* (1996), 45–64, here 49–54.

gunpowder weapons while paying due attention to much older forms of siege equipment. Christine's work included detailed advice on logistics. Neither, however, paid much attention to the engineers who would have had the responsibility for implementing the proposed good practice, although Christine designated how many men were needed to serve the artillery "beside the master", whose presence was therefore just accepted.[15] Of the same period was Taccola's *De machinis*, which contained many different types of engine including military examples. This book may have been completed by 1433 although the manuscript history is complicated. Among other relevant subjects, it described how to measure the height of a tower using a quadrant and how to sink piles and build aqueducts.[16]

Production of texts with the same function continued in Muslim territories. A remarkably detailed work on archery composed in 1368 revealed a deep practical knowledge of the subject. The text described both hand bows and crossbows and their construction, their qualities and requirements, discussed the whole range of ways in which they could be used including in siege warfare, and made the interesting comment on the work of al-Tarsūsī that "his [own] method was eclectic", borne out by the acknowledgement in the earlier writer's text of where he had obtained his information. Before the establishment of the Mamluk regime in Egypt and Syria for which this text was compiled, the 'Ayyubids made use of a treatise by Abu'l Hassan 'Ali al Harawi which covered the employment of masons, sappers and *mandjanik* operators as essential members of the sultans' armies.[17] The tone of the message did not suggest a particularly high status for these men. However, that engineers were respected elsewhere in the Muslim world was suggested by an incidental reference in a history of the Delhi sultan Ala-u'Din to his employment of "skilful engineers" as he prepared his defences against a Mongol invasion in 1300.[18] Among the few surviving general manuals on warfare in Arabic, that of al-Ansari (who died in 1408) suggested no innovations. It repeated traditional advice on how to carry out the attack or defence of fortified places, recommended that the

15 Christine de Pisan, *The book of deeds of arms and of chivalry*, ed./trans. S and C C Willard (1999), 124. This is based on a near-contemporary English translation; Konrad Kyeser, *Bellifortis*, ed. U Lindgren et al.

16 M Taccola, *De machinis. The engineering treatise of 1449*, ed. G Scaglia, 2 vols (1971). L Thorndike, 'Marianus Jacobus Taccola', *Archives internationales d'histoire des sciences* XXXIV, year 8, no. 30, 1–3/1955 (7–26), analyses the issues around its composition. Taccola is further discussed in chapter seven below.

17 J Latham, W Patterson (eds), *Saracen archery. An English version and exposition of a Mamluk work on archery (ca. AD 1368)* (1970). The author was Ṭaybughā Ashrafī l-Baklamishū I-Yūnanī; Nicolle, *Saracen strongholds 1100–1500*, 49. The work of Al Harawi was translated by J Sourdel-Thomine, 'Les conseils du Sayh Al –Harawi à un prince Ayyubide', *Bulletin d'études orientales* XVII, 1962, at 227–8.

18 Zíáu-d Dín Barní, 'Táríkh-í Fíroz Sháhí', in H Elliot, J Dowson, *History of India* (1871), III, 191.

commander himself was familiar with siege engines to the extent of supervising their placement, and left out any mention of engineers, reducing their role to that of the craftsmen whose job was to assemble the engines. That it was based on earlier manuals is suggested by the failure to mention gunpowder weapons, which had certainly reached the Middle East by that date.[19]

Another writer with no interest in matters military casts light on levels of knowledge among surveyors. Bertrand Boysset was a "middle-ranking" bourgeois from Arles, the owner of vineyards and other commercial operations; he also had a financial interest in the fish taken from water for which he may have designed the system of dikes and a complex lock. His main claim to fame lay in two books frequently recopied and used for more than a century after his death (c.1416) on surveying and the measurement of land and boundaries. There is evidence in Arlesian records of the practical work done by Boysset and he therefore represented a very rare example of someone writing on the subject who also did actual surveys. While his modern editor confirmed that he used archaic arithmetical and geometric methods and continued to work with Roman rather than Arabic numerals, recommending the abacus to make the calculations (suggesting that he was working from an earlier text as an exemplar), some of his writing was original and the authority for this was his own experience. If Boysset's urging that surveyors needed to be familiar with a long list of subjects beyond arithmetic and geometry was probably rhetorical, his explanation as to why they needed to be able to read and write was justified by the need to record the findings of a survey. It may therefore be concluded that the various people mentioned in texts as being sent to lay out new towns (as suggested in chapter five), as well as those who took charge of preparing the ground for works on castles and fortifications, very likely had a level of education, and that the prescriptions always laid down in the many texts entitled "practical geometry" did indeed reflect the learning required of surveyors, even if they did not acquire it from these books.[20]

ARSENALS

Rulers continued to maintain arsenals on a large scale, providing them with facilities to construct and maintain their weapons. While storing armour and hand weapons, bows and crossbows was a primary concern, having somewhere to locate

19 Umar ibn Ibrahim al-awsi al-Ansari, *A Muslim manual of war*, trans. G Scanlon (1961), books 19, 20 (114–21).

20 P Portet, *Bertrand Boysset. La vie et les oeuvres techniques d'un arpenteur médiéval (v.1355–v.1416)*, 2 vols (2004). Here, vol. 1, particularly 74–6, 111–16, 144, 171, 208, 269–71. The original texts of *La siensa de destrar* and *La siensa d'atermenor*, in Provençal, are in volume 2.

engines was also important – the care taken by rulers from the outset to have somewhere secure to maintain (and to protect from the elements) these treasured items has been observed already. Such arsenals were necessarily the base for many of the engineers involved in the production of these weapons.

For Edward III of England, the Tower of London was most important: materials for his many wars were both manufactured and stored there. Alongside vast quantities of weapons, the payroll included armourers, carpenters, smiths, bowyers, fletchers and the *attiliator* (in charge of the large crossbows and springalds), who was provided with a house within the Tower. Research by Hewitt has uncovered the names of Nicolas Coard in this role in 1338, and that he was followed by William de Rothwell. Robert de Mildenhall was ordered to repair and despatch guns, pellets and 100 'ribalds' (probably a word with the same source as *ribaudequin* and signifying small guns) for the siege of Calais in 1346. But not everything was stored at the Tower: preparations for the invasion of France in that year included long siege ladders ordered to Sandwich from three southern counties.[21] Other royal castles also served. Inventories of the contents of the stores at Wallingford before and after the deposition of Edward II show that it contained, alongside many crossbows and bolts, two-footed crossbows, a springald and a trebuchet, and royal *attiliatores* had been sent to inspect and repair as necessary.[22]

The port city of Southampton was also a major arsenal for English expeditions as well as becoming an early example of a city fortified to use, and resist, gunpowder weapons. Following a French raid in 1338 the king sent springalds and crossbows, and an inventory of 1353 identified two "great mangonels" and smaller "tripogets" as well as twelve springalds in the town's armoury. A mangonel was apparently sited outside the city's west gate. The first reference to there being gunpowder weapons in Southampton was in 1382. The last reference to the mangonel, surprisingly, was as late as 1454.[23] Perhaps this was actually a trebuchet, since the last references to that weapon's use occur around this period. The presence of these weapons presupposes the existence of people to operate and maintain them.

In Paris, the major royal arsenal in the Louvre may have occupied most of the lower court, although there is now no evidence of exactly where it stood. Houses outside were allocated to the *artilleur* and to other masters. In 1391, the house "where the artillery was made" measured 14 toises by 4 (1 toise = 1.94 metres),

21 Hewitt, *Organisation*, 68, 85.

22 N Christie, K Keats-Rohan, D Roffe (eds), *Wallingford: the castle and the town in context*, BAR British Series 621 (2015), 96–8, 110.

23 R Moffett, 'Military equipment in the town of Southampton during the fourteenth and fifteenth centuries', *Journal of Medieval Military History* IX, 2011 (167–99), here 170–3, 183. This late reference to a now obsolete weapon may suggest that the word was being used to describe a trebuchet, since there is some evidence that it was still occasionally called for as late as this period (see p. 222).

while in 1412, on the same site, there was a large hall that was home to guns and gunpowder along with various hand weapons.[24]

Medieval Venice had possessed an arsenal at least since the fourth "crusade" (1204) when the doge ordered hundreds of masters and workers to construct the fleet that would convey the crusaders and a large quantity of siege artillery. A new arsenal had been built at the end of the thirteenth century on a different site with a range of officials in charge of various aspects. The arsenal became the location for large-scale military functions including shipbuilding and weapons and, eventually, gunpowder. A skilled workforce working under experienced masters was essential to the growing power of the republic.[25]

NAMED ENGINEERS OF THE FOURTEENTH CENTURY

The large scale of the engineering carried on at these arsenals, as well as on campaign, implies that a substantial number of engineers of the rank of master were employed in these occupations, and the example of Venice confirms this – it also shows that such people did not allow themselves the luxury of loyalty to a single employer.

Domenico di Benintendi exemplifies both the high reputation and the range of skills that such engineers could acquire. Fleeing a death sentence imposed by his native Florence in 1391, he worked on the cathedral of Milan for Duke Gian Galeazzo Visconti (1378–1402). When war broke out with Mantua, he worked to divert the river Mincio using a large dike, described by the chronicler as "doing the impossible", and then constructed a bridge. Other architectural works included the great tower of the Scaliger castle as well as the cathedral at Pavia (1396), and the capture and destruction of a defended bridge across the Po using a boat full of gunpowder (one of the earliest examples of such a use). With the return of peace, his activity then included canal building, but after being employed by the Carrara family, his military skills were called for again to construct engines to attack Vicenza and he worked on the defences of Padua against Venice. On 3 October 1404 he was captured during an attack on a fortification by the Venetians, who ten days later employed him at 100 gold ducats a month. He was put in charge of siege works against his former employer, Padua, in 1405, which included castellated "cats". More hydraulic and military work followed until he was killed by a gun shot in 1409 at Reggio. Alongside Domenico di Benintendi, an equally multi-skilled but anonymous German engineer, who wrote a text about both hydraulic and military engineering including the construction of bridges, siege ladders, fire weapons and

24 A Berty, 'Topographie historique du vieux Paris: Louvre et Tuileries, I', in Haussmann et al., *Histoire générale de Paris. Collection des documents* (1866), 160–3.

25 E Concina, *L'arsenale della republica di Venezia* (1894), 10–17, 25.

fire ships (citing as a case study Domenico's successful operation), served the pope for seven years as well as being employed at various times by a number of Italian cities.[26]

The employment of engineers for military purposes by cities elsewhere also becomes evident in the early fourteenth century. In the regional war of 1303 against a lord who broke the recent peace involving Swiss towns, Strasbourg and the Habsburg family, the city of Bern used a *Werkmeister* named Burkhardt and another named Rudolf Rieder, each responsible for a piece of siege equipment that succeeded in taking their enemy's town, though not the castle. Again in 1324, in another regional conflict, a (this time unnamed) *Werkmeister* was employed by the city to construct cats, this time unsuccessfully. The incidents were recorded in the city's later chronicle.[27] The trading cities of northern Germany who, with their allies, constituted the Hanseatic League were no different. One of the largest, Lübeck, employed a *Blidemeister* and a *balistarius*. In 1364, on the eve of a major war against the king of Denmark Valdemar "Atterdag", a contract was made with Johann Stoke to supply artillery for three years, while other cities made similar arrangements to provide altogether a large force of up to 8000 fighting men and fifty-two ships, including several to transport a siege train comprising three of the largest stone throwers with five *blide* built by an *ingenieur*. Some of the engines were to be mounted for use from the decks since conflict with Valdemar's fleet was anticipated. This war ended without success for the Hanse, but a second war followed and in 1369 another army was gathered, this one accompanied by siege artillery supplied by Lübeck, Rostock, Wismar, Stettin and Greifswald, making a formidable array of *blide*. All of these engines were built by engineers and operated by skilled people, even if the surviving records do not identify them.[28] The first reference to gunpowder weapons in use in this region occurred in city records in 1372.

Named engineers now appeared as well in the service of rulers of lower rank, and although this may represent no more than the chance survival of records it may

26 G Fasolo, 'Domenico di Benintendo da Firenze. Ingegnere del secolo XIV', *Archivo Veneto* ser. 5, I, 1927, 145–80, with numerous primary sources cited; D Lohrmann, 'Zwei Ingenieure beim Untergang des Hauses Carrara', in M Arnoux, P Monnet (eds), *Le technicien dans la cité en Europe occidentale 1250–1650* (2004), 25–41; C Promis, 'Biografie di ingegneri militari italiani del secolo XIV alla meta del XVIII', *Miscellanea di Storia Italiana* XIV, 1874, 24–30. The references are from the Annales Estenses and the chronicle of Sanuto.

27 *Conrad Justingers Berner Chronik von Anfang der Statt Bern bis in das Jahr 1421*, ed. E Stierlin, J Muss (1819), 53, 74–5.

28 D Schäfer, *Die Hansestädte und König Waldemar von Dänemark, Hansische Geschichte bis 1376*, 295–305; 46–8. The second war has a modern description by A Reinart, 'The siege and storm of Lindholmen during the second Hanseatic war (1368–9)', in R Atzbach et al. (eds), *Castles at war* (2015), 204–15.

also signify not just recognition of the importance of employing such experts, but also their greater availability. Savoy benefitted from having a long line of intelligent and skilful counts who shared an objective of expanding the lands of their house. Originally located in the commercially profitable territory of modern French Savoie, where they controlled traffic across several Alpine passes, this dynasty engaged in a centuries-long process of expansion, first to control their own domestic nobility, then to extend their sway into neighbouring lands now in Switzerland and Italy; the counts of Geneva to the north and the lords of the Dauphinois to the south and west were their frequent enemies. Alongside a successful divide-and-rule diplomacy, the Savoyards proved equally adept at warfare, in which the siege played a major part and where military engineers, and miners, were prominent.

The count's army was made up of local noble cavalry and infantry from the communes, but the engineers had to be hired and from their names they seem to have come from across the region. Surviving muster rolls have permitted detailed study of these forces. In Amadeus V's siege of Gex (Ain) in 1321, alongside the army of feudatories and 3700 communal infantry were the people to operate engines listed as *couillards* (see below), sows, mangonels, bricolas, trebuchets, perriers and belfries (siege towers). The logistical back up included boats to transport engines across Lake Geneva from Evian to Geneva including one carrying parts of the belfry: an example of a pre-constructed engine awaiting self-assembly. A chief engineer named Bret was responsible for checking the parts on arrival and directing carpenters to assemble it. The roll named other masters in charge of engines, distinguishing them from noblemen who also supplied machines. John of Morithey was in charge of a large engine, Gauthier of a smaller one, and Michael de Benevis and Jaquet Malliet were also on the payroll. The army included a team of miners from what is now German-speaking Switzerland. Command of the weapons in the siege was in the hands, however, of noble captains. Gex surrendered after five weeks.[29] In 1355, a large Savoyard army completed the submission of the noble family controlling Faucigny (Haute Savoie) by capturing Hermance, and the muster included seventeen carpenters brought from Geneva under the control of engineer Master John from Fribourg to prepare the siege engines. They were paid 3d a day, while the engineer received 6d. A team of German miners also played a major role. In this campaign were to be found noblemen who did not think that being in charge of a siege engine was beneath their dignity: Aimon di Challant, the castellan of Lanzo (in Piedmont), was in charge of a large sow.[30]

29 All from L Ménabréa, 'De l'organisation militaire au moyen âge d'après des documents inédits', *Mémoires de l'Academie Royale de Savoie*, ser. 2, vol. 1, 1851, 179–226, here at 182–91. The castle of Gex is now lost under the later fort of L'Ecluse.

30 Ménabréa, De l'organisation militaire au moyen âge', 192–9; L Cibrario, *Cronaca d'Usseglio ricavata da documenti autentici degli archivii regii*, 11.

Count Gaston of Foix (styled 'Fébus' or Phoebus) (1331–91) displayed a similar combination of diplomatic and military skills to extend his reach in the lands north of the Pyrenees in south-west France, playing off his larger neighbours and exploiting the economically advantageous location of his territory which controlled mountain passes and a wealth-producing iron industry that also provided Gaston with teams of miners for military purposes. Again, surviving muster rolls have been analysed and confirm that alongside a force totalling 1,200 foot soldiers and nearly as many cavalry was a substantial body of artisans, miners and engineers deployed for a campaign in 1376, many of them from his Pyrenean territory of Béarn. This time, the expert in charge was a nobleman, Sicard, named after his castle of Lordat in the county of Foix. He was responsible for a train including sows, a *brida* and four mangonels as well as a number of guns, and named the five captains in charge of these engines and cannon. Sicard was also skilled in building fortifications, including new towers and the donjon at Pau and new castles erected at Morlanne and Montaner.[31]

These examples show some people of noble rank reputed for their skills in (at least) directing engineering activities, and the availability of a substantial cohort of non-noble masters with highly valued multiple skills. Other people found that their skills could rescue them from a potentially dangerous turn of events, as had so often occurred before. Such a man was the Fleming John Crab. In 1318 he was employed by Robert Bruce, king of Scotland (1306–29), to help defend newly captured Berwick against the English. He was reputed as a designer of siege equipment, including building springalds and using "Greek Fire". Berwick was assaulted by an English army, who deployed among other devices a bridge lowered from a ship, another unusual engineering achievement, although this one failed and was burnt and the unnamed engineer was captured and forced to serve the garrison as it resisted further attacks. Now with two skilled engineers, the Scots not only succeeded in disabling a sow being used to cover a sap of the walls, when Crab constructed a crane-like machine to push it over, but also by their accurate shooting with stone throwers sank several of the English ships. John Crab was captured by the English, who promptly re-employed him and fifteen years later he was to be found literally on the other side of the wall he had so successfully protected in 1318. The English deployed two large stone throwers built on the spot under Crab's direction and caused significant damage (though surrender only came following the Scottish defeat at the battle of Halidon Hill). Crab continued in English service and was in charge of the engines at the siege of Dunbar in 1338,

31 P Raymond, *Rôles de l'armée de Gaston Phoebus, comte de Foix et seigneur de Béarn 1376–78*, 132, 145–8; P Tucoo-Chala, *Gaston Fébus et la vicomté de Béarn 1343–91*, 159; and see now R Vernier, *Lord of the Pyrenees. Gaston Fébus, count of Foix 1331–91*, 18, 95–7.

when another engineer named Nicholas was also identified in the accounts: he had been responsible for the construction of thirty springalds in 1336.[32]

Sometimes the possession of engineering skill alone did not save a master from an unpleasant fate. A chronicle account of the capture of Schwanau (in Alsace) described not only how the sixty rebellious defenders were all executed, but also that the engineer who had been in charge of their engine was eviscerated and his remains hurled back into the place.[33] Evidently his association with a "robber baron" outweighed any possible usefulness to the cities that had sent the army to suppress him.

These examples all relate to armies mobilised to carry out siege operations. Not all armies had this purpose, of course, and sometimes the engineers were left at home. Regular examples were provided by English armies during the Hundred Years War, because when the objective was to destroy French willingness to resist by fire, slaughter and the destruction of villages, towns and the countryside, or to compel the enemy to fight an unwanted battle, as it was in many campaigns, there was no call for a siege train and no need for engineers, although miners might still be enrolled: there were twenty, with four masters, recruited for Henry of Lancaster's 1345 campaign.[34] The muster roll for the English army led by the earl of Arundel in 1378 showed a force of cavalry (knights and bannerets), archers and crossbowmen. This particular expedition differed from many in that it did attempt a siege of St Malo and had some miners with it, although on this occasion the defenders caught the English off guard and destroyed the mine.[35]

It is also clear that a degree of specialisation was continuing to develop. From England in the fourteenth century there are exceptional examples of highly skilled architects who, unlike the anonymous engineers and Benintendo of Florence, devoted their lives to building: one such is perhaps the best architect of this or many other centuries, Henry Yevele, while working in the north of the country another expert with a distinct style was John Lewyn. Both undertook numerous commissions for work on castles, cathedrals and churches and other buildings for a range of high-ranking clients, but there is no record that they did anything

32 Barbour, *The Bruce*, book XVII, ll 1–204, 235–51; Hewitt, *Organisation*, 72; *Calendar of Close Rolls*, Edward III, 1333–37, 724 and 1339–41, 11. There are many detailed accounts of the sieges of Berwick during these years.

33 'Chronica Iohannis Vitadurani', *MGH* SrG NS III, ed. F Baethgen (1924), 112.

34 Gribit, *Henry of Lancaster's expedition*, 67. These miners were from the Forest of Dean and the Peak District.

35 A Bell, *War and the soldier in the fourteenth century* (2004),16 (citing Froissart's chronicle), 52 (muster roll); A Ayton, 'English armies in the fourteenth century', in A Curry, M Hughes (eds), *Arms, armies and fortifications in the Hundred Years War* (1984), 21–38. (I would question Bell's translation of "alblasters" as gunners (p. 52)).

other than building.[36] Their roles had by now evolved into something recognisably similar to that of a modern architect, as they appear to have been responsible for both the overall design of the buildings and oversight of the construction by other craftsmen. Both seem to have begun as master masons. Their buildings reflect continued development of architectural styles ("English perpendicular" being their particular style but similar processes were taking place everywhere with "late Gothic" being a generic description) as well as the evolution of more refined techniques calling for new levels of skill.

The skills of such builders were available elsewhere. Although all the documentation has been lost, the work of completing the construction of the immense palace-castle of Vincennes for the French king Charles V (1364–80) was achieved in a remarkably short space of time under the direction of master of the works Raymond du Temple. The visitor is likely to be struck both by the great size of the castle (the total outer wall length of 1100 metres is little less than the entire city of Carcassonne and the towers stand 40 metres high; the donjon stands six storeys to 52 metres) and by the symmetry of the design. This may conceal the immense work required to achieve such an outcome on a sloping site, which had to be levelled first, and took into account the existence of an earlier royal site and the first levels of the donjon begun under his predecessor. The symmetry is the result of the careful deployment of geometric proportion using the traditional multiples of 7, 10 and 12 to locate the donjon and its own enclosure within the envelope of the outer wall and to determine the dimensions of towers of four different designs. Modern calculation confirmed that the proportionality actually achieved was almost perfect.[37]

People possessing the skills to achieve such constructions were widely available. Not all are known, but the description of the new fortifications begun by the Florentines to help protect against multiple threats in 1355 described the digging first of the ditches in August and then in September the foundations for the walls were laid, with a barbican, two gates, towers 50 feet apart, and the walls were provided with internal passages. The whole operation was completed in twelve months, and the large total cost ("35,000 gold florins") must have included payments to surveyors and master masons able to plan such a work and successfully transfer such a plan onto the ground.[38]

36 The classic study remains J Harvey, *English medieval architects. A biographical dictionary down to 1550*, revised edn (1987), 181–4 (Lewyn), 358–66 (Yevele). Harvey also wrote a separate biography, *Henry Yevele c.1320–1400. The life of an English architect* (1944).

37 J Chapelot, 'Le château de Vincennes (1372–80)', in H Mouillebouche (ed.), *Châteaux et Mesures. Actes des 17e journées de Castellelogie de Bourgogne 2010* (2011), 100–23.

38 *Cronica di Matteo Villani (1348–1363)*, ed. A Racheli, 181.

TECHNICAL IMPROVEMENT OF SIEGE ARTILLERY

The underlying technology used by these masters remained little changed; the same methods were used, the same materials and the same tools. Nonetheless, there is some evidence that these skills were being put to new use and the example for which there is clearest evidence is the progressive improvement in the design of the heavy artillery represented by the trebuchet.

It is much easier to make a general statement on this subject than it is to propose details as to when and where the evolution of the counterweight stone thrower took a particular form, and it is impossible to know who may have first suggested changes in the design to their employer, to obtain the means for putting their ideas into practice. It is nonetheless certain that this was what happened on a number of occasions during the century or so after the first appearance of this weapon. The recognition of the underlying principle – that replacing human muscle power with the consistent power generated by using a material counterweight would not only ensure a previously lacking consistency in delivery but would also permit a substantial increase in the power of the engine – was relatively straightforward and had been adduced by al-Tarsūsī. Moving from that idea to practical application required another step. It must be expected that a considerable amount of trial and error went into developing the first designs, given (as I pointed out above) that getting the proportions of the various component parts wrong could be catastrophic. By the time of Villard de Honnecourt, it would seem from his drawing, engineers had concluded satisfactorily the critical measurements of what was possibly the first standard design, a single-box counterweight suspended from the end of the throwing arm. Whether these early experiments included assessing the relative benefits of attaching the box directly to the arm or of having it suspended so as to increase the momentum of the downward movement cannot be known now. Quite quickly, however, variants to this design appeared. Giles (Aegidius) had already identified that there were different designs of counterweight stone thrower available to rulers by the third quarter of the thirteenth century. He wrote that the name "trebuchet" traditionally applied to the model with a fixed counterweight and that the range was modified by changing the length of the sling or the weight of the shot. A second type with a hanging counterweight was called a *biffa*. The third variant was called a *tripantum* and had two counterweights.[39] Medieval as well as modern writers loved to organise their subject matter into lists and it is not necessary to believe that Giles' categories necessarily reflected the complexities of reality but they do confirm the development of different designs over a relatively short period of time: they must have already been in use when they were written about, meaning that not only had they been devised, but awareness of the technology had spread.

39 Aegidii Romani, *Libellus de re militari*, ed. Hahn, book III, chapter 18, 50–2.

Illustrations also confirm that there were changes in the way in which the throwing arm was reset after each shot, including the use of treadwheels such as those on the reconstructions at the Danish Middelaldercentret and Warwick castle. As the winding down of the arm to reload the weapon was a critical part of the entire operation, and would largely determine its rate of shooting, finding ways to achieve this formidable challenge (with counterweights of perhaps 20 tons having to be raised well above the height of the machine) must have exercised the minds of engineers from whenever the really large machines first appeared. Another method that appears in illustrations such as those of Konrad Kyeser was a winch with a turning wheel at each end set with handles to enable two people to wind down the arm: it must be hoped that some kind of braking device was included in the design as the consequences of one of the winders slipping would have been grim.

Alongside these developments in the form of the weapon there were also changes in the dimensions, to allow a commander a variety of options when attempting a siege or resisting one. The largest weapons, able to propel the stone shot weighing 100 kilograms a long distance and that could inflict significant damage on stone walls and towers, had the disadvantage of taking a long time to reload. The smaller and lighter the weapon, the quicker it could be brought back into use after each shot, and accounts of sieges where it is recorded that large numbers of weapons were deployed suggest that a variety of sizes best met the needs of the commanders – the description of the 1291 conquest of Acre by the Mamluks specified that the battery *included* a number of the very largest, the implication being that most of the weapons were smaller.[40] Rate of shooting was a particularly important requirement if the bombardment was to achieve the purpose of inflicting damage both physically and psychologically on the defenders.

Illustrations in manuscripts continue to be the only way (with the sole exception of Villard's incomplete drawing) to reconstruct what the counterweight engine looked like, and such illustrations cannot convey the actual dimensions or the details of construction that determined whether the engine worked. Editions of manuscripts of Kyeser's *Bellifortis* (1405), for example, have drawings of counterweight engines with single counterweights attached directly to the arm and suspended from it, and double counterweights, one hanging either side of the arm (and giving the weapon its French name of *couillard* for its evident reminder of male sexual organs).[41] These are not scale drawings (as discussed above) but they do suggest the components needed for the engine to work if built to the

40 For a summary account and list of all the contemporary sources including translations of key Arabic accounts, see Purton, *HLMS*, 61–2 and n. 1.

41 U Lindgren et al. (eds), *Konrad Kyeser Bellifortis. Bayerische Staatsbibliothek Clm 30150*, 2v, 88–9; R Cermann (ed.), *Der 'Bellifortis' des Konrad Kyeser* (2013). See also R Beffeyte, *L'art de la guerre au moyen âge* (2015), 81–3, suggesting the various rates of

correct scale. That Kyeser knew of what he wrote was also implied in his statement that the various designs he illustrated came partly from the German and partly from the Italian tradition, suggesting particular lines of development in different geographical locations, as was only to be expected.

A trebuchet of standard design, but fulfilling a requirement that it could be directed rapidly in any direction, featured in an account from the town of Martel (Lot) in 1356. No doubt in response to the perilous situation in which French towns found themselves, fearing rampaging English armies across the land, the town government commissioned the construction of a machine that would be swivelled using something akin to a millstone set in a masonry base. The master mason took twenty-five days to construct the base, four masons took another day to install the millstone, and then eight carpenters took another nine days to build the trebuchet on top. Timber, metal pins, hemp, leather, oils, tallow, resin and ropes were purchased at a total cost of nearly 20,000 deniers. It took fourteen days for mules to convey the stone and earth to be placed in the counterweight. The completed engine was still in place three years later, although unfortunately it is not known how well it worked – perhaps the town had the good fortune never to need to use it. The account suggests that one small town did not have the resources to construct more than one piece of heavy artillery but decided on having a weapon that could be aimed in any direction, and therefore expended a considerable quantity of money and labour to build a machine of an unusual design.[42] The name of the designer remains unknown.

A similar approach was applied in a fort at Luisandre (Ain), built on the orders of Amadeus V as part of the Savoyard count's conflict with the dauphin and to control a trade route, from 1313. Provided with springalds to protect the builders during construction, part of the armament of the fortification included a bricola on a pivot, alongside a trebuchet. Records also confirm that many stones were quarried and shaped to provide the ammunition, and spare slings were laid in for the trebuchet. The design included an esplanade on which the artillery was stationed. The engines were used to launch an attack on the neighbouring dauphinois fort at Allymes, a short distance away.[43] The engineer was not named but again a range of skills and experience would have been needed to enable this project to be completed successfully. It seems from these examples that the engineer responsible for constructing each real engine knew how to do so, where getting the proportions

---

shooting expected of different types of stone-throwing engine. He dates the *couillard* to the fourteenth century, which seems reasonable.

42 The story is told by Nicolas Savy, who discovered the accounts, and is published in Beffeyte, *L'art de guerre*, 70.

43 A Kersuzan, *La bâtie de Luisandre (Ain). Histoire et archéologie d'une fortification savoyarde de frontière au XIVe siècle*, 24–5, 48–50, 61–2 with reproduction of the original accounts for 1312–13 (85).

right was critical to the weapons actually working. Apparently they did (although there may be good reason why the failures did not get mentioned).

The long survival of variants of the trebuchet design as valued components of a ruler's arsenal has been noted. A frequently cited example was the large engine called "Ruin" that was in the charge of Peter Masuer, "master of artillery" to the (now) duke of Savoy in 1433/37. There is an inventory of all the parts required for the engine, which must have been stored in a dismantled condition. The engine required treadwheels to wind it down. The beams on which they rested were 14 feet long (4.3 metres). The inventory also reported the lengths of all the many timber components, the throwing arm, the parts of the structure holding it and the channel for the sling to rest in prior to shooting. Three different types of timber were called for: oak for the largest pieces that required strength and solidity, deal used where light and elastic functions were called for, and (with no specified usage) elm. The builder of this machine was aware of the different qualities of the timbers and what they were best used for. Following the description of the wooden parts there was an even longer and no less detailed list of all the iron parts needed (*ferramenta*) which ran to more than twenty differently named items, not all of which can now be understood. But the picture that emerges is of significance not only because of its complexity, perhaps to be expected of a heavyweight piece of artillery, showing the knowledge required of its master, but also because of the awareness of the characteristics of the materials.[44] Note, too, the date: large guns were already in regular service.

MILITARY MINING: OLD AND NEW PROVIDERS

A number of references have been made to mines alongside other forms of attack in the siege warfare of the fourteenth century. Technology evolved slowly in this industry but there is evidence that new development took place during the thirteenth and especially the early fourteenth centuries, as the exhaustion of surface deposits drove miners and mine owners to dig deeper and deeper, forcing them to deal with growing problems, in particular flooding. The challenge of creating drainage from hundreds of metres down inside mountains was met; for example, two drainage channels and six water wheels were constructed at the silver mines of Iglau in Bohemia in 1315. Deciding where to dig the channels required skills in geometry, in particular being able to triangulate. Since the results were

44 Ménabréa, 'De l'organisation militaire au moyen âge', 221–4; D Dubois, 'La charpente militaire. Un aspect des engins à jet à contrepoids à la fin du moyen âge', in X Barral y Altet (ed.), *Artistes, artisans et production artistique au moyen âge* (1983), II, 403–8, the inventory at 406–8. The same document has also been published elsewhere.

successful, it is evident that such skills were available to mine owners.[45] We have already observed that hydraulic engineering was among the skills possessed by many engineers in Italy.

Rulers continued to employ such miners in military service. The apparent growth in numbers is possibly only the result of the greater survival of records, but it may also be associated with the expansion of mining to new sites, creating demand for the specialist skills required of underground mining as well as for the knowledge of how to turn the ores into something useful, and thus also increased opportunities for rulers to add a new resource to their armouries. In Durham, for example, where the mining rights belonged to the bishop, a manager was appointed who in turn paid John Gyll, a smith, to clear the ground for a new forge as well as visiting another site to instruct the craftsmen so they "could better inform themselves about the building of it" (this being in 1408).[46]

The frequency of warfare undertaken by the English kings meant a relentless demand for miners to serve in their armies. Time and again it was the miners of the Forest of Dean who were enlisted. The mechanism for doing this appeared in records. The constable of Bristol castle was paid expenses from the royal treasury for travelling to St Briavels, the royal castle in the Forest, and selecting fifty miners who were to be sent to London, their pay being 5d a day, along with hiring a horse to carry their tools (hammers and wedges of iron) (1300–03).[47]

Government records show the reliance of Edward III on these miners. As he prepared his expedition to France in 1346, Guido de Brian, custodian of the Forest was ordered to select forty miners from there and the surrounding county, including four masters, and send them to Portsmouth where the host was gathering. For the expedition of 1359 a similar order was issued, this time specifying four smiths and forty miners, "the best and most apt of the county". This concern with quality, which was probably formulaic, was repeated in the order to find twelve miners to be shipped to Aquitaine in 1365 and again, with the same number, to be delivered to Edward Despenser for overseas service in 1375. Although they could not be used at the long siege of Calais following the victory at Crécy in 1346 on account of the marshy terrain, these miners were put to use in other sieges. 120 miners were used at the siege of Berwick in 1355–56, and while most came from Dean, the chronicler Avebury also specified that some came from elsewhere. Perhaps the need for such large numbers exceeded supply and made it necessary to broaden the recruitment, and the local mining industry in northern England would have been an obvious source. Miners breached the walls of the Great Tower of the castle of Cormicy during the otherwise unsuccessful

[45] Cited in C Bartels, 'Der Bergbau im Zentrum das Silber', in U Lindgren (ed.), *Europäische Technik im Mittelalter 800 bis zum 1400* (1996), 235–48, here at 241–2.

[46] *English Historical Documents IV (1327–1485)*, ed. A R Myers (1969), 1005–6.

[47] M Sharp (ed.), *Accounts of the Constables of Bristol Castle* (1982), 64–5.

attack on Reims in 1359, when forty-three men had been indentured for service. They set fire to the posts to complete the task and the tower "cracked and split into two parts". Froissart's long description of Edward III's wars is replete with references to the key role of miners in capturing towns and castles.[48] It was not only the king's army that use these miners: they were to be found supporting other expeditions. In 1355, miners were responsible for breaching the walls of Limoges during the Black Prince's campaign. In 1369, the duke of Lancaster's army of around 2000 men was accompanied by forty miners. Lancaster was also the author of a religious text written as an allegory but including a reference to the role of miners in "breaking through the walls at their foot and thereby entering as simply as through the doors".[49]

The role of English miners naturally diminished during the decades that witnessed the recovery by the French of much of their lost territory in the later decades of the fourteenth century, but they were found in service again with Henry V's army in 1415, where their part in the siege of Harfleur was given much later prominence through Shakespeare's play. The lie of the land meant that the duke of Gloucester had to begin the digging from within sight of the defenders who were thus able to carry out successful countermining; Harfleur eventually fell as a result of the king's artillery barrage and impending assault on the other side of the town. One chronicler explicitly stated that the location of the mines was *contra doctrinam magistri Egidii,* indicating awareness of the teachings of Giles of Rome. Other commentators noted that the French had ensured there was a very deep ditch to prevent mining.[50] The muster rolls for this campaign (which culminated in the battle of Agincourt) showed that among the specialist companies accompanying the archers and men at arms were 120 miners under the command of John Greyndon, knight, and four master gunners, smiths and carpenters.[51]

One of the chroniclers remarked that the English had forgotten the art of mining for warfare as a result of the long peace,[52] and it is not impossible that the men who had developed particular skills required for this work in conditions of

48 Avebury, *De gestis mirabilibus regis Edwardi Tertii,* 450–1; T Rymer, *Foedera,* ed. G Holmes (1825), III, pt 1, 78, 417; III pt 2, 762, 1021; J Froissart, *Chronique,* ed. Diller, c. 618 (III, 223–8) (Cormicy).

49 *Le livre de seyntz medecines. Unpublished devotional treatise of Henry of Lancaster,* ed. J Arnold, 71.

50 *Gesta Henrici Quinti,* ed./trans. F Taylor, J Roskell (1975), 40–1; 'Elmhami liber metricus de Henrico V', in C Cole (ed.), *Memorials of Henry the Fifth, king of England* (1858), 110–11.

51 H Nicolas, *History of the battle of Agincourt and of the expedition of Henry the Fifth into France in 1415* (2nd edn, 1832), Appendix. A Curry, *The battle of Agincourt. Sources and interpretations* (2009), 421.

52 Pseudo-Elmham, *Vita et gesta Henrici Quinti* (1727), 45.

armed struggle were now gone. While there is no suggestion in the contemporary accounts that it was incompetence on behalf of the miners that led to the failure of this attack at Harfleur, what is clear is that when Henry returned to France in 1417, he did not call on the miners of Dean, but instead recruited masters and teams of miners from Liège: wages were paid from the treasury to a team of sixteen miners and to John Lowart for his miners and money spent for "divers miners from Liège".[53] The reason why the king abandoned the tradition of hiring miners from the Forest of Dean is lost to history.

Whereas miners from the Forest of Dean dug iron, the miners of Liège made their living excavating coal. Histories suggest that the earliest discoveries of the stuff were made around the year 1200 and that the potential profit to be made was quickly recognised. A whole community of coal miners became established during the thirteenth century, with the prince-bishop naturally taking a particular interest.[54] It is perhaps surprising that there is no record that they were called upon for military service until the following century, although there was no shortage of opportunity as the Low Countries were made up of small, competing lordships, counties, bishops with territories to rule, dynastic dukedoms and the lands of distant monarchs. The first time the chronicle of Dynter reported a military role was when miners brought down the castle of the lord of Faucquemont in 1329, following a fierce siege during which the defenders had countermined. In 1346, Liègeois miners were used to take the castle of Arkenteel. The next year, the coal miners went to work against the castle of Argenteau, set fire to the props and brought down a section of wall, allowing for a successful storming. In 1388, the duchess of Brabant summoned miners and diggers from Liège to assist in her siege of the lordship castle of Gasebeek, which surrendered before they set to work. In 1397, however, they were responsible for demolishing 30 feet (9 metres) of the wall of the city of Roermond in a war between the duchess and the duke of Guelders. The region was partly under the sway of the dukes of Burgundy by this time and in 1413 a company of miners was deployed to destroy the Great Tower of the castle called "Mersi" near Verdun. Evidently, like their fellow miners from Gloucestershire, Liègeois miners knew how to operate both in peace and in war and they had by this time acquired sufficient reputation to be in high demand by rulers from further afield.[55]

53 F Devon, *Issues of the Exchequer, being a collection of payments made out of His Majesty's revenue from King Henry III to King Henry VI inclusive* (1837), 351–2.

54 J Chapeauville, *Gesta pontificum Leodiensium scripserunt auctores praecipii* (1613), II, 192; H Kranz, *Quellen zum Lütticher Steinkohlenbergbau im Mittelalter*, 13–14.

55 Dynter, *Chronica nobilissimorum ducum Lotharingiae et Brabantiae ac regum Francorum*, ed. de Ram, II, 783–5; 820–1; III, 623–4; 807–8; Jean d'Outremeuse, 'Chronique abrégée (1341–1400)', *Chroniques Liègeoises*, ed. Balau, II, 179, 232.

It seems that those who commanded them were also aware of the circumstances where mining could succeed and of the good advice available in the standard textbooks in which they had probably been educated. Giles of Rome reproduced much of the good advice of Vegetius including on the subject of mining and countermining, and it has already been shown how the work of Christine de Pisan echoed this. Her words on mining were those of Vegetius, but brought up to date with the recommendation that the sound of the digging should be concealed under the noise of a continuous assault by guns and bombards.[56]

Having access to a community of mineworkers was a critical advantage for a ruler wishing to use this advice. Henry V may have decided to replace his domestic miners by paying for Netherlanders, but the counts and dukes of Savoy had their own resources. Records show the licensing of mining for iron and silver in the valleys and forests of Piedmont from 1316. The mines were exploited for the manufacture of iron weapons, and it is evident that when the count (in this case Amadeus VI in 1351) confirmed the privileges of the iron and coal miners of Lanzo he had wider potential uses in mind. When the castellan of Lanzo was engaged at the siege of Gex in 1355, the miners seem to have been Swiss and German. The next year saw the capture of Balangero where a full artillery train was deployed (including three *trabucchi*) but victory was achieved by mining. The conquest of the marquisate of Saluzzo was achieved in 1363 with the successful siege of castles and finally the main town itself, which was faced with a formidable attack shooting 300 stones a day from the artillery (both stone throwers and "bombarde"), but the surrender was linked to the arrival in the Savoyard camp of a team of miners from Lanzo. They were rewarded with 10 gold florins per month. Nine years later, a team comprising four knights and thirty-five miners was sent to demolish Salto.[57]

The references to the Savoyards hiring Swiss and German professional miners even when there was expertise available on the far side of their Alpine realm reinforce the argument for there being a shortage of these skills and confirm that those who did know this trade were likely to find employers in many lands if they were prepared to move.[58]

56 Christine de Pisan, *The book of Deeds of armes and of chivalry*, ed. Willard, 116.

57 Cibrario, *Cronaca d'Usseglio ricavata da documenti autentici*, 8, 11, 14–16; L Usseglio, *Lanzo. Studio storico* (1887), 158–9, 197, points out that many Swiss and German miners migrated to take advantage of the start of the industry at Lanzo; F Gabotto, 'L'eta del Conte Verde in Piemonte secundo nuovi documenti (1350–1383)', *Miscellanea di Storia Italiano* ser. 3, II, 1895, 74–333; G della Chiesa, 'Cronaca di Saluzzo', *Monumenta Historiae Patriae*, Scriptorum 3, 1848, 1006–7.

58 Hesse, 'Artistes, artisans ou prolétaires?', 446–7.

MINING (AND OTHER ENGINEERING SKILLS) UNDER TIMUR

A move from the local to the supranational is involved in noting how the revived Mongol threat posed by Timur was executed and how it replicated features of previous Mongol conquerors. Having achieved extraordinary success in re-establishing a single empire (with his own capital in Samarkand) by the 1370s, Timur followed the example of his predecessors in securing the skills his nomad followers lacked by forcibly enlisting artists, construction workers and engineers from the urban centres he had conquered and continuing to rely on the educated elite from Iran for administration as he extended his empire to south, east and west over the following decades. Only death (followed by civil war) halted his planned invasion of Ming China in 1405. Although accounts of Timur's many victories are often lacking in detail, where this is provided they give evidence for there being engineers with his armies who were responsible for the construction of military equipment – they created the battering rams to take Terdniz (1381) and knew how to drain the wet ditches there.

That he used artillery is confirmed by accounts in Georgian and Armenian sources of his brutal conquests of those regions, but the first explicit reference to miners related to his capture of Tikrit (in Iraq) in 1393–94, where no fewer than 2000 diggers were apparently sent in. Mining from all sides of the town was used again to capture Batnur in India in 1398. When Delhi fell to the Mongol in 1399 without a fight, Timur spared the artisans and master mechanics from his massacre of the inhabitants, this time so they could be sent to Samarkand to work on the beautification of his capital. At Mirat, where the mining operation was overtaken by a successful escalade, the tunnels were used to demolish the fortifications. A "honeycomb" of tunnels was dug under the towers of Behesna in Syria in 1400 and the great citadel of Aleppo was taken by the same means within two or three days. The massive rectangular towers and ramparts built to protect and glorify the citadel of Damascus were cracked open by Timur's miners in 1401. The only failure of this method of attack was recorded in the abandoned siege of the fort above the Syrian city of Mardin, where the rock was so tough that "the hoes were worn away and the hammers blunted". He left this minor site to march on and destroy Baghdad, which had rebelled. Time and again, Timur's large invading army was accompanied by a full corps of artisans and military engineers as well as a substantial number of miners: while it is likely that the muscle power was provided by soldiers or pressed local labour, those planning the operations must have been skilled and those carrying out the underground work mentioned so often in the sources must have been professional miners. Who they were and where they came from is not revealed, but the history confirms that such military engineering skills were in use across the whole known world.[59]

59 For the contemporary evidence drawn from the accounts of ibn Arabshah, *Tamurlane or Timur the Great*, trans. J H Sanders, Sharaf al-Din, *The history of Timur Bec*, trans.

GUNPOWDER ARTILLERY AND THE GUN MASTER

One feature of Timurid siege warfare is that there is no evidence that he or his commanders used gunpowder weapons despite their growing use in Europe from seven or eight decades before his death. Had Timur actually carried out his assault on Ming China, he would have encountered gunpowder. By the end of the fourteenth century, the emperors of China did have guns that appear to have resembled their western counterparts of the same date, but evidence has not been found to confirm which came first.[60] It has long been assumed that awareness of gunpowder was transmitted from China to the west sometime in the second half of the thirteenth century, whether or not one believes in the existence of the mysterious German monk "Berthold Schwarz" alleged (by German writers) to have "invented" it, or the more reliable evidence from the *Opus tertium* of Roger Bacon suggesting that the inquisitive friar had come across the recipe and had tried it out (see chapter five).

Reliable evidence for the first use of guns is not always available, and a number of early sightings have been challenged, but the general picture is clear. From the first surviving manuscript illustration (bound with the Milemete manuscript dated to 1326) to the unquestionable evidence of growing numbers of references in chronicle accounts and government and town records to guns, gunners and gun makers, it is unarguable that – whether or not the references to the use of guns at such actions as the sieges of Faucquemont (1329), Cividale (1331), or Castle Eltz (1331–36) are reliable – by the 1340s guns were in regular use. Not having any datable surviving weapons from this early period (the metal was always liable to be reused) has been an obstacle to grasping their performance and inevitably it has been necessary to rely on reconstructions. A replica of the "Loshult gun" found in 1861, preserved in Stockholm and just 30 centimetres long, shot gun arrows 500 metres and lead balls 1200 metres, and subsequent tests using different formulae for the powder managed even greater ranges [Fig. 22]. Although it was very inaccurate, this range, and the ability to pierce armour (albeit only at closer range), would explain why so many rulers immediately turned to the new technology. Unfortunately, the Loshult gun itself has been ascribed an early date only because of its similarity in appearance to the Milemete illustration and not through any scientific test.[61]

J Darby, and Khandamir, *Life of Tamerlane*, see Purton, *HLMS*, 176–94 where full references are given.

60 See the discussions by Zhou Jiahua, 'Gunpowder and firearms', 183–91; Needham, *The guns of Kaifêng-Fu*.

61 The tests were carried out by P Hansen, reported in *Journal of the Royal Ordnance Society* 14, 2002, 5–16 and R D Smith for the Medieval Gunpowder Research Group, available on https://www.middelaldercentret.dk. For a summary of the various records and accounts covering the first use in Europe of gunpowder artillery, see Purton,

Fig. 22 Early gun (barely 30 centimetres long) reconstructed at Middelaldercentret, Nykobing Falster, Denmark. (Photo: author)

The use and impact of guns continued to grow during the remainder of the fourteenth century until a combination of factors – in particular, technical improvements in metallurgy and experiments leading to greater effectiveness of the powder itself – made the new artillery increasingly effective. But the transformational impact of these processes had to wait for many decades from the first recorded appearance. Meanwhile, it is also clear that the enthusiasm of warmongering rulers (and of those determined to defend themselves by keeping up with the arms race) was matched by the appearance of a new category of military engineer, the gun master (in French *maître d'artillerie*, or *de canons*, and in German *Büchsenmeister*), and a new industry, racing to fill the gap in the market. Regardless of who had the dubious honour of being the first to use the new means of destruction, people across Europe hurried to acquire the skills needed to manufacture the weapons and the gunpowder or to offer their services as the master gunner the modern ruler needed to employ. The question of exactly how these men acquired the requisite knowledge cannot now be answered. By the end of the century there existed handbooks (such as the famous *Feuerwerkbuch*) and

HLMS, 116–18. For a good modern summary, L White, R D Brown, *Medieval Military Technology*, 2nd edn, 137–63.

the writers of advice to princes (such as Christine de Pisan) were incorporating wise words on modern technology, but no such manuals survive from the earlier period. There must have been a great deal of rapid transmission from the early masters to their successors, certainly aided by the continuing geographical mobility of the engineer, but even the identification from records of the first generation of gunpowder and gun masters and gun founders gives no insight into how they acquired the knowledge.

The interaction between technological progress and the demands of the military cannot be clearly demonstrated in the way it can for later times, but it is certainly likely that the requirements for ever-larger artillery drove the search for improvements in metallurgy. Fourteenth-century gun founders certainly moved around to where they could sell their services at the best rate, in the process spreading their knowledge, and no doubt a reputation as the best in the business would attract offers of employment, as suggested by the 1392 agreement of an alliance between the dukes of Lorraine and Bar and the city and bishop of Metz to pay for foreign "artilleurs, mineurs, tarrillons, maistres de canons engeigneaulx et teille manière d'ouvriers d'artillerie". Evidence of the first blast furnaces appeared in Liège, a centre of metalworking, from around 1340 and in Sweden from 1360, and if these developments had a much slower impact on gun making than on smaller operations (such as plate armour), they represented technological progress.[62]

Sometimes the people making the guns and operating them were the same; sometimes it is not possible to tell. By the fifteenth century, the trade was practised by hundreds of manufacturers across Europe, but the number of names recorded during the earlier decades was very much smaller. There was a foundry at the Tower of London, consistent with it being a major arsenal for the kings of England; John Byker was listed as "artilliator regis" in 1354 and (the presumably related) William Byker was there between 1382 and 1388. Though the title could also apply to a maker of springalds, records show that the Tower did have a stock of twenty-nine iron guns as well as saltpetre, sulphur and other requisites in 1370–74. William of Aldgate made four copper guns for the king in 1353. John of Cornwall supplied guns in 1361. John Brazier provided guns for the duke of Clarence in the same year. Also identified as founders by their names or the record were William the Founder (London 1385), who supplied twelve guns to Dover castle, William Woodward (London 1382–88), Stephen atte Maistre (Tower of London, 1385) and John Mollyng (London 1382–96), who made brass guns for the king. What of the people who manufactured the very first guns, from at least two decades before the earliest of these? All that can be said is that enterprising metalworkers

62 A Girardot, 'Fondeurs d'artillerie et siderurgistes', in P Benoit, P Braunstein (eds), *Mines, carrières et metallurgie dans la France médiévale* (1983), 67–76 (quotation at 67); R Forbes, 'Metallurgy and technology in the middle ages', *Centaurus* 3, 1983, 49–57; Williams, *The knight and the blast furnace*, 881–3.

must have decided to try their hand at the new inventions. Having been hired as gun founders, they would also have been the only people with the skill to use the weapons so must also have served them in war, along with a team of other workers to provide ancillary service, and from amongst whom, possibly, more gun masters would have been trained.[63] While a certain precocity may be allowed to England in this questionable league table, there is also evidence from other parts of Europe for the early manufacture and use of guns. The same lists identify manufacturers who, while living in other lands, were selling to the English, such as Jean and Philippe Chautdouvrier and Bernat de Montserrat in Caen (1374–75) and Arnaud Naup in Bordeaux (1398). Joseph Collard was providing both bells and guns for the duke of Burgundy (1380–90), a not uncommon combination of skills, while the dukes of Savoy were obtaining theirs from a local source, the brothers Anne and Pierre Dorier of Bourg-en-Bresse (in both 1386 and 1418 – a long-lasting business). A certain Guillaume was active in Laon in 1358.

The urbanised Low Countries were not surprisingly also at the forefront. In Ghent, Claus van Lit was appointed master of artillery (*engienmeester*) at 20 sous per day for the war that broke out with the count of Flanders in 1380. With two valets he took part in several military expeditions and was exiled following the defeat of the citizens. The first recorded use of guns in the modern Netherlands dated to the siege of the castle of Rozenburg in 1351. The nearby town of Deventer bought three guns (*donderbussen*) from a manufacturer (Master Herve of Campen) in 1348 and further purchases were recorded in subsequent years. The city of Nijmegen's partial surviving accounts show the first reference to gunpowder in 1382.[64] The earliest reference to the use of guns in the kingdom of Denmark dated to 1372, when a citizen of Ribe was executed for delivering gunpowder to the enemy.[65]

In German-speaking lands a thriving industry rapidly developed and gun founders and gun masters of German origin were among the first to diffuse the knowledge more widely. The city of Strasbourg was the site of gun manufacture from the second half of the fourteenth century. Foundries were set up in the city,

63 These names have been extracted from the invaluable lists of A Kennard, *Gunfounding and gunfounders. A directory of cannon founders from earliest times to 1850* (1986) and C Ffoulkes, *The gunfounders of England, with a list of English and continental gunfounders from the XIV to the XIX centuries* (2nd edn, 1969). See discussion in Hewitt, *Organisation*, 72–3.

64 N de Pauw, 'Claus van Lit, Maître d'artillerie de Gand de XIV siècle', *Bulletin de la Commission Royale d'histoire* LXXXIII, 1920 (233–9); F Hoefer, 'De artillerie in de Rekeningen der Stad Nymegen', *Gelre. Bijdragen en Mededelingen* XXIV, 1921 (77–100).

65 Tøjhusmuseet, *The cannon hall*, trans. E Christensen, 6. The first datable guns in Denmark were found in a wreck near Anholt and have been ascribed to around 1400: Tøjhusmuseet, *Billedkatalog for den permanente udstilling*, 12–13; A Hoff, 'The Töjhusmuseum, Copenhagen', *Armi antiche*, 1963, 53–68, at 54.

although it did not become a large-scale industry until the following century – but guns featured prominently in the records of Strasbourg's campaigns at this time. Jean de Troy was appointed as *Büchsenmeister*. Presumably because the local foundry was not yet functioning, Dietherick Kessler and another gun maker were brought in from Basel. Despite this apparent resource, other Swiss cities and cantons seem initially to have relied on northern Italy, with the substantial iron-making tradition of Lombardy, until it became cheaper to import what they needed from south German cities, where big industries became established in Augsburg, Nürnburg and in Thuringia during the fifteenth century. Before this, it seems that the first named supplier of guns in western Germany hailed from the Low Countries: Walter of Arle (Arlon) may have made the first guns for Cologne (1377) and Frankfurt am Main, Augsburg, Nürnburg and Passau (1378–79). In 1382, a letter concerning a feud over possession of Berchtesgaden named Walther as the *Büchsenmeister* of Arnolt of Swiken and this may be the same Walter of Arle. However, locals were also available. Jakob von Toran was serving the archbishop of Salzburg in 1378; he was committed to teaching his skills to the archbishop for an annual salary of 70 pounds of Vienna and had a deputy, a horse and lodgings. He was therefore ranked with a household knight. The archbishop also paid another master gunner, Hermann, and a third named Georg who received 100 gulden, a horse and accommodation and when he left in 1391 was pledged not to serve against his former master. These were the master gunners, whose pay and rank were equivalent to those previously provided for engineers working pre-gunpowder artillery. As was to be expected, there were also men of lower rank, such as Georg of Friedburg, paid 20 pounds for his craft, who served from 1382 to 1395, Heinrich from Nürnburg, paid 40 gulden in 1389, and several others. The siege of Mühldorf by Duke Stephan of Bavaria in 1364 is the first recorded use of guns in south-east Germany but it seems that there was no looking back from then on, with the archbishop of Salzburg, Pilgrim (died 1396), paying particular attention to providing himself with a whole team of experienced gunners as well as a gun founder.[66]

The earliest references to guns in the kingdom of Poland dated from 1383 and in Russia from the 1380s, during the defence of Moscow against the (Mongol) Golden Horde, and it seems all were introduced from further west, there being as yet no indigenous industry. Chroniclers referred to guns being imported. The Teutonic Knights' regular and brutal warfare against their eastern neighbours (Lithuania) involved a wide range of military equipment and they seem to have possessed access to engineering skills, reflected in the recorded employment of a brother knight named Marquadus, a master carpenter, and Matthias, a carpenter from Königsberg, to construct wall-breaking types of battering ram to break through the outer walls of Lithuanian Kauve, and then repeat the exercise with the inner

66 This paragraph from H Klein, 'Die Salzburgischen Büchsenmeister des 14 Jahrhunderts', *Zeitschrift für Historische Waffenkunde*, new series VI, 7, 1938, 141–7.

defences, in 1362. But there was no clear reference to the use of guns until they were deployed against Vilnius in 1383, which is consistent with the other dates for this region, and again in 1390. On this occasion, one of the visiting crusaders was Henry Bolingbroke, the future King Henry IV of England, whose touring party included "diversis engynours".[67] In the Balkans, it is suggested that the earliest guns are of similar date although the first reference in this region dates from 1351.[68] There is no suggestion that the Muslim great powers of the fourteenth century made use of gunpowder artillery until several decades after its adoption by Christian states, although they cannot but have become familiar with it as a result of continuing confrontations in Iberia, North Africa and the Balkans. The means to transfer awareness of the new technology into practical results may not have been available at this time.

Thus it seems that a new breed of engineer had taken his place alongside the traditional master of siege equipment or bridge building, who had been routinely drawn from the ranks of those trained in carpentry. The generations of master gunners, many of whom may indeed have doubled up as gun founders – at least at the beginning – had different, metalworking skills. Nonetheless, for most of the fourteenth century, there was too little technological progress for the master gunner to become the dominant figure among engineers. For most of the century, these limitations were reflected in the very slow process of modifying the design of fortifications either to combat or to mount (at gun ports) the new weapons. During the last quarter of the century, however, guns became larger as manufacturing improved, allowing for more powerful charges that did not risk exploding the barrel, and powder became more efficient. Shot size was steadily increasing, and chroniclers noted the greater size and impact of the projectiles deployed at the siege of Chioggia in 1380, when the Venetians finally achieved a crushing victory over their historic rival, Genoa. Among the weapons they used were two large guns made in Treviso capable of hurling shot weighing 88.5 kilograms and 63.5 kilograms that brought down whole lengths of the wall of a monastery held by their enemies.[69] In this war as in many still to come, both stone throwers and guns were now deployed side by side. But it was evident which way the future lay and the much more rapid evolution of gunnery in the century ahead changed the face of warfare and of military engineering.

67 The evidence from contemporary chronicles is summarised in Purton, *HLMS*, 155–60. For detail of the acquisition of guns by the Teutonic Knights, see G Zabiński, 'Ways of acquisition of firearms and related equipment in the state of the Teutonic Order in Prussia', *Acta Militaria Medievalia* X, 2014, 119–42.

68 W Wilinbachow, 'Początkowy okres rozwoju broni palnej', *Kwartalnik Historii Nauki i Techniki* 8/2, 1963, 213–35, English summary, 234–5.

69 The contemporary sources for the siege of Chioggia are discussed in Purton, *HLMS*, 170–5.

7

Polymaths and Specialists in the Fifteenth Century

The fifteenth century in European history is traditionally seen as marking the end of medieval times and the dawn of a new age, "the renaissance". The "the" is important, because other "renaissances", such as that dubbed Carolingian and that ascribed to the twelfth century, are distinguished by a temporal qualifier not apparently needed by their later incarnation. The significance of the last of the renaissances, however, is that it was accompanied by the significant changes in economy and society that truly mark out new times. The scientific and artistic progress recorded in each renaissance was based on recovering the lost secrets of the ancients, and fifteenth-century humanists were no less anxious than Charlemagne that their work should be understood in this light, not as a revolution that marked the beginning of a new age. This is ironic in light of the massive transformation getting underway during these decades. By the sixteenth century, world-changing events included the first overseas expansion of Christian states, led by Portugal (which had started early in the fifteenth century) and then by newly unified Spain, leading to the extortion of vast wealth from the Americas at the cost of the virtual annihilation of the native populations. While the "universal" authority of the Church now reached new continents, in Europe it was forever overturned by the protestant reformation, which was also deployed as cover by many rulers to reinforce their efforts to establish independent states. The biggest changes simultaneously occurred in society and the economy, where feudal social structures gave way to the progressive enrichment of traders, merchants and manufacturers and the continuing development of capitalism. This new class was achieving a weight in society that would in due course assert a political voice. Gutenberg's invention of the printing press brought mass production of reading material to Europe from the middle of the fifteenth century with growing social impact. Some of the military engineering texts discussed here were among that material. Countries that did not develop this way (even if printing was already available, as in China) would be overtaken in economic, scientific and technological progress by Europe for the first time, Ming China being a classic example once the dynasty resolved to close itself off from the outside world in mid-century.

Despite developments in knowledge during the European renaissance, there was no sudden significant technological progress during the fifteenth century: for the most part, traditional forms of industrial and agricultural production continued. Those improvements that did occur were important, however, both in their own right and in their application to the military sphere, and in the related area of shipbuilding. Progress was continuous in both the quality and quantity of gunpowder weapons, and the design and construction of fortresses evolved in parallel. There were engineers capable of working on both the weapons of attack and ways to resist them, but alongside these men there was some suggestion of increasing specialisation that would in due course separate the profession of master of artillery entirely from that of builder of fortresses. No less significant (in the longer term) was the growing understanding of the need for renaissance rulers to have more effectively organised and administered armies (even if the necessary expertise was contracted in for the duration rather than sustained as a standing force) and that meant applying the same principle to military engineering. It will be seen how the French kingdom established itself as the leading power of Europe by the second half of the fifteenth century, accompanying and building on the final expulsion of the English by creating a standing army in which professional organisation of the artillery train was a vital component. Other states, such as newly unified Spain and the rich and powerful duchy of Burgundy, would do the same. These states continued to rely on other engineering skills such as bridge building and mining, but only the arrival of the carrack at sea matched the drive to professionalise the artillery in terms of enduring significance.

SHIPBUILDING

Making it possible for wooden sailing ships to navigate across open oceans was the basis for the voyages of discovery that opened up Africa and the Americas to European exploitation. The technical knowledge needed (including navigational expertise, based on astronomy and instrumentation[1]) was developed over a relatively short period and while records sometimes make it possible to identify the physical process of construction, and marine archaeology can sometimes confirm the changes from analysis of physical remains long buried, it is still uncertain how the technological improvements took place, who may have been responsible and how physically they were implemented. The same questions that arise time and again in understanding how medieval engineers and builders decided on dimensions and materials, and how these decisions were carried through in practice, apply as well

1 For a good summary of progress, F Maddison, 'A consequence of discovery: astronomical navigation in fifteenth century Portugal', in T Earle, S Parkinson (eds), *Studies in Portuguese Discoveries*, I (1990), 71–110.

to shipbuilders. It is again necessary to propose hypotheses that fit the evidence rather than the proof provided by a master's own writings or technical drawings. The handful of exceptions are intriguing but do not resolve these questions.

It has been convincingly demonstrated that the illustrations of ships in medieval manuscripts offer powerful evidence of technical progress in that they usually show not just the overall design but also the details of medieval vessels with remarkable accuracy. The critical development in this period was that of the three-masted carrack or caravel, a shape and design of such lasting impact that its main features, it has been pointed out, would have been familiar to a sailor on a "ship of the line" of the eighteenth century. Its precise evolution remains uncertain and it seems to represent the coming together of a number of shipbuilding traditions from both the Mediterranean and north and western Europe. The finished product was built around a heavy keel and close-fitting frames with a curved stem and raked sternpost, around which the hull timbers were added. Developments in shape and dimensions and additional masts accompanied improvements in key elements of the design, such as the size and shape of sails and rigging and changes to the location of the rudder(s).[2] The appearance of such ships in the English navy is documented and helps suggest what was happening but without explaining the "how".

Maintaining ships of war during periods of peace was an expense that cash-strapped monarchs could not afford, so renewals of hostility were usually the occasion of creating a navy anew either by arresting merchant ships (exercising an age-old right) or building new vessels. When England's Henry V (1413–22) decided to reassert his claim to France he resolved to create a new fleet and decided that Southampton was an appropriate base. Wealthy merchant William Soper was appointed to oversee the work of converting a captured Spanish ship for war and also of building new ships. One of these was (in 1418) the *Grace Dieu*. It was possibly the largest ever clinker-built ship and its remains, discovered and excavated only during the lowest tides, suggested an original length of 200 feet (62 metres) and beam of 48 feet (15 metres). A Florentine visitor in 1430 attested to its immense size and beauty. It is known who built it: accounts show John Hogekyn as the original master carpenter and that Robert Beerd was appointed as surveyor (at a higher rate of pay), with a team of carpenters, joiners and labourers. Beerd's role may have been broader. He was later paid to make "trepogets", guns and ladders (1421). William Bachyne was employed as shipwright when Hogekyn fell ill. Thomas Crane was another master carpenter described as a shipwright building another royal ship at the same daily rate, and so were William Kyff and William Savage, who was the only man called a "carrek carpenter" "because of his own skill in the

2 See J Flatman, *Ships and shipbuilding in medieval manuscripts*, 77–8, 91–7; I Friel, *The good ship. Ships, ship-building and technology in England 1200–1520* (1995), 101–3, 126–7, 157–80.

work". Hogekyn continued in royal service and was rewarded with a pension (4d a day), paid until (presumably) his death in 1439.[3] Beerd's range of work suggests an unusual overlap between maritime and land-based military engineering tasks. The inspiration may well have been three Genoese carracks that had been captured in battle by the duke of Bedford at the (French) siege of Harfleur in 1416 as it appears that the group of four great ships that were subsequently built were based on the same design but were almost certainly all new in having at least two masts.[4]

It is confirmation that this was new technology that not long after starting work on the new ships it was recognised that native shipwrights, apparently unfamiliar with the carrack, could not complete the task and a shipwright was brought in from Bayonne. That city also provides documentary evidence of the process of ship design and building related to Henry's commission of another "great ship" to be built there in 1419. Bayonne had maintained its long history as a commercial maritime city but it also met royal demands when required. This time the ship was to be 186 feet long (57 metres) with an inside beam of 46 feet (14 metres), making it not much smaller than the *Grace Dieu*. While it appears that local reluctance to pay up and the death of the king meant that it was never completed,[5] the other significant conclusion to be drawn from the project was the claim that its design reflected the use of particular ratios to establish the vessel's dimensions in a way familiar from that of the approach used by master masons in deciding the proportions of their buildings. The naval historian Carr Laughton argued that dimensions of the keel and beam were used to determine the length of the raked sections of the hull fore and aft as well as other dimensions and he justified this conclusion by using a study of a captured sixteenth-century galleon of which all dimensions were recorded. It is not surprising that shipbuilders must have had a method of deciding the most efficient proportions of their creations that was more reliable than their eyes, especially when the vessel was of such size and where its seaworthiness would depend on getting the design right. Susan Rose, a leading

3 Rose, *The navy of the Lancastrian kings*, 75, 127, 211, 230–1, 242.

4 S Rose, 'Southampton and the navy in the age of Henry V', *Hampshire Papers* 14, 1998, 1–14; *Calendar of State Papers, Henry V, I: 1413–1416*, 178 (Commission to William Soper); I Friel, 'Documentary evidence for ship-building in England 1294–1500', in C Villain-Gandossi et al. (eds.), *Medieval ships and the birth of technological societies*, I (1989), 139–49; S Rose, *England's medieval navy 1066–1509*, 81–4, 87. In 1436, a poet wrote in the 'Libelle of Englyshe Polycy' that Henry V had "made the great dromons ... the Trinity, the Grace Dieu, the Holy Ghoste ..." (after) Bedford destroyed "the carrikkys grete and stoute" off Harfleur: Hattendorf et al., *British Naval Documents 1204–1960*, 12–13. It was reported by Ian Friel that another wreck that may be that of another of this squadron, the *Holy Ghost*, had been discovered near Southampton (*British Archaeology* January/February 2016, p. 11).

5 A letter to Henry V, giving the key dimensions, and putting the blame for lack of progress on the city was reproduced in *English Historical Documents* IV, 222–3.

modern historian, has concluded that "while hands-on experience was essential for shipwrights, those at the highest level most probably were also literate and numerate". On the other hand, wages paid to shipbuilders were sometimes lower than those of other craftsmen, so the ordinary skills were not recognised as of high status.[6] It is not possible to generalise from the evidence of wage rates that survives beyond suggesting that supply and demand would have enhanced the value of the highly skilled masters of ship design, especially when they were needed urgently, whereas those who made their living from everyday work on the thousands of ships plying their trade could expect much less for their efforts.

The question of skill levels and transfers is also highlighted in the evidence from the shipyard originally created by the king of France to build galleys at Rouen on the river Seine, which was eventually destroyed by fire in 1419. Ships were built here using carvel construction but the comprehensive surviving records show that those doing this work were brought from the Mediterranean, because Norman ship workers were used to clinker building. Enrico Marchese, Albertino Spinola and Lanfranc Tartaro were from Genoa. Throughout the fourteenth century the yard employed shipbuilders from the south and they were paid more. Those in charge of caulking the ships to make them watertight were particularly skilled and served long apprenticeships. They were rewarded by pay of 6s 8d compared to just 2s 6d for their local co-workers.[7] For whatever reason, there was no transfer of skills over many decades. In the Mediterranean the reputation of such Italian masters was high. In 1444 the humanist scholar Cardinal Bessarion (1403–72) encouraged the Byzantines to send their young men not to Constantinople but to Italy to learn "arts and techniques, mechanics, metallurgy, arms manufacture, naval construction; send them there where timber is cut by automatic saws and mills turn very fast ... on their return, they will instruct others".[8] Although perhaps telling more of the catastrophic loss of resources of all kinds that afflicted the soon-to-be-extinct empire than of anything else, the words nonetheless confirm distinctly contrasting levels of skills and technical knowledge.

The survival of a manuscript containing the book of Michael of Rhodes offers insight into the level of knowledge possessed by shipbuilders while also leaving some questions unanswered. Michael had begun his career as an oarsman on a

6 L Carr Laughton, 'The great ship of 1419', *Mariner's Mirror* IX, 1923, 83–7; Letter of John Alectre to King Henry V in *English Historical Documents* IV, 222–3; Rose, 'Bayonne and the King's ships', in Rose, *Medieval ships and warfare*, 69–76 (originally *Mariner's Mirror* 86, 2000 (140–7)); Rose, *England's medieval navy*, 89; Friel, 'Documentary evidence', 143–4.

7 E Rieth, 'Le clos des galées de Rouen, lieu de construction naval à clin et à carvel 1293–1419', in Villani-Gandossi et al., *Medieval ships*, I, 71–7.

8 Quoted in H Antoniadis-Bibicou, *Études d'histoire maritime de Byzance* (1996), 22 (my translation).

Venetian galley in 1401, had travelled the Mediterranean and visited northern waters on both commercial and military duty on numerous occasions. He began his book in 1434. The work demonstrates an intense interest in mathematics as it was regularly taught in the schools of Italy, well beyond what was required to accomplish maritime duties. He also offered very detailed instructions on how to build five different types of ship, and his modern editors state that while this is the earliest extant treatise on shipbuilding, parts of it appear to have been copied, indicating earlier lost sources. However, among the lists of dimensions, which state relative proportions starting with the size of the keel and deducing the rest from that figure, analysis shows an absence of advanced geometrical knowledge and that the dimensions that follow, not being based on any meaningful or memorable proportions, must have been written down, a conclusion seemingly confirmed by the fact that this was the practice in the arsenal of Venice where all the republic's ships were built.[9] The many associated drawings in the manuscript, it is suggested, were evidence that the author (who had risen to the rank of adjutant to a captain) was seeking to impress people of higher rank. The implication of the analysis is that shipbuilders did not have knowledge of geometric proportions, but it may also reflect the particular way of working in the Venetian arsenal. Another surviving text of similar date, written by Giorgio Timbotta, who was a Venetian merchant, covered much of the same ground. Giving dimensions and some not-to-scale illustrations for the construction of eight different types of ship as well as galleys, the instructions do make frequent references to the deployment of proportions for the ships' fittings – the mast needed to be three and a half times the beam, for example. Analysis of surviving notebooks confirms that shipwrights did not need to know geometry, but they did know relative proportions of critical parts of the vessels they were constructing.[10]

The general establishment of the new carrack design in northern and western Europe took place over the coming decades, although the shortage of archaeological evidence makes it difficult to define it tightly. The ship discovered at Newport (Wales) has been dated to 1448+. A three-master, what was uncovered of the building technique led the excavators to conclude that "the builders were incredibly skilful" and it is suggested that it may be the same as the vessel that Richard, earl of Warwick, a renowned naval commander, ordered to be made good by John Colt,

9 Chapters by P O Long, R Franci and D M McGee in P O Long et al. (eds.), *The Book of Michael of Rhodes. A fifteenth-century maritime manuscript* (2009), vol. 3, 1–4, 116–46, 211–41.

10 R Anderson, 'Italian naval architecture about 1445', *Mariner's Mirror* XI, 1925, 135–63. The example quoted is at 153; M Bondidi, 'The art of designing and building Venetian galleys', in C Beltrano (ed.) *Boats, ships and shipyards* (2007), 222–7.

Fig. 23 The 'Newport ship' reconstructed in the painting *Final Resting Place* by David Jordan. (Copyright Newport City Council and Friends of the Newport Ship, reproduced with permission.)

Fig. 24 The 'Newport ship': clinker planking, nails, caulking, reconstructed in the Newport ship centre, Newport, Wales. (Photo: author)

one of his officers, at Newport at the cost of £10 on 22 November 1469 [Figs 23 and 24].[11]

Among the many artefacts recovered from the Newport ship were small stone shot with diameters of 2.5, 3.5 and 4 inches respectively (65, 88 and 102 millimetres) that would probably have been fired from wrought-iron guns. Armaments required for the *Grace Dieu* included three guns and it would be some time before large numbers of gunpowder weapons were used on board ships – lost in 1479, the *George* carried sixteen bombards while the *Edward* had fifteen. Royal ships, however, quickly took on much more substantial armaments by the end of the fifteenth century. Three ships recorded in 1485 carried 30, 48 and 70 guns respectively, while by 1497 Henry VII's *Sovereign* carried 141 and the *Regent* no fewer than 181.[12] What was lacking at this time was a reliable way to mount the guns to make them effective weapons for fighting at sea and it would be the following century that witnessed the development of ports in the side of warships through which heavier guns could shoot at enemy ships (as opposed to the smaller anti-personnel weapons that had to be operated from the deck and the fore and aft superstructures). The overloading and sinking of Henry VIII's great *Mary Rose* suggested that the technology of combining large ships with low-sited gun ports remained imperfect into the 1530s.

That aside, however, ships demonstrated a formidable technological evolution during the fifteenth century which signified the presence of those capable of creating new designs and able to make them real, and of masters who continued to develop the skills. Maritime technology did not of course evolve in a bubble, but overlapped with developments elsewhere. One evident requirement for shipbuilding and ship use was the availability of facilities for the construction and repair of vessels, and for loading and unloading ever-larger examples. The English ships described already were usually constructed in one-off docks dug beside the sea and protected against pilfering by fences: it would be late in the century before specific dockyards were built for the English navy, such as the dry dock at Portsmouth in 1495. The French had already established their facility of Rouen. Where there were harbours, facilities had to be created to accommodate and process ships' cargos. The fifteenth century witnessed further technological progress in the building of cranes suitable for dealing with a new generation of vessels. If the underlying principles behind the limited range of possible designs

11 R Trett (ed.), *The Newport medieval ship* (2010), 13–16; updated by the Friends of the Newport Ship newsletter 21 (2013).

12 Trett, *The Newport medieval ship*, 20; Rose, 'Southampton', 8; Friel, *Good ship*, 152, 202; F Howard, *Sailing ships of war* (1979), 38–9, provided a breakdown of the size of guns carried by ships across the fifteenth century, noting the significant increase in quantity but also the use of some much heavier weapons suggested from a mid-century Danish shipwreck with a gun barrel 10 feet (3 metres) long.

FIG. 25 The medieval crane tower at Gdansk, Poland, from a pre-war postcard kindly provided by the Narodowe Muzeum Morskie w Gdansku.

had been well established for centuries and ultimately reached back to Vitruvius, a new scale of instrument was now needed. The Baltic provided examples of many where the detail is known and that at Gdańsk is still preserved [Fig. 25]. With complex timber moving parts, and designed to rotate, the famous crane here dates from 1443 and was housed inside a stone tower. It was operated by two 6.5-metre treadwheels and with these paired teams of eight workers could lift weights of 2000 kilograms to a height of 27 metres.[13] This engine represented a culmination of long-existing technology rather than anything revolutionary, however.

ENGINEERS

What is known, then, of the engineers themselves? The men who were hired to carry out these roles represented a continuation of the tradition of skilled people of non-noble social background who needed to earn their living. There are sufficient surviving records from the fifteenth century to follow some of them in more detail than is possible for many of their forebears, and there is no question that a much larger number of names is known than previously, almost certainly confirming that there was indeed a larger group of people matching a broad definition of engineer.

Much is known about the lives of the polymaths, multiply talented individuals such as Felipe Brunelleschi (1377–1446) or Leon Battista Alberti (1404–72), the evidence of whose skills across many areas of art, architecture and engineering remains with us today. In the next generation, the many talents of Leonardo da Vinci (1452–1519) included being able to apply his inventiveness to military engineering. Leonardo's military drawings were often futuristic imaginings, such as the well-known proto-tank not capable of being realised with the technology of the time, or else were not actually particularly innovatory (such as the designs for fortifications). But other contemporaries such as Francesco Giorgio di Martini (1439–1502), whose work included civil and military architecture and hydraulic engineering as well as art and sculpture, were also employed to design eminently practical fortresses (see below). One aspect of all of their lives that appears new is the high level of literacy they seem to have shared, demonstrated not only in the depth of their reading but also in their own numerous writings.

It still remains frustratingly unclear how far this literacy extended beyond such people who were educated in the classics. To what extent could master masons and carpenters who would sometimes have been employed for military purposes, and gun masters, read – and if they could, was it just in their own language? It is still possible that everything they knew had been acquired through experience rather than by reference to any texts, that the administration of their projects was executed by hired scribes who would read out letters received and write responses and by those who could count (and account for) money received and paid. It remains impossible to decide whether a builder could reproduce geometrical ratios by reference to long-established practice and drawings in the sand rather than measurement or really did understand some geometry. The existence of textbooks

13 D Ellmers, 'Development and usage of harbour cranes', in Villain-Gandossi et al., *Medieval ships*, 43–70.

shows that some could both read and write, and the books' reproduction suggests that others could read them. The architectural design books of German master masons Roriczer and Hans Schuttermayer offer some clues. The Roriczers were a family based in Regensburg and three successive generations were master masons working on the cathedral there, the oldest from 1415–19 and the last sufficiently famous to commission a portrait by the elder Holbein (c.1490). Schuttermayer was a goldsmith from Nürnberg (c.1433–1518). Both Mathias Roriczer and Hans Schuttermayer published printed works addressed to master masons and which called for knowledge of geometry – right angles, pentagons, heptagons, finding the circumference and centres of circles, areas of triangles and so forth – and advised on the use of levels, set squares and compasses among other instruments.[14] These late-fifteenth-century booklets were the first of this kind to survive since Villard de Honnecourt's, although it is widely accepted that lodge books (such as Villard's) may have been commonplace but were retained in particular lodges and have not survived.

Among so many masters across continents it is impossible to reach a conclusion except that the gap between university education and practical knowledge remained largely unbridged, but master craftsmen shared a common level of knowledge of their subjects, acquired in different ways. This knowledge could embrace a wide range of distinct skills. While detailed knowledge of metals explained why people skilled in clockmaking could also turn their hand to the manufacture of guns, as records show, and master carpenters familiar with the qualities of different timbers and the design of wooden buildings could readily switch to erecting a trebuchet, the mechanisms by which engineers combined skills normally related to distinct areas of activity cannot now be recovered, unless they were gained by straightforward curiosity and the willingness of one master to share knowledge with another from a different craft.

DEVELOPMENTS IN METALWORK AND GUNPOWDER ARTILLERY

During the fourteenth century, gunpowder weapons had arrived in Europe and had slowly taken their place alongside traditional devices as part of the arsenal available to rulers and generals. Towards the end of the century there had been a substantial

14 L Shelby (ed./trans), *Gothic design techniques. The fifteenth-century design booklets of Mathias Roriczer and Hans Schuttermayer* (1977), esp. 3–38, 54–65; the same, 'The education of medieval master masons', *Medieval Studies* XXXII, 1970, 1–26. Harvey had previously argued from the existence of such pattern and text books for a level of literacy among master craftsmen: *Gothic England*, 21–2. The trend was also discussed by Bucher, *Architector. The lodge books and sketchbooks of medieval architects*, I, 10–13.

improvement in metalworking technology making possible the construction of ever-larger weapons, projecting ever-heavier shot for greater distances and with greater ballistic force. Alongside refinements in the making of gunpowder that delivered a more consistent result, the impact and potential of the new weapon were transformed over a few decades. This would lead to increased demand for guns, and a large growth in the number of gun foundries and gun makers that continued throughout the fifteenth century. In due course these processes would lead to changes in the organisation of this element of military engineering.

Very few fourteenth- or fifteenth-century guns have survived because they were melted down when they became redundant, but one early exception was the 1375 bombard found at Boxted (Suffolk) in England. It was made of iron staves bound with iron hoops, 2.38 metres long with a bore of 343 millimetres, with its powder chamber much narrower than the barrel of the gun. This example is evidence of the growth in size of such weapons, suggested also from chronicle accounts of growing destructive impact (as noted in the last chapter). Such weapons could project shot that would cause serious damage to defences and represented a big change from the noisy but unreliable and (for siege purposes) ineffective weapons of the type shown in the Milemete illustration. At the other end of the scale, handguns were now also appearing, the earliest examples dating from the 1390s [Fig. 26].[15] Evidently, gun makers were discovering how to create an ever-greater variety of weapons to meet a diverse set of needs that were themselves evolving. Christine de Pisan early in the fifteenth century had discussed different types of gun and what they could be best used for. Her recommendations specified in particular their value in destroying a besieger's own siege artillery, and went into detail on the quantities of powder and lead for making pellets, while the besieger was advised to make use of "great guns" able to shoot missiles weighing 100, 200 and 400 pounds alongside smaller guns shooting lead bullets.[16]

There must have been interaction between maker and user, since it is unlikely that makers would have gone to the trouble of building a new size of weapon without having a customer waiting, but evidence for such discussions is lacking. Historians have noticed the main trends in design: the technical improvements discussed below allowed for the construction of ever-larger guns, culminating in a group of monster weapons some of which survive to this day, such as *Dulle Griet* in Ghent, *Mons Meg* in Edinburgh (both of Netherlandish origin) and the sultan's great guns in Istanbul, while others are known only from drawings or

15 H Blackmore, 'The Boxted bombard', *Antiquaries Journal* 67, 1987, 86–96; R D Smith, 'Artillery in the 100 year's war', in Curry and Hughes (eds), *Armies and fortifications*, 154–6 for the general development of artillery. For discussion of the examples and analyses of the early handguns, Purton, *HLMS*, 174–5.

16 Christine de Pisan, *The book of deeds of arms and of chivalry*, ed. / trans Willard, 108, 111, 117–18.

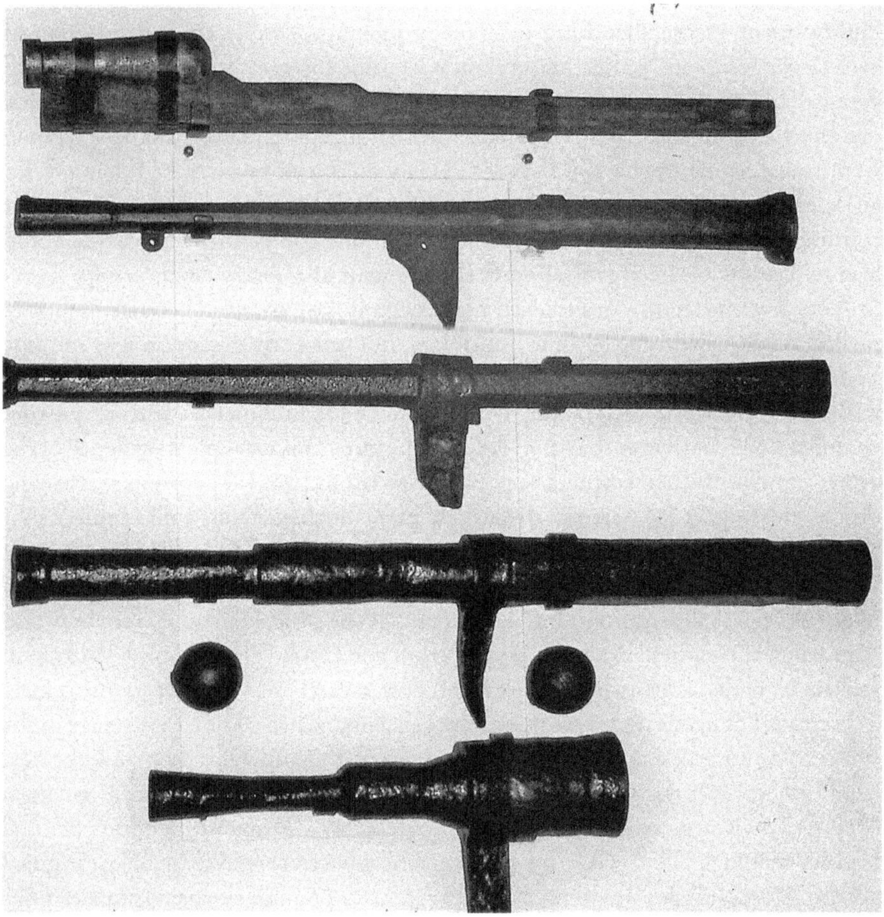

FIG. 26 Small fifteenth-century handguns displayed at Eltz Castle, Rheinland-Pfalz. (Photo: author. With the permission of Count Eltz)

records. Some were of iron, some of bronze. The earliest "monster gun" appears (on uncertain evidence) to date from the first years of the fifteenth century and the youngest was built in 1464. Shared characteristics included the use of enormous quantities of metal, the capability to shoot the largest shot to a range of well over a kilometre, and the reality that they were almost never actually employed, the best known cases being their use by the Ottoman army to blast the ancient walls of Constantinople in 1453 [Fig. 27] and three recorded uses in war of *Mons Meg* (in 1451, 1489 and 1497). *Faule Mette*, made at Brunswick (Braunschweig) in 1411, was shot twelve times in a life of 376 years, only five of these for war. The effort of moving and operating such guns and their ammunition was not worthwhile compared with the military benefits of manufacturing more mobile and efficient

Fig. 27 A great bombard used in the siege of Constantinople, 1453. Army Museum, Istanbul, Turkey. (Photo: Dick Osseman)

weapons.[17] Nonetheless, perhaps comparable to more recent vanity projects, they served to demonstrate the virility of their owners but also required their makers to achieve feats of design and construction called for by working on a previously unimagined scale.

Metalworking techniques were themselves stimulated by the demands for guns, with progress in achieving more effective manufacture of cast iron and bronze and the spread of more efficient furnaces across Europe from the fourteenth century, accompanied by the development of precision tools and machinery.[18] Significant limitation had been placed on earlier ironwork by the inability to achieve sufficient

17 There is an enormous literature devoted to these giant weapons quite disproportionate to their actual significance. For a good summary, V Schmidtchen, 'Riesengeschichte des 15. Jahrhunderts, technische Höchsleistungen ihrer Zeit', *Technikgeschichte* 44, 1977 (153–230). Evidence for the dates and technical details provided by records is collected by A Essenwein, *Quellen zur Geschichte der Feuerwaffen*.

18 Forbes, 'Metallurgy and technology in the Middle Ages', 49–57, and 'Metallurgy' in C Singer et al., *A history of technology* (1956, reprinted 1967), II, 73–7; Girardot, 'Fondeurs d'artillerie et siderurgistes', 67–76. Summarised in E Whitney, *Medieval science and technology*, 118–19. A similar process was taking place in India, long a source of major metalworking industries, during the same centuries: S Mahmoud, *Metal technology in medieval India* (1988), 46–8.

temperature in its manufacture, but this problem was now being overcome. Progressively larger and taller furnaces had been developed across the continent, with Styria a leading base for innovation. The introduction of water-driven bellows in the fourteenth century had helped make bigger lumps of "pig iron" and water-driven tilt hammers had begun to turn this into wrought iron in larger quantities. Now, the higher temperatures achieved by taller ovens meant retaining sufficient carbon in the iron to enable liquefaction and hence casting using much larger quantities of the metal. By the early fifteenth century, it was practical to cast directly from the furnace into moulds. Guns could now be cast, and were, in bronze and then iron. Knowledge of the techniques spread across Germany then to northern Italy and into north-east France and the Low Countries. Imported German iron makers will have spread the technique. It is indicative of the arrival of the new technology that those who made the guns became known as "gun founders" in the records. Figures for the number of smiths and armourers in the city of Bruges are indicative of the significant increase in the trade, where metalworkers made up five percent of the population by the fourteenth century. The number of armourers rose from 25 to 73 by the following century, of smiths from 21 to 75 and of metal founders from 20 to 66. At Rouen, a gun foundry was established in the market space at the start of the fifteenth century and a new one was ordered by the English king Henry V following his capture of the city in 1419, using metals that had to be imported from England. In Orléans, soon to become a centre of French resistance to the conquering English, a gun foundry was set up around 1411 and the son of a clockmaker from Tours, Naudin Bouchard, was employed to adapt the town's fortifications for artillery. He gained considerable wealth from his continuing efforts to strengthen the defences. Nearly 40 per cent of Orléans' revenue was spent on adaptation of the defences between 1415 and 1419.[19]

In independent Brittany, regularly involved in the Hundred Years War, records show towns and cities investing in the construction of guns and defences, totalling hundreds of weapons of various sizes commissioned from the mid fourteenth century through to the late fifteenth. The records confirm the technical changes taking place: the switch from forging to casting iron, movement from iron to bronze, both being more complex tasks and often calling for the importation of skilled workers to compensate for the absence of native skills. To use the finished articles, a number of Breton cities had *maîtres d'artillerie*, while the rulers appointed

19 J-P Sosson, 'Metallurgies urbaines en Flandre et en Brabant', in P Benoit, C Cailleaux (eds), *Hommes et travail du metal dans les villes médiévales* (1988), 163–72; M Monteillard, 'Artisans et artisanat du metal à Rouen à la fin du moyen âge', in same vol., 109–26; F Michard-Frejaville, 'Fèvres et forgerons d'Orléans 1399–1430', in same vol., 69–82. For the siege of Rouen in 1418–19, where old and new artillery was widely used and where the English threw a bridge across the river Seine, see the sources cited in Purton, *HLMS*, 253–4.

cannoniers du duc in different cities but kept them on the ducal payroll. Some of these people were of high rank. Pietres d'Alemagne (presumably of German origin) was the commander of 80 culverins at Nantes and Bastien (from either Crécy or Quercy) was in charge of 12 of these guns. No fewer than 217 men were named as cannoniers, makers or gunners, with the earliest being from 1418 and the majority from the 1440s onwards.[20] This is firm evidence of the big growth of works devoted to making guns, a growing number of gun founders employed in the trade, along with the fact that sometimes the skill had to be bought in from abroad before sufficient local craftsmen were available. Alongside this process, and sometimes as a part of it because the men were the same, those then given command of the use of the completed arsenal represented a new group of military engineers. While some of those who were appointed to such a rank were themselves gun makers, others were not. J Larchie, 'bombardier', had been a nail maker (1411), while Didier Morel (also "bombardier") had been a carpenter (1471). Another group of masters was appointed when the artillery of the duke of Lorraine was reorganised after 1473.[21]

There were other consequences of the industry observable from the records of cities in the Low Countries. These cities saw the creation of guilds of gunners, culveriners and hackbuteers defined by the weapons they used for the defence of the commune, but who were not themselves involved in the manufacture of the weapons, which by the fifteenth century involved large numbers of other workers. These records also indicate how much time and how many workers were required to manufacture the whole range of weapons from the smallest handgun (at Mons in 1489, one worker took one week) to the largest giant bombard named *Bourgogne* (two cannoniers and six valets took seven months and eight days at St Quentin in 1436–37).[22]

Evidence of similar developments across Europe could be multiplied. Reference has already been made to the lists compiled from records of the names of gun makers and the growing number of centres where the manufacture was carried out. What is startling about the fifteenth century is the sheer number of those taking up the work in response to an apparently relentless demand. From the small number of men identified as being active in making guns during the fourteenth century, Kennard's directory of cannon founders listed well over a hundred during the fifteenth. A high proportion were north Italian, south German or Tirolean, or Netherlandish. A Genoese (Paulo Bosio) was hired to make guns for Ivan III

20 J-P Leguay, 'Un aspect du travail dans les villes armoricaines au moyen âge: la fabrication des canons et armes blanches', in Benoit, Cailleux (eds), *Hommes et travail du métal dans les villes médiévales*, 185–226.

21 Girardot, 'Fondeurs d'artillerie et sideurgistes', 68.

22 C Gaier, 'L'industrie et le commerce des armes dans les anciennes principautés belges du XIIIe à la fin du XVe siècle', *Bibliotheque de la Faculté de Philosophie et Lettres de l'université de Liège* CCII, 1973, 136–7, 241 (guilds), 287–9.

(1462–1505), grand prince of Moscow, and Alberto Fioraventi from Bologna set up a foundry there in 1480. Gun maker Michelazzi Michelozzo from Florence (1396–1472), who was actually French, was also an architect and sculptor. Hans Reichart, identified as a master gunner, made weapons at both ends of the scale (1474–78): two large guns and 130 hackbuts for Salzburg. Others were recorded as specialist makers of handguns, but Hildebrand Tuphingiesser (his name itself indicative of his being a caster of iron (*giesser*)) provided four (big) *steinbüchsen* for Mühlhausen in 1430 when large guns were still most favoured. At least two (probably German from their names) were to be found delivering their expertise to the Ottoman Turks. The variety of work carried out was confirmed by Antoine d'Arnelle, who also cast a bell for Metz cathedral in 1438. The 2000 pounds of iron left over was not wasted but was used to make a bombard. As already demonstrated, guns were also increasingly needed for use at sea now that ships had developed to the point at which the new weapon could be mounted on them: a Philip Loker was paid to manufacture eight serpentines for ship use in 1485.[23] The trend continued to grow in the sixteenth century and countries where gunpowder weapons were introduced later than elsewhere – such as Scotland, where the records show expenditure on the manufacture of guns and the equipment and ammunition for *Mons Meg* in the 1450s – imported foreign craftsmen to make up for the absence of local skills in a number of areas, including gun making.[24]

Many of the makers were family businesses, creating dynasties of iron workers who included gun making among their skills or who rapidly moved into this promising new area of work. The Ferrarese family of the Alberghetti first served that city in 1484 as casters, but also did work for Florence and Modena. One son served Florence as a bombardier, the other was appointed gun founder for Venice in 1487. Here, the state had been quick to make use of gunpowder artillery but there is no official record of investment until 1440. Master Ferlino had been recruited in 1452 following his capture in war. Then in 1463 Bartolomeo da Cremona served as Sigismondo Alberghetti's predecessor as chief gunner and gun founder.[25] Peter Layminger was originally a clockmaker but made guns for Archduke Sigismund and Emperor Frederick III (1440–93) from 1486 and was elevated to the nobility three years later. His sons continued the business. The records of the emperor Maximilian (1493–1519) show the continuous construction of large guns as well

23 Kennard, *Gunfounding and gunfounders*, 33, 74, 105, 115, 131, 146, 147.

24 *The Exchequer Rolls of Scotland, VI (1455–60)*, ed. G Burnett (1880), 119, 161, 201–2, 209, 294, 436, 455; M Pearce, 'Craftsmen and accounts', in Historic Scotland, *Stirling Palace, Conference paper 2012*, 7, referring to "George the Almain the gunner" who served Henry VIII of England in 1513, had worked on the fortifications of Boulogne for Louis XII in 1510 and probably served James IV of Scotland in 1508 to cast guns at Stirling.

25 M Mallett, J Hale, *The military organisation of a renaissance state: Venice c.1400 to 1607* (1984), 81–4.

as serpentines and newer weapons such as mortars. One of the most renowned makers was Martin Merz (1440–1501) of Amberg in the Bavarian Palatinate whose lavish tomb records him as an honoured "buchssenmeister in der kunst mathematica buchssenschissens vor anders berümbt" (gun master famous above all others in mathematical arts and gun shooting). He also wrote a handbook (see below).[26]

The Tirol became renowned as a base of the industry after the area came under the rule of the Habsburg dynasty in 1386: conflict with the neighbouring Swiss Confederation and then with the rival Luxembourg family led to an increase in warfare and the appearance of the first *Büchsenmeister* in 1400, Heinrich Ratolf and Hans Spielhammer. The province's silver mines financed gun making and its copper mines were used to help manufacture bronze. Improvements in bronze casting techniques around this time enabled the manufacture of much bigger guns. The duke (Frederick I) employed salaried gun masters and created an artillery park. The industry continued to expand, producing many of the gun founders mentioned above, and also exporting copper to neighbouring centres such as Nürnburg, which in turn became centres of large-scale manufacture where new dynasties set up, and Augsburg. Surviving inventories record the large store of guns in the duke's arsenal at Innsbruck.[27] In the newly independent Swiss cantons, the weaponry was quickly taken up but had to be imported from these same south German cities until foreign masters could be brought in to establish a native industry during the following century.[28] Nürnburger gun founders Hans Widerstein and Claus Schoner were induced to move to Strasbourg to enable that city to build its own weapons as an earlier effort by Rhenish masters Hans and Ulrich seems not to have met the need – the city was one of those allied against Charles the Bold of Burgundy and was in need of modern artillery, although it seems to have relied on continuing to buy guns or, indeed, to make use of captured weapons until the end of the century.[29]

Ability to manufacture larger guns was lacking in Bohemia but this did not prevent the emergence of an independent state during the third decade of the fifteenth century following the execution of religious reformer Jan Hus at the Council of Constance (the main role of which was to have ended the "great schism" in the Church (1415)). The subsequent uprising of his followers succeeded, remarkably, in repelling repeated invasions from German "crusaders" as well as defeating native Czech opponents, loyal to their faith and to the ruling house

26 All from W Boeheim, *Meister der Waffenschmiedekunst von XIV bis ins XVIII Jahrhundert* (1897), 6–7, 113–15, 135–6; and E Egg, 'Der Tiroler Geschützguss 1400–1600', *Tiroler Wirtschaftstudien* 9, 1961, 28.

27 Egg, 'Der Tiroler Geschützguss 1400–1600', 11–29, 45.

28 H Schneider, 'Schweizerische Büchsenmacherei', *Armi Antiche*, 1963, 91–106.

29 P Martin, 'L'artillerie et la fonderie de canons à Strasbourg du XIV au XVIII siècle', *Armi Antiche*, 1967, 71–102.

of Sigismund (king of Bohemia, Holy Roman emperor 1410–37, and king of Hungary) with an army composed largely of peasants; much of the Bohemian nobility remained Catholic and royalist. The Hussites succeeded in part because of the inspired leadership and generalship of Jan Žižka and Prokop. They were able to rely on the local manufacture of thousands of handguns using readily available iron and charcoal, and for larger guns, used in their frequent successful siege operations (albeit against unmodernised castles) they soon had a plentiful supply left on the battlefields by their defeated enemies. The iron handguns were small enough to be constructed manually using a forge, and surviving examples include weapons made of plates bent around a mandrel and (later) forged rings shrunk over a tube. A metre long, and weighing not much more than 10 kilograms, they proved very effective. Apart from unusual reliance on this weapon, the only other technical advantage gained by the Hussites over their more traditionally armed foes was the introduction of defensive "wagon fortresses", in which ordinary farm carts were adapted by providing them with timber shields pierced for guns and crossbows with entry through a side door. Time and again, the peasant armies used a large number of these simple devices drawn up together to disrupt and defeat the invaders, who for a long time failed to come up with a tactical solution. Where the idea came from remains the subject of speculation. However, whereas once the builders of the Hussite stronghold of Tábor were suggested as the first designers of projecting gun bastions, more recent scholarship has dismissed this dating.[30]

Gunpowder weapons technology was also taken up early in Asia and across the Muslim world. It seems probable that the origins lay in Europe, however. The Ottomans had begun their expansion into the Balkans and this is thought likely to have been where they first encountered the means of constructing guns, which were known in Serbia and Bosnia from the late fourteenth century and Dubrovnik was a centre of manufacture. The first Turkish use is disputed and most of the accounts were not contemporary, but it seems certain that the Ottoman armies used guns at their (failed) sieges of Constantinople in 1394 and 1402 as well as the capture of Thessaloniki in 1430. Handguns (called *tüfenks*) appeared in registers from the 1420s. The earliest identification of the engineer responsible was of Ömer, a cannonier (*topçu*) under Bayezid I (1389–1402). It appears that one of the earliest corps of artillery may have been established under Sultan Murad II (1421–51) and

30 D Braid, 'Ordnance and freedom of thought: the development of gun-making in Bohemia 1350–1450', *Journal of the Ordnance Society* 5, 1993, 75–94. Hussite tactics and the "wagenburg" and illustrations of models in Czech museums are well shown in S Turnbull, A McBride, *The Hussite Wars 1419–36* (2004). There are many discussions of the fortifications of Tábor: see Purton, *HLMS*, 235–9, 404, but the claims by Czech castellologists Menclova and Durdik have been rejected more recently by J Varhanik in a series of articles re-examining fortifications of this period. Tábor's bastions have been redated to c.1500, which certainly seems more likely (pers. comm).

the deep interest of his successor, Mehmet II (1451–81), is well known. Mehmet ensured a supply of expertise by either importing European artillerists or using captured ones (as the example of Jörg will show), but this was no different from the practice already identified across the continent, as well as being consistent with the Ottoman practice of employing smiths, carpenters and masons from among the peoples they conquered. It is also suggested that there was native talent available. Copper for bronze guns came from the mines at Kastamonou (securely conquered in 1417), while Mehmet ordered the establishment of both gun foundries and gunpowder factories in newly conquered Istanbul, where it has been speculated that he may have taken over existing Genoese works. Prior to the conquest of Constantinople his guns had been constructed at Edirne in Thrace.

There is still debate on the take up of guns by the Mamluk rulers of Egypt and Syria, not helped by the use of the same words to describe gunpowder weapons and the much older incendiaries (*naft*). Additionally, the sultans had no access to copper for making bronze and could only obtain iron and timber from Anatolia. Nonetheless, analysis of the evidence suggests the adoption of guns from the 1360s and widespread use of handguns by the early fifteenth century. An account of the sieges of rebel cities in 1417–19 talked of guns being set up to bombard the citadel but victory being achieved by the miners, and also referred to the continued use of stone throwers.[31] What is not known is who was constructing or operating the artillery, but it might be assumed that initially at least the expertise had to be imported.

We noted above an absence of evidence of gunpowder weapons in Timur's army. In central Asia, one of Timur's successors, Shahrukh, was manufacturing guns in the 1430s where a gun cast by a founder named Farruk shot a missile weighing 320 kilograms: clearly the craze for giant guns was not geographically limited. It seems too that native expertise was available to rulers even if sometimes they had to import it either to pass on the skills or to make up the required numbers.[32]

31 R Irwin, 'Gunpowder and firearms in the Mamluk sultanate reconsidered', in M Winter, A Levanoni (eds), *The Mamluks in Egyptian and Syrian politics and society* (2004), 117–39; Abu'l Mahasin ibn Taghri Birdi, *History of Egypt 1382–1469AD*, ed./trans. W Popper (1957–60), 47–9, 69. Irwin re-evaluated the evidence cited by D Ayalon, *Gunpowder and firearms in the Mamluk kingdom* (1956).

32 G Agoston, *Guns for the sultan. Military power and the weapons industry in the Ottoman empire*, 15 (Shahrukh, with the reference to a source in Russian), 17–29, 46, 129, 167–80; the same, 'Ottoman artillery and European military technology in the 15th and 17th centuries', *Acta Orientalia Academiae Scientarum Hungaricae* XLVII, 1994, 16–46. See also C Heywood, 'Notes on the production of fifteenth-century Ottoman cannon', *Proceedings of the International Symposium on Islam and Science*, reprinted in the same, *Writing Ottoman history* (2002), 3–13.

The demand for gunpowder weapons increased the demand for gunpowder and therefore also saltpetre. The first European reference to how to obtain it (awareness had been long established in China) was written down in Konrad Kyeser's *Bellifortis* and English documents preserve evidence of ever-larger amounts being procured. The first reference to its manufacture in England, however, is from 1450 when a clerk was ordered to provide a site for the "manufacture of saltpetre for the king's ordnance" which required equipment for boiling "nitrous earth" and crystallising it as saltpetre. It seems that England had to obtain most of its supplies from abroad well into the following century, however.[33] Illustrations of the equipment required for the crystallisation process appeared in a number of manuscripts contemporary with Kyeser's and followed a period of around twenty years when the price fell dramatically, initially in German cities, another indication of the technical progress that helped further proliferation of gunpowder weaponry.[34]

ADVICE AND INFORMATION FOR THE NEW AGE OF GUNS

If there was a relationship between the greatly expanded industry dedicated to the manufacture of the new weapons and the availability of suitable metals, it is also interesting to identify how awareness of the necessary skills was spread. There is no evidence to confirm whether those who set up and ran foundries in the many cities cited were themselves literate and numerate, although some no doubt were. Traditional routes of knowledge transmission undoubtedly continued. Observing the highly skilled craftsmen engaged in building furnaces for metalworking in modern India, one expert has written that "it has often surprised me to see men who are unquestionably ignorant of the principle construct them with precision" using centuries-old methods of measurement based on the thickness of the finger and maintaining constant ratios as required.[35]

However, for those who could read, written material became available. More than one of those whose trade it was decided to put their knowledge into writing, the best known being the *Feuerwerkbuch* (literally, *Firework book*) dated to around

33 All from A R Williams, 'The production of saltpetre in the Middle Ages', *Ambix* 22, 1975, 125–33.

34 On the evidence for this, Partington, *A history of Greek Fire and gunpowder* (1999), 314–15. Modern analysis of medieval recipes for gunpowder, and in particular how saltpetre was extracted and prepared, are being carried out by the Medieval Gunpowder Research Group whose reports are at http://www.leeds.ac.uk/ims/fieldsofconflict/report. Purchases of large amounts of saltpetre by the Teutonic Knights were analysed by Zabiński, 'Ways of acquisition', 123.

35 Mahmoud, *Metal technology*, 28–9.

1420. The author remains anonymous and the original attribution to Abraham of Memmingen, who was *Büchsenmeister* to Duke Frederick of the Tirol, has been challenged. Certainly, there were at least twenty manuscripts copied before 1450 and fifty-eight altogether survive, a considerable diffusion. Describing it as a guide both to the manufacture of all the parts of the gun and its use to defeat an enemy town, fortress or castle, the author laid out the correct ingredients of gunpowder through a traditional question-and-answer teaching method. He stated that 1 pound of powder will propel a 9-pound shot. The advice also covered different types of powder. Shooting stones, arrows and multiple missiles were also explained. This book, then, was written for practitioners not princes, implying the existence of a literate audience of masters who would be expected to carry out the many functions laid out.[36]

Other similar texts have been found. The earliest (found in Vienna) has been dated to 1411, a purely technical description of the use and manufacture of gunpowder and guns. Next in date was the first accompanied by illustrations (two manuscripts in Munich), dated 1415–18. Roughly contemporary with these and the famous *Feuerwerkbuch* was another text (from Vienna) describing the techniques of manufacture, including of fire arrows, and set out in the same question-and-answer style. Johannes Formschneider of Nürnburg was a gun master for thirty years and seems to have made a collection of these manuscripts. Johannes Hartlieb's *Buch der Kriegskunst* of 1436 contained illustrations showing handguns operated by a crew of two using a large wooden screen to protect themselves, which suggests familiarity with actual practice. Martin Merz's work consisted of practical information on measurement, how to make quadrants and how to use guns against strong houses, towers and walls, echoing the approach of other writers. Merz offered early evidence of the use of the matchlock for firing the gun.[37] He was succeeded in his post of *Büchsenmeister* in the Palatinate by Philipp Mönch, who also left a collection of such works, the earliest dated 1479, while his practical experience was suggested by his inclusion of illustrations of Elector Frederick's sieges of Schauenburg and Dossenheim in 1460. Among the writings preserved was another called the "anonymous of the Hussite wars", now dated to around 1430 though preserved in a later copy, which differed in that it copied earlier drawings of a wide range of siege engines (including trebuchets and siege towers) but was brought up to date with illustrations of cannons and fascinating if probably exaggerated portrayals of the famous Hussite war wagons, showing them as veritable mobile fortresses bristling with guns.[38]

36 W Hassenstein (ed.), *Das Feuerwerkbuch von 1420* (1941). Translated into English by G Kramer, 'The Firework book: gunpowder in Medieval Germany', *Journal of the Arms and Armour Society* XVII, 2001.

37 A Hoff, *Feuerwaffen* (1969), I, 12–13.

38 R Leng, 'Ars belli. Deutsche taktische und Kriegstechnisische Bilderhandschriften und Traktate im 15. und 16. Jahrhundert', *Imagines Medii Aevi* 12 (1,2), 2002; the same

A successor in the same genre of writing as the "firework" books of the 1420s and 1430s was the copiously illustrated volume written by Johannes Bengedans, who had served the kings of Denmark and was seeking employment with the master of the Teutonic Knights – his letters were found in their archives. The work was written around 1450 and its contents reveal a sophisticated knowledge of contemporary science and technology. It offered detailed advice on making gunpowder but also on how to aim and shoot guns and how to use them to destroy walls and towers. His advice on sighting confirmed that use of quadrant and scales was current, and the purpose was to keep hitting the same spot, low down on the target. Also of great interest in that it reveals an evidently common skill of the time was his preoccupation with incendiaries: he gave advice on no fewer than thirteen types of fire arrows or bombs to be propelled either from crossbows or from guns.[39] That fire arrows were in regular use was confirmed by a letter from the commander of the Teutonic Knight garrison of the town of Lobau as it resisted a siege by the king of Poland in 1422, describing that the Poles shot *fuer pfeylen* into the town, but noting they did little damage.[40] Bengedans evidently expected, like the master military engineers of earlier ages, that he could sell his wide range of skills to a prince in need of someone to bring him the latest military engineering skills.

Of approximately the same date was the treatise *De machinis* by Mariano Jacobus (known by his family nickname Taccola) of Siena (1382–?1458), originally written around 1430 as *De ingeneis*. It had a wide circulation. Military engineering featured among many other devices in this text, presented as advice to commanders and describing how to use various types of gun. Taccola's work included advice on how to destroy bombards by allowing humidity to occur in the barrel, or by inserting a plug made of hard wood, or by mixing camphor with the gunpowder.[41] His work also illustrated by now traditional versions of bridges of boats, scaling ladders and large winch-wound crossbows. It has been pointed out how the style of illustration changed between the two versions of his book to one where the drawing was presented such that an engineer could derive what he needed to know to reproduce the machine. Taccola came from a family of wine dealers, became an artist working

(ed.), 'Anleitung Schiesspulver zu bereiten, Büchsen zu laden und zu beschiessen. Eine Kriegstechnische Bilderhandschrift im cgm 600 der Bayerischen Staatsbibliothek München', *Imagines Medii Aevi* 5, 2000. B S Hall, *The technological illustrations of the so-called 'Anonymous' of the Hussite wars, Codex Latinus Monacensis 197* (1979).

39 J Bengedans, *Krigskunst og Kanoner. Kriegskunst und Kanonen*, ed. J Blosen, R A Olsen et al. (2006). Fire arrows were reconstructed and tested by the Medieval Gunpowder Research Group, loc. cit, report 9.

40 Quoted in M Burleigh, *Prussian society and the German Order. An aristocratic corporation in crisis c.1410–61* (1984), 84–5.

41 Mariano Taccola, *De machinis. The engineering treatise of 1449*, ed. G Scaglia, 2 vols (1971), here 69–70.

on the cathedral (as a sculptor in 1408) and then obtained public office as a notary (1424). His first claim to fame as an engineer was his proposal on harbour works such as cranes (1427) related to his city's plans to develop its maritime capacity. His notebook confirmed that he had met and conversed with Brunelleschi and that he was admitted as a "familiar" of King Sigismund. Analysis of his drawings and text identified that his knowledge was rooted in practicality and efforts to improve on existing technology (especially in the area of hydraulics) but that he was sometimes wrong. He appears to have been aware of ancient knowledge available in Siena at his time and his military engineering, according to modern editors, was based on "sober realism" and actual testing of his designs.[42] The element of his book related to mining is discussed below.

It is not known just how many engineers also wrote down their skills although later references confirm that those who did were consulted as authorities. The book written by Konrad Grüter of Werden, discovered in the Vatican in 1998, was primarily a detailed technical treatise in Latin on hydraulics – techniques of raising and moving water, mills of various kinds and related devices – but it contained a section on military equipment including, relevant to his main subject, designs for portable bridges, as well as information about both gunpowder weapons and older artillery. Unlike some of the material on water engineering, that on military topics was not original. Grüter was the son of a brewer who served the abbey of Werden in Germany; he had become an expert in water engineering in northern Italy and possibly hoped to attract royal patronage from King Erik VII of Denmark, who was on pilgrimage in 1424, to when the book has been dated. The *De machinis et rebus mechanicis* was also interesting in making frequent acknowledgement of whence Grüter's designs came. Classical writers featured, but also the work of recent or contemporary engineers, including Taccola's *De ingeniis*, Bartolinus of Novara, famous as the chief architect responsible for castles in Ferrara and Mantua alongside churches, canals and mills during the late fourteenth century, and Johannes de Livello who built walls and bridges at Verona. The most frequently recognised, however, was Domenico of Florence, whose wide-ranging skills and works for, among others, the dukes of Milan were discussed in the last chapter. Reflecting that there was a pool of engineering talent available, Grüter also named eight or nine other contemporary engineers specialising in water works for north Italian cities but also responsible for buildings, not least of them Johannes Fontana (c.1395–1455) of Venice, who also produced books which alongside civil engineering proposed designs for torpedoes and rockets.[43] The specific statements that this or

42 These are the interpretations of F Prager and D Scaglia, *Mariano Taccola and his book De Ingeniis* (1972), 153–9. On the drawing styles, McGee, 'The origins of early modern machine design', 73–4.

43 D Lohrmann, 'Das Maschinenbuch des Konrad Grüter für Erich VII, König von Dänemark (1424)', *Deutsche Archiv für Erforschung des Mittelalters* 63, 2007, 71–92.

that design was based on another engineer's work in a particular city confirm that these were examples which had been put into practice. They also demonstrate that experts were not working in isolation but were familiar with each other's works (not surprising given their employment in the same cities, but also suggesting active exchange of information).

One conclusion from all of these texts and the analysis by their editors is that unlike many of the works of previous ages, most of these works were entirely practical. They were not written for the education of princes or other nobles who would be expected to lead military campaigns, but represented a sharing of knowledge among the growing fraternity of men associated with the manufacture, and the use, of guns. The relatively large number of surviving manuscripts and the indication that some master gunners had assembled collections of them are testimony to both the literacy and the technical knowledge possessed by these masters. The illustrations sometimes provided were not presented as scale drawings to guide a novice, but contained sufficient information that would have been well enough understood by other practitioners. They demonstrated as well, if looked at in chronological order and if the occasional anachronism is taken into account, a record of progressive improvements in the design of guns across the fifteenth century, with visual evidence of such developments as the use of trunnions to assist the fitting of the gun into a timber frame and the adoption by mid-century of wheeled gun carriages, a step that helped greatly in making the increasingly efficient artillery suitable for use in the field as well as in its long-established role as a siege weapon. While it cannot now be known if any one among the named and anonymous writers was responsible for first identifying these technical improvements, it is at least a reasonable guess that they were.[44]

In contrast to these practical guides stood the genre of texts that followed the earlier approach of re-presenting the classics. An outstanding example was the *De re militari* of Roberto Valturio (1405–75), commissioned by the condottiere Sigismondo Pandolfo Malatesta of Rimini and completed by 1455. The author was a well-educated man of letters and 'doctor artium' who had served popes. For modern students, most interest attaches to the more than fifty illustrations. The artist was possibly Giovanni Giocondo Veronese who also illustrated Francisco di Giorgio Martini's treatise of 1492. Unlike the drawings in the texts discussed above, these were copies of illustrations in previous manuscripts and some were evidently based on ancient survivals such as Trajan's column in Rome. The drawings of

For the full texts, D Lohrmann, H Kranz, U Alertz (eds), 'Konrad Gruter von Werden De machinis et rebus mechanicis', *Studi e Testi* 428–9, 2006. Analysis of the acknowledgements of other engineers is in 428 at 99ff. The pontoon bridges are in 429 at 188–202.

44 On this see Leng, 'Social character, pictorial style and the grammar of technical illustration', 89–100.

trebuchets, for example, were excellent reproductions of weapons dating from more than a century before, but certainly not in regular use in 1450. Some were copied from Guido da Vigevano or Konrad Kyeser. Nonetheless, Valturio had naturally identified the appearance of gunpowder weapons and the artist illustrated several, including drawings of mechanisms for elevating gun barrels. Again, however, they were no longer contemporary. Valturio's work was copied repeatedly and may have been the first military treatise to be printed, but it was not of practical value, nor was it intended to be.[45] One of those whose own military drawings owed much to Valturio (as well as to Kyeser and Taccola) was Leonardo da Vinci, whose work in this area was far from original.

A NEW BREED OF ENGINEERS

The rapid spread of gunpowder artillery generated both a new industry to supply the weapons and associated supplies like gunpowder, and the people able to meet the insistent calls of princes, kings, bishops and city corporations for their own arsenal of weapons. As the quality of guns improved, new and better weapons were also continually called for. But most of the men so far mentioned were the manufacturers of the gun rather than its operators, and the exceptions who performed both roles have been few. Having paid someone to create a battery of guns, the commander then needed people who could be relied on to make effective use of them in the field, or more often during a siege. Who were these men? The fifteenth century saw an interesting range of people in this role about whom usually little is known, with occasional exceptions.

Sometimes this information can be gathered (as in the past) only from royal accounts of what they cost. Such was the apparently Flemish master Nicholas who was paid more than £580 for his service with guns, engines and other warlike equipment in Scotland in 1430. Gun making was still a skill that had to be imported from abroad. It is not clear whether he operated the guns as well as constructing them but this is likely if native skills were lacking. At least he was named, unlike the anonymous Germans paid for serving as gunners during the siege of Roxburgh (1436), no doubt additionally not pleased that they did not receive payment until two years later.[46]

One sign of the reality that greater value was now usually attached to these men was the citing of their names in descriptions of events: in ages past they would probably have been left anonymous. The chronicler Georges Chastellain,

45 S Ricotta et al. (eds), *Le macchine di Valturio nei documenti dell'archivio storico AMMA* (1988), essays by F Gambaruto, P Bassignana and V Marchis.

46 *Exchequer Rolls of Scotland*, ed. Burnett, IV, 677; V, 32; Fordun, *Scotichronicon*, book 16, chapter XVI, 1287.

describing the Burgundian siege of Compiègne in 1430 (where Joan of Arc was captured), highlighted the part played by a Franciscan from Valenciennes named Noirouflé whose culverins were responsible for killing "hundreds" of besiegers.[47] Another example was the master gunner of the duke of Burgundy, Philippe de Vaudrey, who was recognised in the memoirs of Olivier de la Marche, a knight who clung to ancient traditions of chivalry, who served in Burgundy's many wars, and who noted the key role of this master in setting up the batteries of bombards and serpentines that brought the surrender of the town of Villy in Luxembourg in 1443. The account continued to talk about the significant role of guns (including mortars) in other sieges in the Netherlands and also confirmed the construction of bridges of boats to cross rivers during the bloody suppression of the revolt of Ghent, all without identifying the engineers who must have been responsible, but a later holder of the office of master of the duke's artillery, Pierre de Hacquemberg, made an appearance with his guns to help suppress a rebellion of the Liègeois in 1464.[48] Accounts of the same events by the chronicler Wavrin simply report the names of the noble commanders of the army, leaving the masters in their more accustomed anonymity.

These men are known only from references by others, but another master left his own account of his deeds. Jörg, of Nürnburg, wrote his account probably during the 1480s. It is the only evidence for his story. Serving as the papal "Büchsenmayster", he was sent to Bosnia to cast guns for Duke Stephen the Great of Moldavia in 1456. He was captured by the Ottoman Turks in 1459 and (following traditional practice) was employed by them. He described his role in frequent conflicts between his enforced employers and the Venetians as well as the usually unsuccessful attempts to defeat the Albanian rebellion led by Skanderbeg during the 1460s and '70s. Finally, he was sent to Alexandria, whence he escaped in 1480 and returned to serve Pope Sixtus IV in his original function. This remarkable story confirms much about the skills (including literacy), employability and the artillery in use: Jörg described culverins, smaller weapons and mortars (*mörser*), indicating the range of his own talents.[49]

Another man who served in Italy, Francesco d'Angelo, called "Il Cecca", was known from the substantial account of his life given by Vasari in his *Lives of the artists*. Il Cecca began as a carpenter, but Vasari described him as concentrating on the construction of engines, scaling ladders and battering rams. The city of Florence appointed him as inspector of its towns and defences at times of war but his expertise involved both attack and defence. In the Florentine siege of Piancaldoli, he enabled capture of the town "without striking a blow" by mining

47 G Chastellain, *Chronique*, ed. Kervyn de Lettenhove (1863–66), book II, 53.

48 *Les mémoires de Messire Olivier de la Marche*, ed. M Petitot (1825), IX, 369–71, X, 257.

49 Jörg von Nürnburg, *Nachrichten von den Turken*, an unpaginated printed edition of 1500; A Vasiliev, 'Jörg of Nuremberg', *Byzantion* 10, 1935, 205–8.

under the walls to allow soldiers to enter unobserved. He met an early death as a result of exercising his obviously substantial talent when he was killed by a bolt in 1488, at the age of forty-two: he put his head above the rampart to use a plumb line to help measure distances and was shot by a priest serving the other side "who feared the genius of Il Cecca more than the might of the whole camp".[50]

Also working across subject boundaries was a man chiefly remembered as an architect, Francisco di Giorgio Martini, born in Siena like Taccola in 1439 (he died in 1501). Like Grüter, his employment included working on his city's water supply (in 1469). He was involved as an engineer in military operations in southern Italy for a number of years around 1480 but also wrote an important treatise on military architecture, which presented models of new fortresses specifically designed to counter the threat posed by the growing ability of gunpowder artillery to overcome more traditional defences. His book (and illustrations) argued for the creation of ravelins in front of double lines of walls, for wide and deep ditches, including with caponiers to enable defenders to cover the ditch, and he included strongly projecting bastions. The design had to be consciously devoted to resisting attack by guns; and he presented what he proposed in considerable detail. The work discussed and illustrated older forms of siege equipment but was clear about what mattered most in modern times. It was not the blueprint that foreshadowed the bastion traces of the following centuries – for example, he retained traditional elements of medieval castle design such as a donjon without defensive function – but it was a contribution to an evolving process of new thinking. Once again, it is evident that much of Leonardo da Vinci's work on fortress design (he spent part of his early career inspecting and overseeing the repair of his various masters' fortifications) derived from the work of Martini.

USE OF GUNPOWDER IN SIEGE MINES

Experts argue about whether Martini's *Trattati* were also the first to talk about and illustrate the practical use of gunpowder for mining, or whether that dubious honour belonged to the works of Taccola,[51] whose words and illustrations suggested that he had proposed using gunpowder to blow a mine under a fortress as a particularly effective method of destroying the walls. Different manuscripts offer slightly different wordings but the clearest suggest that he was proposing using miners to dig deep under the target then to place three of four barrels of

50 Vasari, *Lives of the Artists*, trans. A de Angelis, at http://members.cfn.org/~acd.vite/vasaricecca.html (2011).

51 Francisco di Giorgio Martini, *Trattati di architectura ingegneria e arte militare*, ed. C Maltese, 2 vols (1967), esp. I, 3–16, and illustrations from ff. 4, 5 and 6 (fortress plans) and 55 (powder mine).

gunpowder there, with thin cord coated in sulphur (*sulfuratus*) laid from the barrels to the entrance to the tunnels. A strong, mortared wall should then be built at the entrance to the tunnel before the cord was set alight [Fig. 28]. The problem is that there was no evidence that the feat had actually succeeded for several decades after his books appeared. There were evident practical problems in achieving this perhaps obvious application of gunpowder to siege mining, one of which was the serious danger that the explosive would blow back through the mine rather than upwards, and the few chronicle accounts that appear to state that this form of attack was used (such as an account of a mine at the Florentine siege of Pisa in 1403 where the engineer was named as Domenico di Matteo) are rarely sufficiently unambiguous to be definitive that it was tried.

Taccola's proposal was not apparently attempted, maybe because constructing such a wall would have signalled too obviously what was happening, perhaps because the arrangements for the "fuse" were not practical. The "correct" solution (the one adopted in practice) was to make the mine tunnel zig-zag so as to prevent blow back and send the explosion in the right direction; the problem with this was that once miners started digging away from a straight line, they might end up in the wrong place, and this was possibly what happened in one of the examples of mines attributed (falsely) to Martini himself (the Genoese siege of Sarzanello in 1487, where the explosion took place short of the desired spot). This objection itself is questionable, however, given that magnets could be used to preserve the correct orientation of the mine tunnel, and Taccola's text stated this. Examples multiplied towards the end of the century during the Spanish conquest of Granada and then during the Franco-Spanish wars in Italy and the Pyrenees. The first printed descriptions appeared in the first half of the sixteenth century when the techniques had been mastered but this represented very slow progress (see pp. 275–6). The safest conclusion is that Taccola and Martini identified the possibilities alongside military engineers who may well have attempted to execute the task but in doing so had encountered practical difficulties that were not solved properly for many decades. Taccola was certainly never called upon to deliver what he had presented in writing but Martini probably was. The story is a reminder that translating theory into practice was no automatic process and it may well have been the advice of practitioners – engineers and miners – that highlighted the risks and delayed the widespread adoption of the practice for nearly a century.[52]

52 Partington, *A history of Greek Fire and gunpowder*, 171–5; J B Bury, 'The early history of the explosive mine', *Fort* 10, 1982 (23–30) and S Pepper, 'The underground siege', same journal, 31–8. All discuss and explore the evidence for use of explosive mines and reproduce the illustrations in Taccola and Martini. The clearest illustration of Taccola's proposal is in the Paris manuscript of his work: M Taccola, *De rebus militaribus. De machinis 1449*, ed. E Knobloch (1984), 257 (fol. 48vn of (Paris) cod. Lat. 7239).

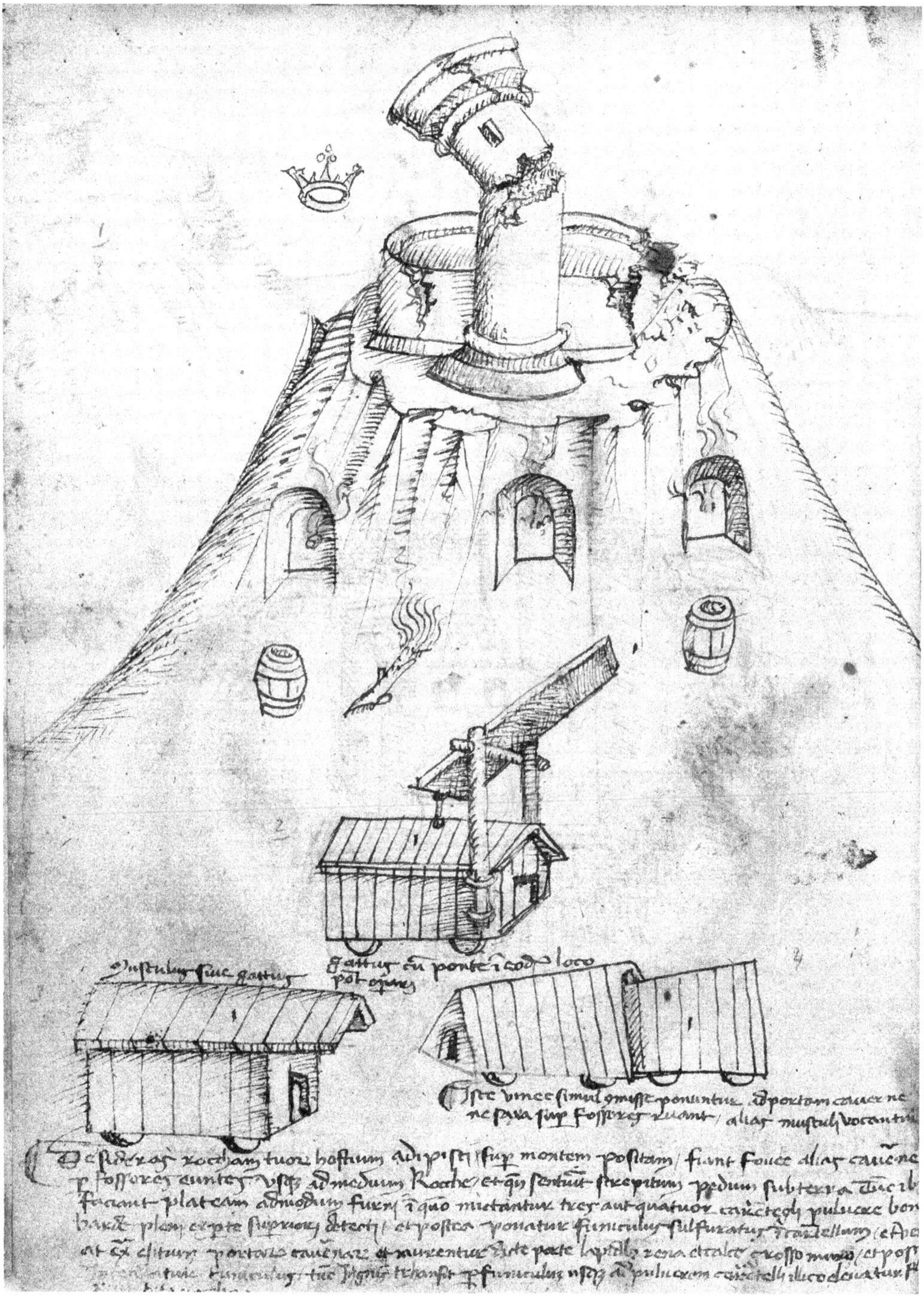

FIG. 28 A gunpowder mine imagined in Taccola's *De re militaribus*. (München Bayerische Staatsbibliothek, clm 28800, fol. 48v, with permission)

OTHER VARIED CAREERS

A man with no theoretical training and only his experience to draw upon but who nonetheless used this successfully was François de Surienne (d. 1462), known as "L'Aragonnais", who served as master of artillery for the duke of Burgundy as well as later fighting for the English. Like Martini, he developed a particular interest in modern fortification, and his varied career included writing a proposal for the refortification of the city of Dijon in 1461 based on updating it to resist powerful artillery. This advice was not followed (it was probably too expensive) but Surienne played a prominent part in the wars of the middle of the century and became an expert in gunpowder weaponry.[53] He was an adventurer, but like others he was able to use his acquired knowledge to secure employment from rival rulers. His ideas were not just a repetition of old formulae, but an original attempt, as with Martini, to address new circumstances.

Il Cecca and Martini stand out as fitting better the earlier model of engineers capable of a multitude of different roles than most of their contemporaries, where serving the artillery needs of their rulers was creating the largest demands. The role of master of artillery was increasingly reflected in records, if not (unlike in the Burgundian cases cited) in the chronicles. England was riven following the final collapse of its French possession by the dynastic "wars of the roses", and guns were used in these wars, which unusually were dominated by field battles. Thomas Vaughn was "master of the King's ordnance" in 1460 when he was commissioned to identify bombards, cannons, serpentines, fowlers (the French *veuglaires*), culverins and gunpowder, sulphur, saltpetre, lead, iron, bows and crossbows belonging to the king (Henry VI), and to find out where they were, with a view to making repairs and storing them in a safe place.[54]

During what turned out to be the last decades of the Hundred Years War, the English artillery in France had been organised into companies of ordnance. The masters here had a more clearly administrative role and nine in all were appointed in Normandy between 1423 and 1448. Typically, a company comprised a master forger (with an aide), a master carpenter (ditto), a master mason, a master gunner and a carter. Philipot Lorin was master of artillery for three years and John Potel served at several different times. The others were Robert Cottes (1423–24), John Harbottel (1424), Philibert de Moulant (1425–27, still serving in 1431), William Appleby (1428 and possibly also 1439), William of Gloucester (1431, 1434), John

---

Interestingly, Taccola had offered a detailed version of the traditional siege mine reliant on burning timber props on the previous page of this version of his text (p. 257).

53 J Richard, 'Quelques idées de François de Surienne sur la dèfense des villes à propos de la fortification de Dijon (1461)', *Annales de Bourgogne* 16, 1944, 36–43, for the advice to the city.

54 *Calendar of Patent Rolls Henry VI*, 14 August 1460, VI, 612.

Banaster (1433–35), William Forster (serving the earl of Warwick in 1438), William Hicklyng ("controller of king's ordnance" in 1441) and Henry Gryffyth (from 1441). The records confirm that the artillery was centrally controlled by the crown and that it saw frequent military service. Some of the master gunners had previously been archers and there was a lack of specialisation which possibly reflected that it was easy to learn the skills of a gunner, while it is possible that with relatively low wages, the career itself was both dangerous and unrewarding, hence the rapid turnover.[55]

The only remaining presence in continental Europe of English fighters after 1453 was in the garrison of Calais and its closely surrounding forts. Analysis of the records shows continuing royal attention paid to the provision of artillery there with new weapons ordered and sent to the town in substantial numbers to ensure a supply of the most modern weaponry and gunners being paid among the defenders, both under Edward IV (1461–83) and under Henry VII (1485–1509). In 1467, master gunner Robert Potte was sent to Brussels to test guns ordered from there. A foundry was established to melt down older guns and manufacture new. Under Edward, the arquebus made its appearance, as well as iron gunshot. William Rose was controller of the ordnance in 1474, and Master Giles van Rasingham (evidently not English) assembled the artillery for the king's invasion of France the following year when there were 150 large guns at Calais. At this outpost, English government maintained a substantial artillery and corps of artillerists as up to date as any of its neighbours during this time.[56]

ARTILLERY ORGANISATION IN FRANCE, BURGUNDY AND THE SPANISH PENINSULA

The most significant steps towards a recognisably modern organisation of artillery have always been identified with the transformation of the army of France during and after the final defeat of the English and in the subsequent wars of Louis XI and his successors that carried French power eastwards and southwards as well. These changes are closely connected with the names of the two brothers Bureau,

55 All from A King, 'Gunners, aides and archers: the personnel of the English ordnance companies in Normandy in the fifteenth century', *Journal of Medieval Military History* IX, 2011, 65–75, and C Allmand, 'L'artillerie de l'armée anglaise et son organisation à l'époque de Jeanne d'Arc', in *Jeanne d'Arc. Une époque, un rayonnement* (1982), 73–83 (list of names at 82–3).

56 All from D Grummit's works, 'The defence of Calais and the development of gunpowder weaponry in England in the late fifteenth century', *War in history* 7/3, 2000, 253–72, and *The Calais garrison. War and military service in England 1436–1558* (2008), 59–60, 123–39.

Jean (1387/93–1463) and Gaspard (1393/1400–1469), with the elder more the administrator and the younger more the technician. They were appointed centrally as masters of artillery in France (Pierre Bessineau had been a previous appointee from 1436 and Gerard was his deputy from 1440, taking over in 1442). Dispute over their precise role is unresolved for lack of evidence but it was during their time that a significant number of improvements took place. Technically, these included the move to cast guns, further improvements in the quality of gunpowder, development of muzzle-loading one-piece guns in place of the traditional separate chamber for the charge from around 1460, the first genuinely mobile guns, trunnions to simplify changing the elevation of gun barrels – a development also (questionably) attributed to Surienne – wheeled gun carriages and iron shot: all happened during the years when the brothers were in service. The precise timing is as unclear as where the credit is due, but what is certain is that under Charles VII (1422–61) and his son Louis XI (1461–83) the artillery of France was transformed in quantity and quality, a result both of technical progress and centralisation of organisation. It is most likely that these changes were progressive and cumulative rather than revolutionary but the consequence was dramatic. French guns, famously, were decisive for the first time in open battle at the French victory over the English at Castillon in 1453.

If it is true that no evidence remains for the actual organisation of French artillery under Charles, it is clear that there had been change and that this continued after the deaths of the Bureaus with several successive masters including Gobert Cadiot and Louis de Crussel. Guillaume Bournel was appointed by the king in 1473 with wide powers and a deputy went to campaign in Roussillon as master of artillery. In the war that saw the destruction of the duchy of Burgundy following the death of Duke Charles the Bold at the battle of Nancy in 1477, for the first time the French artillery was organised into three "bandes", the first containing ten of the largest guns and the most experienced gunners. The second was commanded by the "renowned cannonier" Giraud de Samain, a man with long service who was succeeded by his brother Bertrand. The third was led by Thomas Bachelier, who had previously served in the same role with his two sons and had in 1471 been "canonnier et maistre des oeuvres de maconnerie" at Rouen – an interesting coupling of different functions. Giraud's career confirmed the value attached to his skills: he was captured by the Burgundians at the battle of Montlhéry in 1465 and then served them. He was later re-employed by the French in Normandy. Presumably he had been taken when the artillery fell into enemy hands: a record by Gaspard Bureau showed the loss of a big cast culverin and four falcons and five culverins at the battle. One of the guns was later recovered during a sortie at the Burgundian siege of Beauvais in 1472.[57] Artillery, like its masters, knew no nationality.

57 H Dubled, 'L'artillerie royale française à l'époque de Charles VII et au début du règne de Louis XI (1437–1469). Les frères Bureau', *Mémorial de l'artillerie française* 50, 1976,

Two other west European states stand out in the development of effective artillery and growing specialisation in the role of the masters of guns. In Iberia, the final expulsion of Muslim rule from the peninsula owed much to the power of the Castilian and Aragonese artillery. Burgundy under its ambitious and expansionist fifteenth-century dukes had become a substantial regional power and maintaining military might was central to their rule. Reference has already been made several times to their masters, who were of sufficient reputation to be named in chronicles. What of their organisation? The Burgundian dukes had understood the importance of the gun early on, and evidence of organisation dates from early in the century. Masters of the duke's artillery have been identified from the archives from 1413; the role evolved to include responsibility for materials, organisation of transport, recruitment of personnel and directing the batteries in use. Surienne held the post in 1440. A separate role of controller was created to take charge of the supplies in around 1430 and increasing specialisation took place during the century. The widely dispersed character of the duke's realms required a decentralised approach to the storage of the ever-growing number of guns being manufactured, inventories showing hundreds of weapons of various sizes in the arsenals of fifteen towns and cities. The critical role was that of the *canonnier*, and evidence shows that many were still recruited abroad in the 1460s. In a plan for a crusade against the Ottomans prepared in 1454, the duke's requirements included 500–600 gunners along with carpenters, smiths, pioneers and miners. At the battle of Montlhèry (1465) the Burgundians used a substantial artillery train, and although there continues to be dispute over who won, the guns do seem to have made a significant contribution. Charles the Bold (1467–77), in parallel with his nearest enemies (the French), established a standing army from 1467 and a properly organised artillery train was an essential component, along with the master gunners who organised and operated it. For the most part, however, these masters remain unnamed.[58]

The armies of Ferdinand of Aragon (1479–1516) and Isabella of Castile (1474–1504) completed the unification of the Spanish kingdoms by destroying the last remaining Muslim state in Iberia, Granada, over a conquest that took several years of hard fighting to take well-fortified cities one by one. They relied on a range of siege skills. Guns had been in use in Iberia since the middle of the fourteenth century and by the start of the fifteenth the whole wide range of different sizes and calibres of weapons was available. It appears that Ferdinand took an interest and may have

---

557–603; E Perroy, 'L'artillerie royale à la bataille de Montlhéry (10 juillet 1465)', *Revue historique* 49, 1925, 187–9; the same, 'L'artillerie de Louis XI dans la campagne d'Artois', *Revue du Nord* 26 (no.103), 1943, 171–96.

58 C Brusten, *L'armée Bourguignonne de 1465 à 1468*, 6–10, 12–20, 35 (arsenals), 126–7 (crusade); R Smith, K DeVries, *The artillery of the dukes of Burgundy 1363–1477*, 124 (crusade), 144 (Montlhéry); N Michael, G Embleton, *Armies of medieval Burgundy 1364–1477*, 10ff.

brought in French and German master gunners to improve his capability, including the introduction of casting. Previously, the great lords of Spain had maintained their own artillery trains and gunners. The title of *capitan de la artilleria* appeared in 1465 and a permanent royal corps was established by Ferdinand in 1489 – an inventory showed it containing 180 guns in 1495. The earliest captain of artillery in Castile was Diego Rodriguez de Zapata (1406). Fernan Rodriguez de Monroy held the post in 1429 and a number of other *maestros bombarderos* appeared in the records. In the war of Granada the captain of artillery was Francisco Ramirez de Madrid who had given long service to King Henry IV (1454–74) and his successor Isabella. This included war against Portugal in 1475. He was appointed as *obrero mayor* of the fortresses of Seville in 1478. He was put in charge of the guns in 1483 and in the siege of Alora (1484–85) employed German and French gunners. He died in 1501. His role in the capture of Velez Malaga in 1487 was noted by chroniclers. At the subsequent attack on Malaga, he was recorded as ordering the digging of a mine under a tower then placing a heavy but squat short-range gun (a *quartadgo*) inside to shoot upwards, destroying the floor, leading to its capture, a most unusual operation. He was knighted in 1487 as a result of his role in the taking of this city, a critical step towards the final conquest of the kingdom. Records show many hundreds of masters, gunners and workers employed with the artillery train, including more than twenty Breton gunners during this year, and batteries of artillery set up to bombard particular angles of the (very strong) defences, evidently by commanders who knew what they were doing. The defenders of Granada were also well provided with artillery and equally skilled in its use and in other engineering roles, although no names of those in charge have been identified. By the turn of the sixteenth century, the joint monarchy had a very powerful artillery train centrally organised and administered and commanded by professionals.[59]

Francisco Ramiro Lopez also served as a master of artillery from 1493, having been presented to the monarchs by his previous employer, the powerful nobleman the duke of Medina Sidonia. He organised the expedition and commanded the guns at the capture of Melilla. He was also responsible for the new fortress constructed at Salses (Pyrénées Orientales), now in France but then in Spanish-controlled Roussillon, started in 1497 during the war with France. This new construction was one of a number of novel designs attempted as military engineers strove to find the best defence against modern artillery [Fig. 29]. A regular rectangle, squat and low-lying with massive walls and projecting circular bastions designed for artillery,

59 W Irving, *A chronicle of the conquest of Granada (from the MSS of fray Antonio Agapida)* (reprinted 1986), 106–9; M Ladero Quesada, *Castilla y la conquista del reino de Granada* (1967), 117–27; F del Pulgar, *Crónica de los Reyes Catolicos*, ed. J de Mota (1943), II, 126–8, 164–70, 170–1, 261, 292, 323–5 (on particular references to the roles of artillery and the part played by Francisco Ramirez de Madrid); J Vigon, *Historia de la Artilleria Española*, I, 53–9, 70–5, 107–10.

FIG. 29 Salses, Pyrénées Orientales, built by the Spanish at the end of the fifteenth century, representing a transitional phase in the evolution of artillery fortification. (Photo: author)

along with outworks, the concept behind this fortress demonstrated again (as with Surienne decades earlier and his contemporary Martini) that military engineers of this period were capable of applying their experience on one side of the gun to the lessons learnt about how best to resist them.[60]

THE SIEGE OF NEUSS

Far too little considered by most English-language historians of the fifteenth century,[61] the failed attempt by Charles the Bold to capture the small Rhineland

60 L Bayrou, N Faucherre, R Quatrefages, *La forteresse de Salses*; Purton, *HLMS*, 373–4.

61 The most detailed account, placing the siege in its political and diplomatic context rather than its military, was R Vaughan's *Charles the Bold* (1973, new edn 2002), 319–45. There is a modern account by J Metzdorf,'Bedrangnis, Angst und grosse Mühsal – Die Belagerung von Neuss durch Karl den Kühnen 1474/5', *Beihefte zur Mediavistik*, ed. O Wagener, 7, 2006 (167–88). See Purton, *HLMS*, 316–22.

town of Neuss in 1474–75 can be followed in immense detail as a result of there being numerous contemporary accounts by participants on both sides of the struggle. Unlike the new fortresses being designed by engineers in the second half of the century, Neuss did not enjoy the benefits of modernisation, but nevertheless the evidence means that its siege offers a valuable opportunity to observe engineering skills in operation. The siege of Malaga already mentioned was also recorded in great detail, but from only one side of the battle.

Neuss became a target as a result of conflict within the archbishopric of Cologne when the prelate, deposed by his chapter, enlisted the aid of Duke Charles against the city and Neuss, which was provided initially with a garrison of 1500 Hessian soldiers (later reinforced) led by their landgrave, Hermann. The town itself was small, perhaps 5000 inhabitants, and was protected by a circuit of medieval stone walls with four gates, but more importantly, outer defences of two ditches and a big rampart provided with bastions for artillery had been thrown up, while the eastern side was additionally protected by a branch of the Rhine (now on a different course) which also fed the wet ditch. The besieging army was drawn from all parts of Charles' widespread domains and included some archers sent by his ally Edward IV of England; it numbered upwards of 20,000 at times. Examples of contemporary military engineering were to be found in all aspects of the attack and defence. The initial strengthening of the ramparts was well designed with professional expertise, which was just as well because the Burgundians disposed of a large artillery train that rapidly flattened the old stone walls and gates but was much less effective against the defenders' stone, brick and earthwork bulwarks. Time and again, following bombardments, the Burgundians sought to storm their way in but were driven back in fierce fighting. The garrison also made much use of gunpowder weapons; the town possessed two powder mills but lacked supplies of saltpetre, and so intense was the fighting that shortage of powder threatened an early end of the siege, only solved when 500 men from Cologne somehow slipped through the lines each carrying 10-pound (4.5-kilogram) bags of saltpetre (November 1474).

Between the eastern branch of the river and the main stream stood two islands and the duke decided early on that securing them would be important to cutting off the town from its friends as well as opening up another part of the defences to bombardment. The isle of Waidt was taken on the second attempt by an attack launched with boats and then secured with two bridges across the two arms of river that had to be crossed. It must be assumed that someone was responsible for organising this vital project although no names have been recorded. The people of Cologne attempted to destroy the bridge with a fireship, as a result of which Charles ordered that this channel be blocked with a dyke 9 metres wide. Scaling ladders were prepared for an assault that took place on 10 September, which failed because the ladders proved to be too short. From a different side, the besiegers attempted undermining, but this failed when the defenders successfully counter-mined. It seems that miners from Liège were engaged on both sides. One account suggested that the Burgundian mine was betrayed by a Liègeois to a fellow countryman on the other side.

One chronicler recorded that a Castilian in Charles' camp was familiar with Vegetius and described to the duke a proposal for a device to conquer the town. The duke was convinced by his sketch and ordered it built: a wooden tower on four wheels, 6 metres square and 12 metres tall, but both it and another similar machine ended up stuck in the muddy ground (Neuss was low-lying and the terrain largely flat) and useless. It is strange, and unexplained, that reliance should have been placed on such throwbacks to earlier times. More usefully, devices to shoot fire arrows into the town had considerable effect, causing substantial damage, while a few surviving recovered shot strongly suggest that the Burgundians were also firing chain shot. On 16 February 1475, the besiegers finally took the bulwark in front of the northern ('Rhein') gate but the defenders had erected a temporary rampart immediately behind and their Liègeois miners had helped in digging a deep ditch. During this time the garrison and townsfolk were mightily encouraged when a vast army mobilised (incredibly slowly) by the cash-strapped emperor (Frederick III) arrived on the opposite bank of the Rhine and started to bombard the besiegers (as well as getting further supplies into Neuss), and resorted to the novel tactic of placing messages into hollow cannon balls to communicate with the town – most fell short but some arrived safely and were preserved. Fierce fighting continued with the arrival of spring and the defenders had to sally to prevent a mine being completed by Italian besiegers (8 April). The siege ended officially on 28 May when a truce between Charles and the emperor was agreed.[62]

Engineering had been involved in many of the moves by both attack and defence during the siege, the design of temporary ramparts and bulwarks to mount artillery, mines and countermines, the erection of bridges, significant hydraulic works and the occasional reversion to by now historic creations. The layout of the duke's encampment was remarked upon by contemporaries, too, and it may be that someone had been in charge of surveying it and proposing its plan. None of those

62 This summary drawn from the following accounts: *Chroniques de Jean Molinet*, ed. G Doutrepont, O Jordagne, 3 vols (1935), I, 29–102; *Les memoires de Messire Olivier de la Marche*, ed. Petitot, 294–8; C Wierstraat, 'Die histori des beleegs van Nuis zusamen', ed. C Nörrenberg, *Die Chroniken der Deutschen Städte* XX (1887), which is a long verse chronicle written by the town secretary in 1476; F Gingins la Sarra, *Dépeches des Ambassadeurs Milanais sur les campagnes de Charles le Hardi (1474–77)* (1858), I, 69–159; Johannis Knebel, 'Diarum', *Basler Chroniken*, ed. W Vischer, H Boos (1880), II, 112–235 (reporting letters received in Basle from Cologne recounting the events). The archaeological finds identified by the Neuss museum attributable to the siege (many of them subsequently lost) were analysed by C Pause, 'Weapons and military equipment from the Burgundian siege of Neuss in 1474–5', *Acta Militaria Medievalia* VI, 2010, 121–44.

responsible were named in any of the accounts, and it may be that the commanders on both sides were sufficiently professional to be able to plan the works themselves and trust their ordinary soldiers to execute them: it is, however, unlikely given what is known of the make-up of late-fifteenth-century armies.

By the turn of the fifteenth and sixteenth centuries, books covering the art of work were everywhere. Printing helped their dispersal but manuscript copies continued alongside: it is fitting to conclude this study by reference to a rare manuscript dated to possibly 1483 commissioned by what is assumed to be an "aspiring bourgeois" family in southern Germany. This "housebook" is a mixture of delightful illustrations designed to please its owners, some of which are pictures of significant technical developments (water wheels for pumping out water from mines, spinning wheels, bellows driven by counterweights, pile drivers) and military equipment. The guns shown differ from those illustrated by Konrad Kyeser even though the other engines drawn were otherwise the same. The style of illustration was contemporary, tools, engines and guns drawn with sufficient detail for their construction to be understood by professionals. It is entirely without written text, so the artist must have assumed awareness. This book combined traditional approaches with the most modern technology.[63]

One issue has emerged continuously throughout this book: what link if any can be traced between scientific knowledge and practical activity? The fifteenth century provides the firmest evidence that by this time there was definitely an overlap. One scholar has proposed that people like Alberti "obliterated" the distinction between Latin learning and artistic skill in his work on buildings written in 1452 which also addressed the subjects of shipbuilding and war engines. Even at an individual level, the way in which artisans switched from one career to another, entirely different, and so many decided that guns offered the way to wealth and success indicates a breakdown of strict boundaries between professions. The now famous renaissance polymaths played multiple roles in developing both scientific understanding and technical progress.[64] Men like Leonardo da Vinci were of humble origin but based their own education on reading the works of their forebears and on talking to more knowledgeable contemporaries. The technical progress that was made over these decades was limited – as with so many of Leonardo's famous drawings, the technology did not exist to make them real. But an accelerated process both of inquiry and experiment – I have argued that both must have also occurred to explain progress in earlier centuries (such as with trebuchet design) but where names and dates are lacking – set the scene for the early modern world. References

63 C Graf zu Waldburg Wolfegg, *Venus and Mars. The world of the medieval housebook* (1998).

64 See the brilliant study by P Long, *Artisan/practitioners and the rise of the new sciences 1400–1600* (2011).

to Leonardo in this chapter so far have played down his importance in terms of his military designs, but this is not to discount his importance in advancing a new approach to science rooted in both principles and the observation and recording of experiment, in his case across a wide range of increasingly specialised subjects. His notebooks contain propositions taking forward understanding of mechanics and its practical applications, and the examples he used to illustrate his theses included propulsion from bows, crossbows and catapults.[65] If a thirteenth-century trebuchet builder or operator had reached the same conclusions from watching how far a particular design propelled shot of various weights, Leonardo proposed ways of deriving scientific principles from the observation.

The final decades of the fifteenth century were in many respects little different from the early years of the sixteenth, but the engineering skills required and provided by 1500 were significantly changed from those of 1400. It is only necessary to note the transformation of ships and shipbuilding from coast-hugging or short hops between ports to oceanic voyages, or the design of fortifications where the move towards forts designed specifically for artillery and entirely distinct from the castle designed for show of status and gracious living was accelerating rapidly, to confirm this long-term trend. Responsible for the latter was the great progress in the effectiveness and reliability of gunpowder weapons noted during this century, accompanied by greater centralisation and professionalisation of all the associated processes, from manufacture to administration to deployment. It would still be a long time before rulers recognised the potential of creating similarly distinct bodies of professional military engineers apart from those responsible for the guns, but that separation of function was also now underway. Surveying, designing temporary defences, building bridges, diverting or blocking waterways, transporting, mounting and operating artillery, and mining or countermining all continued to matter a great deal in the military operations of early modern Europe, and it would be fair to conclude that many of these skills had already attained during the fifteenth century the levels they would sustain during the sixteenth. The many named engineers cited in this chapter and the larger number who remain anonymous have the right to be recognised for their achievements. Their willingness to serve any master was not controversial in their day. Their courage in the dangerous lives their commitment involved deserves recognition. Their own understanding was itself based on following a long tradition of military engineering that can be traced back for many centuries. It would be the seventeenth century before more recognisably modern science made an appearance and before the "scientific" fortifications associated chiefly with the name of Vauban became widespread, but these revolutions did not spring from nowhere. They built on a legacy of knowledge and experience inherited from the medieval centuries.

65 Leonardo da Vinci, *Notebooks*, selected by I A Richter, ed. T Wells (revised edn. 2008), 6, 10 (scientific method), 71–4 ("impetus", "percussion").

Postscript

From Medieval to (Early) Modern in the Sixteenth Century

THE LAST CHAPTER DESCRIBED some of the revolutionary changes marking the end of the medieval world. In warfare, the most obvious transformations were the creation of standing armies in which infantry replaced cavalry as the dominant element and the professionalisation of artillery as processes begun in the middle of the fifteenth century took root. In terms of engineering, this continued to have an effect, with gunnery becoming a separate specialism, although – as we will see – the application of scientific knowledge to the skills of the artilleryman took a great deal longer to register. The acceptance that a geometric approach to fortress design also required the integration of both the defensive and offensive use of modern artillery continued to evolve before finally arriving at commonly agreed principles, but not before the middle of the sixteenth century, and throughout this time called upon parallel progress in military engineering aiming to overcome such defences.

In Europe, warfare was more or less continuous in the first half of the new century. Successive kings of France initiated campaign after campaign to make good their claims to the inheritance of the kingdom of Naples and the duchy of Milan, starting with the irresistible invasion of a large army under Charles VIII in 1494. The small city states of Italy, even with their advanced systems of fortifications, proved powerless and it took military intervention from Spain to throw out the invaders. This led to more than sixty years of conflict involving France, Spain, the Holy Roman emperor (the Habsburg Maximilian, succeeded in 1519 by Charles V who combined rule of Spain, the empire and the Low Countries), the papacy and all the states of Italy, including Venice, in a patchwork of changing alliances. Peace was not made until 1559. Charles V was also engaged in conflict with German states that embraced Lutheranism from 1517, and faced the continuing expansion into Europe of the Ottoman Turks (who conquered Hungary in 1526 and besieged Vienna in 1529), while Henry VIII of England (1509–47) had broken with his father's prudent abstention from military adventure to demand his moments of glory.

It was Italy – home of great renaissance thinkers and creators but also now trampled under the devastating mercenary armies of Swiss and Landsknechts hired by foreign invaders – that produced the treatise that made a big impact on contemporary thinking: Niccolò Machiavelli's *The Art of War*, written in 1521. Chiefly a call to arms to Italians (particularly his fellow Florentines) to create citizen armies to assert their ancient glories, the book relied time and again on the history of the great Roman generals of the past for its examples. A whole section was devoted to laying out camps, relying on ancient Roman practice as reflected by Vegetius. Indeed, whole sections are lifted from this writer without acknowledgement. Of course, he sought to make it relevant, talking about Swiss pikemen, while the section on undermining a fortress discussed (briefly) the use of gunpowder. His apparent dismissal of artillery was once believed to represent a weakness in Machiavelli's approach; more recent studies have established the context of his views and, along with work on his other writings, have restored his credentials.[1]

Machiavelli, while lifting from Vegetius on the value of craftsmen such as masons and carpenters to his army, makes no reference to engineers, but in doing so he *was* overlooking a group of people who made an important contribution to all sides in the Italian wars and whose careers are sometimes quite well known. The first indisputable use of a gunpowder mine was in 1495 when the Aragonese deployed one to breach the defences of the Castelnuovo in Naples. The same method was used again in the capture of the same castle, and the nearby Castel d'Ovo, in 1503. The contemporary historian Guiccardini (on whose writing, as well as that of Sanuto, much of the information for the first stages of the Italian wars relies) noted of this form of attack that it was "more frightening, because the defences against them have not yet been thought out".[2] The names of two engineers are associated with these achievements, finally putting into practice the ideas of Taccola. Antonello, from Trani in Italy, was well rewarded by the famous commander of the Spanish army in this campaign, Gonzalo de Córdoba (the 'gran capitan'), with 150 ducats and land for a house. He was later sent to organise the defences of Puglia, where he was rewarded with houses in Trani, and then became an artillery commander in the wars against Venice (1510–12). His work included advice on strengthening fortifications and actual commissions to construct new ones. He died in 1526 (or 1527) and his son Gian Girolamo followed the same

1 I use Machiavelli, *The Art of War*, ed. N Wood. Camps: book six, pp. 150–66; mining: book seven, 199–200. See A Gilbert, 'Machiavelli on fire weapons', *Italica* 23/4, 1946, 275–86; A Ghat, 'Machiavelli and the decline of the classical notion of the lessons on history in the study of war', *Military Affairs* 52/4, 1988, 203–4; B Cassidy, 'Machiavelli and the ideology of the offensive: gunpowder weapons in the *Art of War*', *Journal of Military History* 67/2, 2003 (381–404).

2 Guiccardini, *Storia*, book VI c.1, quoted in Mallett, Shaw, *The Italian wars 1494–1559*, 62–3.

career.[3] His colleague at the Naples operations was Pedro Navarro (1460–1528) from Spain, originally a sailor, subsequently a mercenary serving Genoa, during which time he seems to have operated as a pirate. He enlisted with Gonzalo and tried his hand at mining at Civitavecchia in 1500, without success. But he served with distinction in subsequent campaigns culminating in the successful mining of Castelnuovo and Castel d'Ovo in 1503, for which he seems to have taken much of the credit and become noticed as the premier engineer of the age. He invented a kind of floating battery used to capture Velez de Gómera in North Africa. His fortunes changed when he was captured by the French and imprisoned. The Spanish king refused to ransom him, whereupon, after three years, he offered his services to his captors and served them in more Italian campaigns. In 1528, while devising elaborate siege works (again at Naples) he was recaptured by the imperial forces, but this time Charles V ordered his execution as a traitor (although he died before it could happen). Having been created a count, Pedro is another example of a humbly born engineer who made his fortune through his skills, although how he acquired them is unknown, but one who tried to change sides once too often.[4]

It seems that both of these engineers turned an expert hand to fortress design as well as fortress destruction. The sixteenth century witnessed a definite trend in military architecture. Whereas some of the renaissance writers who devoted time and effort to designing model fortifications did find themselves in a position to put their ideas into practice (see chapter seven), others worked in a more theoretical world, their drawings of ideal shapes taking a greater interest in the harmony of their geometry, perhaps, than in the ability of the result to mount defensive artillery capable of keeping an enemy's guns at a safe distance. Designing bastions proved to be a complex process advancing towards the ideal that became standard across Europe and the Middle East by the second half of the sixteenth century. There were many questions to solve. Issues included how to cover the space in front of or between bastions from close-range attack, how to mount large guns without creating enormous and vulnerable embrasures (the answer was to put them on top, of course, but many designers were obsessed with defending the ditches), and what angles gave the structure the best resistance to bombardment. There is consensus that the people who achieved the right outcome were specialist military architects who were professionals in the service of rulers. One problem, of course, was that they were not working in a world in which it was possible to come up with abstract ideas on which it was possible to experiment and then amend the plans. Rather, the fortifications they planned (or in many cases added to existing towns and castles) were continuously being attacked. Albrecht Dürer's work of 1527

3 He was mentioned in Sanuto's *Diarii* and discussed by G Ceci, 'Un dimenticati ingegnere pugliese del sec. XVI, A da Trani', *Japigia* I, 1930, 54–60.

4 Article in Encyclopedia Britannica online; 'Pedro Navarro, count de Olivetto', accessed 13/1/2017; and Mallett and Shaw, *The Italian wars 1494–1559*, 167.

was the first of the new practical handbooks, but the honour of writing the first professional text on modern fortification probably belonged to a Spaniard, Pedro Luis Escrivá. He was definitely not writing in the abstract but was responsible for the design (finished by others after his death) of two forts, L'Aquila and Sant Elmo, which stand above Naples. Both show two key features: compliance with a set of design principles and also respect for the actual terrain on which they stand. Escrivá followed in a tradition of Spanish engineers begun by Ramiro López with the construction (from scratch) of Salses and works at other Spanish towns now in French Roussillon (Collioure, Perpignan). Antonio de San Martin plied his trade for the crown on Rome's Castel Sant'Angelo or on the fortress of Rhodes (which despite its claim to the most advanced defences of the age fell to the Ottoman Turks in 1522). There were also native Italian engineers working on modern fortifications, probably the most famous being Giuliano and Antonio di San Gallo, who were responsible for the new defences of many towns in Puglia and also the Fortezza da Basso of Florence between the 1510s and 1530s – at the same time as Michelangelo's designs for Florence, which have been described as "opulent fantasies" (J R Hale), and the bastions built there on his design were criticised by Antonio di San Gallo for their military weaknesses.[5]

Italy may have been the focus for much of the new military engineering but the ideas spread rapidly and individuals – in a time-honoured way – took their expertise with them. By the middle of the century, the *trace italienne* was in vogue everywhere. A well-preserved example surrounds the old bastide town of Navarrenx (Pyrenées Atlantiques) [Fig. 30]. The transformation from medieval to modern defences here is startlingly clear. There was once a major castle, but all that now remains apart from a few stones on the ground is an outer wall that has been cannibalised by part of the enceinte of the new fortress.

England offers a partial and time-limited exception to this line of development. Henry VIII had personally seen continental developments but they were then still at the stage of experiment and featured gun towers rather than bastions. Perhaps this was the reason for the design of the artillery forts erected at great expense when Henry fell out with both his powerful neighbours (Francis I and Charles V) and feared invasion. The proceeds of the monasteries liquidated in the English reformation were spent on twenty-four forts built and garrisoned between 1539 and 1540. The forts – many still standing as English Heritage sites open to the public – were based on a circular plan and comprise layers of rounded bastions often around a central round tower; they were designed to accommodate as much

5 On the evolution of military architecture, J R Hale, 'The early development of the bastion trace: an Italian chronology c.1450–1534', in the same, *Renaissance war studies* (1983), 1–30; F Cobos-Guerra, 'Pedro Luis Escrivá and the first treatises on modern fortification. Naples 1538', in A Cámara Muñoz, B Revuelta (eds), *Renaissance engineers* (2016), 25–52

FIG. 30 Navarrenx, Pyrénées Atlantiques, a bastion fortress of the mid sixteenth century. (Photo: author)

firepower as possible in a limited space. The names of two designers are known, Richard Lee and John Rogers, while the Moravian Stephan von Haschenperg was definitely associated with the forts at Sandgate and Camber. Militarily, they would have been extremely vulnerable if attacked by an invading army from the land side, but of course (in defence of their design) their purpose was actually to prevent an enemy invasion in the first place.

That Henry could appreciate the need for modernity was reflected in the range of fortification for which he was responsible after 1540. Having seen the Italian-inspired bastions built by the French at Le Havre and then the pentagonal bastioned fort put up to besiege Boulogne (1545) and listened to the advice of Lee and Rogers who were won over to the Italian style, Henry immediately adopted this style in the planning of the next generation of defences at Berwick, Yarmouth (Isle of Wight), Lindisfarne and forts on the Scottish border. A host of Italian engineers were employed on these works under Lee's direction, although financial crisis often prevented the projects' completion (such as at St Mary's on the Isles of Scilly).[6]

These engineers then were part of a new generation of specialists in military architecture, not generalists turning their hand to whatever function the ruler

6 All from J R Hale, 'Tudor fortifications 1485–1558', in the same, *Renaissance war studies*, 63–98.

would pay them for. This represented the culmination of another trend away from the world of the medieval military engineer. The development of the bastion trace would continue for another century and a half before culminating in the immense works of Vauban for Louis XIV and Coehorn for the Dutch republic and William of Orange.

Now in a separate category, engineers in charge of artillery also continued to evolve as a profession from the beginnings exemplified in the fifteenth century by the Bureau brothers. As guns became more efficient and effective and therefore became a more significant part of a state's armed forces – the Italian wars were marked by a number of bloody pitched battles in which guns sometimes played an important part, as well as being vital for breaching the rapidly developing defences of towns and forts – so also did the need for professional skill and knowledge from the artillerist. New works appeared in which scientists grappled with the principles that gunners would benefit from knowing. The differences between the *Nova Scientia* of Nicola Tartaglia of 1537 and the *Feuerwerkbuch* of a hundred years before were substantial. While both claimed to focus on solving practical problems, Tartaglia sought to apply mathematical science to the calculations needed by the gunner, such as working out the range and the elevation required. His was the first work to do so, although he had no practical experience of gunnery. Fifty years later, Luis Collado de Lechuga attempted the same challenge, this time asserting that he had such experience. However, they were attempting to create a science of ballistics that was fatally rooted in the natural philosophy of Aristotle, and it would not be until the work of Galileo in the seventeenth century that a new science could develop.[7]

A great number of engineers are to be found in the records of the states of the sixteenth century, mostly with clearly defined specialisms, without significant overlap. They represented a logical development of the crafts and knowledge of generations of military engineers slowly progressing through the previous centuries. As societies changed and a new world emerged, they would carry out the same functions as their medieval predecessors, but in a different way. In due course, states would recognise the benefits of establishing a corps of engineers as part of their military establishment, thus at last returning to the model used for so long and so successfully by ancient Rome. But the twentieth-century soldiers recruited from mining communities who mined and countermined on the western front during the First World War were practising the same skills as Vegetius' Bessi and the men from the Forest of Dean or the coal mines of Liège; and just as across the intervening centuries, their aristocratic commanders took the credit for the achievements of ordinary people who acquired engineering skills and showed the courage to put them into practice, whether it was digging mines or building bridges or siege engines, at the risk of their lives, for a few pennies a day.

7 J García Hourcade, 'From Tartaglia to Lechuga. The artillery engineer', in Cámara Muñoz, Revuelta (eds), *Renaissance engineers*, 53–74.

Appendix

Military Engineers and Miners in the Pipe Rolls of the English Exchequer

EXTRACTED BELOW ARE SOME of the references in the published volumes of the Pipe Roll Society (1844–2012) to aspects of engineering with military purposes. The Pipe Rolls (PR) – the accounts of the Exchequer preserved in the Public Record Office (now The National Archives) – have been published from that surviving for 1130 (31st year of Henry I), resuming from the first years of Henry II (1154–89) up to 1224 plus that for 1230, with gaps especially between 1131 and 1155, in 1215 (incomplete) and in 1217–18 (civil war). This table lists year, volume of the published series, name and title of recipient, amount paid and what the amount is recorded as being paid for. Under 1210 is added a record from the Liberate Rolls, published in 1844. Publication references are in the bibliography.

The reigns covered are: Henry I (1099–1130) (last year only), Henry II (1154–89), Richard I (1189–99), John (1199–2016), Henry III (2016–72). No Exchequer rolls survive for the reign of Stephen (1130–54).

Dates	PR volume	Name, title	Amount where given separately	What for
1130–31	31 Henry I (separate volume)[1]	Gaufridus, engineer (ingeniator)	£10 12s 11d	wages
1158–90	every year	Ailnoth, engineer	£10 12s 11d	wages paid every year
(1155–58)	separate[2]	Bertram, arbalistarius	30s 5d	wages
1166–67	XI	Ailric and Robert, carpenters		wages, also paid in subsequent years

Dates	PR volume	Name, title	Amount where given separately	What for
1172–73	XIX	Yvo, engineer		for various works and delivering carpenters to the army for making engines
1173–74	XXI	Yvo, engineer		carpenters to make engines for king at Huntingdon
1173–74, 1175–76, and 1176–77	XXI, XXII, XXV, XXVI	Ailnoth in charge of carpenters and masons …		… to destroy castles at Framlingham, Walton, Benington
1179–85	XXIX–	Helpa, "arbalistarius"	12s	wages
1181–87	XXXI–	Maurice, engineer	£10 16s in 1186–87	wages each year at daily rate of 1s per day
1184–88	XXXIV	Urric, king's engineer and "arbalistarius"	20s	wages
1186–1203	XXXVII–	Elyas, variously master mason, carpenter, engineer		paid for specific jobs
1189–90	XXXIX (NS 1)	Reginald, engineer		owes 6s 8d (repeated in subsequent years)
1191–92	XL (NS 2)	40 miners		wages
		Richard, engineer		5 marks
1193	XLI (NS 3)	1 engineer	14s	"making stockade and bretasches at Doncaster"
		?	43s	"for working one petrary and one mangonel"
		?	£67 2s 6d	"repairing mangonels and ditches"

Dates	PR volume	Name, title	Amount where given separately	What for
1194	NS 5	Albert, John, King's clerics		"for one petrary and one mangonel"
		Roger, "Enganet"	22s 8d	repairing the king's house and transporting bolts and "Greek Fire" to Nottingham
		Elyas, carpenter (Helie)	8s and £4 0s 6d	conveying engines to Nottingham; conveying 1 petrary and 1 mangonel from Windsor to Nottingham
1195	NS 6	Radulf de Planez and Helye, engineers		repairing castles of Hastings and Pevensey
1196	NS 7	14 miners who were at Welshpool castle in service of archbishop of Canterbury and 24 miners in royal service	42s 6d and 22s	expenses
		William, engineer		owes 2 marks
1204	NS 17	Fortinus, engineer	1 mark[3]	for a robe
1205	NS 19	William Báárd, engineer	1 mark	
1206	NS 20	16 miners at Bristol	£15.10s	wages
1207	NS 22	Albert, engineer	£18 13s	
		11 miners and 3 masons		13 at 3d per day, masters at 6d
1210	Rot. Lib.	13 carpenters under master Urric; 13 miners under masters Pinell, Ernulf		wages
1209–11	NS 24–	costs of shipbuilding and dockyard at Portsmouth		see chapter five, p. 179

Dates	PR volume	Name, title	Amount where given separately	What for
1211	NS 28	Master Peter, balistarius	53s 8d for making ballistas	wages 9d per day
1212	NS 30	Peter as above	£14 12s 9d	wages
1214	NS 35	Roger, Nicolas, balistarii	26s 5d	making ballistas
		Pinellus, miner, and 10 associates	£33 13 10d	wages
		Radulf and Nicholas, master carpenters	£38 19s 5d	making 2 "Turkish mangonels" in Nottingham castle
1215	NS 37	Nicholas de Andeli, master carpenter, and 3 others	master 9d per day, others 6d per day	for making 2 petraries and 3 Turkish mangonels at Knaresborough castle
		Master Ernulf, William Lamb and 11 miners	£35 12s 6d	wages at 4s 9d per day for 150 days
		Nicholas balistarius	£4 16s	for ballistas
1219	NS 42	30 miners	31s 6d	for travel from Gloucester to Stamford
1221	NS 48	Thomas, Robert, Burnell, king's carpenters	£31 6s	carrying and preparing 1 petrary and 2 mangonels
		10 miners	£7 4s	expenses to Castle Bytham
		2 master carpenters	36s	bringing 2 mangonels to Bytham
1222	NS 51	40 miners from Forest of Dean		expenses for coming to London
		Philip and Master Guillot, king's balistarii		construction of ballista parts

Dates	PR volume	Name, title	Amount where given separately	What for
1224	NS 54	Arnulf, William son of Lambert, Roger and Ade Bruno, miners; John de Standon, miner		wages at 5d or 4d per day
1230	NS 4	Ernulf, miner	40s 10d	?debt or pension

Notes

1 Published separately as *The Great Roll of the Pipe of 31 Henry I, Michaelmas 1130* (1833) (facsimile edition 1929).

2 Published separately as *The Great Roll of the Pipe for 2nd, 3rd and 4th years of Henry II (1155, 1156, 1157, 1158)*, ed. J Hunter (1844).

3 A mark represented 13s 4d.

Glossary

Abacus: (Latin, from Greek). A counting device often using counters in a wooden frame. Of ancient origin it has continued in use to the present day.

'Abbasid: (Arabic). The 'Abbasid dynasty assumed the Muslim caliphate reigning in Baghdad after their defeat of the 'Ummayads in 750.

'Arrada: (Arabic, from Syriac). A generic term for a large crossbow-type weapon, cf. Ballista.

Alidade: (Arabic – Greek equivalent Dioptra). A measuring device of ancient origin consisting at its simplest of a rod with raised ends placed on a surface through which one could line up distant objects both horizontally and vertically.

Arquebus: Handheld firearm appearing in the late fifteenth century, forerunner of the musket.

Astrolabe: (Greek) Device for measuring the height of the stars, crucial in astronomy and also in determining distances. Possibly invented in Hellenic period, certainly known by the start of the first millennium CE.

Attiliator: (Latin; also Artiliator). Title of a craftsman responsible for making or repairing crossbows, in use from twelfth century.

'Ayyubid: (Arabic). The dynasty ruling in Egypt and Syria during the first half of the thirteenth century, descendants of Saladin.

Ballista: Originally, a Roman torsion-powered stone-throwing siege engine, replaced by the Onager as a stone thrower. The name then became applied to a tension-driven bolt thrower in which form it continued throughout medieval times. The name applied both to large engines requiring mounting on a frame and a crew of two, and to the hand-held crossbow.

Barbican: An outer defence erected in front of a gate to provide an additional layer of protection.

Bastide: (French). Normally describes the large number of planned towns created particularly in S W France/Aquitaine during the thirteenth century. Also occasionally used to describe a military fieldwork.

Bastion: Any form of tower or projecting defensive work, occasionally by transference applied to the whole defence. During early modern times (approximately the sixteenth century onwards) became the term for the geometrically designed artillery platforms projecting from the enceinte of a fortress or city wall.

Belfry: (Latin Berefridum, French Beffroi; Greek helepolis although this was sometimes applied to other weapons). A mobile wooden siege tower.

Bent entrance: Originally Muslim design of a gatehouse in which the passage turns through 90 degrees, requiring entrants to make a sharp turn before entering the space beyond.

Biffa: Type of trebuchet.

Blide/a (German). Siege artillery generically but became applied to the trebuchet.

Bombard: Type of gun, commonly applied to the very large weapons built especially in the fifteenth century.

Büchse: (German). A gun. A master gunner was a Büchsenmeister.

Burh: (Anglo-Saxon). A fortified site, first constructed under Alfred the Great of Wessex.

Capitulary: (from Latin). A legal or administrative instruction issued by Carolingian rulers (eighth and ninth centuries) so named because of its division into chapters.

Carolingian: Refers to the empire of Charlemagne (d.814) and his descendants in (modern) France, Italy and (western) Germany and is attached to many other features of the period (e.g. 'Carolingian renaissance').

Carvel : Ship-building method based on butting the planks forming the hull end to end, the whole structure held together by internal frames. First developed in Iberia in the fourteenth century.

'Castle': A fighting platform built on trading ships (e.g. cogs) to enable archers to shoot at the crews of enemy ships. Occasionally carried artillery.

'Cat': (Latin Gattus). Wooden shelter used in siege operations.

Chelandion (plural chelandia): (Greek). Byzantine ship. Meaning seems to have moved from being a (larger) type of war galley to also describing transport ships able to carry horses.

Chevauchée: (French). The fast-moving raids carried out by usually small, mounted English and allied armies designed to devastate French towns and countryside during the Hundred Years War (1337-1453).

Clinker: Ship-building method relying on nailing overlapping planks together. In regular use in northern Europe (from Viking longships to Hanseatic cogs) from early in the first millennium CE.

Cog: Clinker built sailing ship common in northern Europe and the Baltic from the tenth century.

Countermine: A tunnel dug by defenders of a fortification to intercept a besieger's mine in order to neutralise the threat it posed.

Culverin: Type of gun, applied to various sizes until eventually becoming one of the standard categories of cannon by the sixteenth century.

Dioptra: (Greek). See Alidade.

Dromon: (Greek). Byzantine warship, a galley. Was the mainstay of the Byzantine navy for hundreds of years although changing in size and form over time.

Fabrica: (Latin). Arms factory of the Roman empire (see chapter 2).

Fatimid: A Muslim dynasty that established a rival caliphate to the Abbasids based in North Africa, Egypt and Syria.

Galley: Warship propelled by bank(s) of oars on each side, the standard form of naval vessel in the Mediterranean from classical times until the sixteenth century.

Greek Fire: often used to describe any type of incendiary weapon, the original meant a Byzantine invention of the late seventh century that could burn in water (see chapters 2 and 3).

Hoarding (or Hourd): Timber platform erected from the battlements of a fortification to enable soldiers standing inside to drop missiles on enemies attacking the base of the wall or tower.

Liberal Arts: the seven subjects of the curriculum taught for centuries in church-run schools then in universities, designed to train future clergy. They were divided into three categories, known as the trivium (logic, grammar and rhetoric), while the quadrivium comprised arithmetic, geometry, music and astronomy/astrology. Several of these subjects could have practical significance. There has been a long debate as to how far (or when) this teaching did impact on the world of the 'vulgar arts' (i.e. craft skills like engineering and building which belonged to common people) because the purpose of formal education was to gain 'knowledge' (Latin Scientia) quite separate from practical applications.

Machicolation: A permanent form of hoarding, comprising a stone projection from the top of a wall or tower, with holes for dropping missiles on anyone below. Sometimes continuous, it could also be a single structure (e.g. over a door) when it is called a box machicolation. In later periods they could be added for purely decorative effect.

Mangonel: A stone thrower operated by ropes pulled by teams of people. Became a generic term for all types of catapult. See chapter 2.

Merovingian: The dynasty of Frankish kings descended from Clovis, who established his realm in late Roman Gaul in the early sixth century.

Mine: The application of industrial mining techniques to capturing a fortification by driving a tunnel under the wall/tower, excavating a chamber then filling it with combustible material, before setting it alight to burn the props and (hopefully) collapse the masonry above to create a breach.

Munera publica: (Latin) A service required from all people to provide labour for the state to (for example) maintain bridges or fortifications. In England, for example, it took the form of the 'trinoda necessitas', an obligation on all residents to work on bridges, fortifications and roads (see chapter 2).

Naphtha: (Arabic Naft). Petroleum used as the basis for incendiary weapons (see also Greek Fire).

Onager: (Latin). A torsion driven stone thrower introduced in late Roman times. Its survival into medieval times is hotly disputed.

Petrary: (Greek, Latin). A generic term for a stone throwing catapult.

Ram: (Latin Aries). A battering ram, usually a tree trunk provided with an iron tip and operated by teams of people driving it against a wall or gate.

Sap: Sometimes used as a synonym for mining, strictly means undermining a wall or tower by hacking away at the base above ground level.

Serpentine: Type of gun of medium size that appeared in the second half of the fifteenth century.

Sow: One of several animal names applied to timber shelters for besieging soldiers.

Springald: A torsion-driven bolt-throwing engine that appeared during the thirteenth century (see chapter 5).

Trebuchet: (Origin unknown). The counterweight stone thrower first described in the late twelfth century, which became the standard 'heavy artillery' for more than two centuries (see chapter 5). Confusingly, the name has been applied to manually- operated stone throwers (e.g. mangonel, petrary) as 'traction trebuchet'.

Ummayad: The first dynasty of Muslim caliphs following the death of Muḥammad and his immediate successors. They were overthrown in 750 but one branch continued and established a rival caliphate in al Andalus (Iberia).

Veuglaire: Translates as 'fowler', a type of small gun.

Bibliography of Primary Sources

Abbo: Abbon, *Le siège de Paris par les Normands*, ed./trans. H Waquet, Paris 1942.

Abou Chamah (Abu Sharma), 'Livre des deux jardins. Histoire des deux règnes', trans. A-C Barbier de Reynard, *RHC, Historiens orientaux* IV, Paris 1898.

Adam of Bremen, *History of the archbishops of Hamburg-Bremen*, ed./trans F J Schan; new edn ed. T Reuter, New York 2002.

Adae Murimuth, *Continuatio chronicorum Robertus de Avebury. De gestis mirabilibus regis Edwardi Tertii*, ed. E M Thompson, RS, London 1889

Aegidii Romani, Bituriensis Archiepiscopi, *Libellus de re militari veterum ad mores praesertim medii aevi*, ed. S F Hahn, Brunswick 1724.

Aegidii Romani, *Opera Omnia: I: Catalogo dei manoscritti (1001–1075) De regimine principum*, ed. F Del Punta, C Luna, Firenze 1993.

Agathias, *The histories*, trans. J D Frendo, *Corpus Fontium Historiae Byzantinae* IIa, Berlin/New York 1975.

Agricola, G, *De re metallica*, trans. H C and L H Hoover, New York 1950 (orig. edn 1912).

Albert of Aachen: *History of the journey to Jerusalem*, trans. S B Edgington, 2 vols, Crusade Texts in Translation 24, 25, Farnham 2013.

Alcuin, 'De rhetorica et virtutibus, Opera didascalica', Migne *PL* 101, 1851.

Aldhelm: Aldhelmus, 'De virginitate', ed. R Ehwald, *MGH* AA 15, Berlin 1919.

Alpertus of Metz, *De diversitate temporum libri II*, ed. G H Pertz, *MGH SS* IV, 700-723. Hannover, 1841

Amatus of Monte Cassino, *History of the Normans*, trans. P N Dunbar, ed. G A Loud, Woodbridge 2004.

Amelli, A M (ed.), *Miniature sacre e profane dell'anno 1023 illustrati l'enciclopedia medioevale di Rabano Mauro*, Montecassino 1896.

Amira, K von, *Die Dresdener Bilderhandschrift des Sachsenspiegels*, Leipzig 1902.

Ammianus Marcellinus, *Rerum gestarum libri qui supersunt*, trans. J C Rolfe, London/Cambridge (Mass.) 1938

Anglo-Saxon Chronicle, ed. G N Garmonsway, London/New York 1954.

Anna Comnena: *The Alexiad of Anna Comnena*, trans. E R A Sewter, Harmondsworth 1969.

Annales Cestrienses, or chronicle of the Abbey of S. Werburg at Chester, ed. R Christie, Lancashire and Cheshire Record Society XIV, 1886.

Annals of Fulda, trans. T Reuter, Manchester, 1992.

Annales Marbacenses, ed. G H Pertz, *MGH* SS XVII, Hannover 1861.

Annales Mettenses priores, ed. B von Simson, *MGH* SrG, X, Hannover 1905

Annales Placentini Guelphi, ed. G H Pertz, *MGH* SS XVIII, Hannover 1863.

Annales prioratus de Dunstaplia, *Annales Monastici* III, ed. H Luard, RS, London 1866.

Annales Quedlinburgenses, in *MGH SS* III, 22-90. Ed. G H Pertz, Hannover, 1839.

Annales Wormatienses, ed. G H Pertz, *MGH SS* 17 (19), Hannover 1861.

Annalista Saxona, in G Loud, *Roger II and the Making of the Kingdom of Sicily*, Manchester/New York 2012 (270–8); and ed. K Nass, in *MGH* SS XXXVII, 2006

Anon., 'An account of the seaborne journey of the pilgrims heading to Jerusalem who captured Silves in 1189', in *The Crusade of Frederick Barbarossa*, trans. G A Loud from C W David, 'Narratio', in *Proceedings of the American Philosophical Society* 81, 1939, Crusade Texts in Translation 19, Farnham 2013 (193–208).

Anon., *The crusade of Frederick Barbarossa. The history of the expedition of the Emperor Frederick and related texts*, trans. G A Loud, Crusade Texts in Translation 19, Farnham 2013.

Anon., 'Gesta Berengari imperatoris', ed. P de Winterfeld, *MGH* Poetarum Latinorum IV (1), Berlin 1899.

Anon., 'On Skirmishing', in G T Dennis (ed.), *Three Byzantine military treatises*, Dumbarton Oaks Texts IX, Washington 1985.

Anon., 'The first and second crusades from an Armenian Syriac chronicle', trans. A Tritton, *Journal of the Royal Asiatic Society*, 1933

Anon., 'Vita Aniani Episcopi Aurelianensis', ed. B Krusch, *MGH* SRM IV, Hannover 1902.

Anon., 'Vita Desiderii Cadurcae Urbis Episcopi', ed. B Krusch, in *MGH* SRM IV, Hannover 1902.

Anonymi gesta Francorum et aliorum Hierosolymitanorum, ed. B A Lees, Oxford 1924.

'Anonymi Novocomensis Cumanus sive poema de bello et excidio urbis Comensis ab anno MCXVIII usque ad MCXXVII', ed. J M Stampa, *Muratori*, *RIS* SS V, Milan 1724.

Ansari: Umar ibn Ibrahim al-awsi al-Ansari, *A Muslim manual of war*, trans. G T S Scanlon, Cairo 1961.

Appelt, H (ed.), *Die Urkunden der deutschen Könige und Kaiser. Die Urkunden Friedrichs I 1158–67*, *MGH* Diplomata regum et imperatorem Germaniae 10, pt 2, Hannover 1979.

Attaleiates, M, *The history*, trans. A Kaldellis, D Krallis, Dumbarton Oaks Medieval Library, Cambridge (Mass.)/London 2012.

Avebury *see* Adae Murimuth

Bacon, Roger, 'Communica Mathematica', ed. R Stede, *Opera hactenis inedita Rogerii Baconi*, fasc. XVI, Oxford 1940.

Bacon, Roger, 'Opus tertium', in A G Little (ed.), 'Part of the opus tertium of Roger Bacon including a fragment now printed for the first time', *British Society of Franciscan Studies* IV, Aberdeen 1912.

Bahā' al-Dīn Ibn Shaddād, *The rare and excellent history of Saladin*, trans. D S Richards, Crusade Texts in Translation, Aldershot 2002.

Al Baihaḳī, Abu'l Faẓl, *Tārīkhu-s Subuktigīn*.In *The history of India, as told by its own historians*, II, 53-154. Ed. H M Elliot and J Dowson. London, 1867.

Bain, J (ed.), *Calendar of documents relating to Scotland in the PRO*, 4 vols, Edinburgh 1884–87.

Baladhuri: Abu-l Abbas Ahmad ibn Jabir al-Baladhuri, *The origins of the Islamic state (Kitāb Futūh al Buldān)*, trans. PK Hitti, Piscataway (NJ), reprint 2002.

Barbarossa in Italy, translation of the *Carmen de gestis Frederici I imperatoris in Lombardia* by T Carson, New York 1994.

Barbour, J, *The Bruce*, ed./trans. A A M Duncan, Edinburgh 1999

Bardonnet, A (ed.), *Comptes d'Alfonse de Poitiers (1243–47)*, Archives Historiques de Poitou IV, Poitiers 1875 (1–234).

Barni, Zia'd-al Din, 'Tarikh-l Firuz Shahi', in H M Elliot, J Dawson, *The history of India* III, London 1871 (146–233).

Bartholomew: 'Bartholomei scribae Annales', ed. G H Pertz, *MGH* SS XVIII, Hannover 1863.

Bates, D (ed.), *Regesta regum Anglo-Normannorum. The Acta of William I (1066–87)*, Oxford 1998.

Bede, 'Vita beatorum Abbatum Benedicti, Ceolfridi, Eosterwini, Sigfridi atque Hwaetberhti', in C Plummer (ed.), *Venerabilis Baedae Opera Historica*, vol.1, Oxford 1896.

Belgrano, L T (ed.), 'Une charte de nolis de S. Louis', *Archives de l'orient Latin* II, 1884.

Bengedans, Johannes, *Krigskunt og Kanoner. Kriegskunst und Kanonen*, ed. H Blosen, R A Olsen, 2 vols, Aarhus 2006.

Bernoulli, A (ed.), 'Chronikalien der Rathsbücher 1356–1548', *Basler Chroniken* IV, Leipzig 1890.

Biruni: Al-Bīrūnī: Jamil Ali (trans.), *The determination of the coordinates of positions for the correction of distances between cities. Translation of al-Bīrūnī's Kitāb taḥdīd Nihāyāt al-Amākin Litaṣḥīḥ Masāfāt al-Masākin*, Beirut 1967.

Bothmer, H-C Graf von, 'Architekturbilder im Koran – eine Prachthandschrift der Umayyadenzeit aus dem Yemen', *Pantheon* XLV, 1987 (4–20).

Brooks, E W, 'The campaign of 716-18 from Arabic sources', *Journal of Hellenic Studies*, 19, 1899, 19-31.

Caffaro di Rustico da Caschifellone et al., *Annali genovese di Caffaro e de' suoi continuatori, dal MXCIX al MCCXCIII*, ed. L T Belgrano, C I di Sant'Angelo, 5 vols, Fonti per la storia d'Italia, 11–14bis, Genoa and Rome 1890–1929; and M Hall, J Phillips (ed./trans), *Caffaro, Genoa and the twelfth-century crusades*, Crusade Texts in Translation 24, Farnham/Burlington, 2013.

Cahen, C, 'Un traité d'armurie composé pour Saladin', *Bulletin d'études orientales* XII, 1947–48.

Calendar of Close Rolls, Public Record Office: 47 volumes, organised by reign, London, 1892–1963.

Calendar of Liberate Rolls, Public Record Office: Henry III, 6 vols London, 1916–64.

Calendar of Patent Rolls: Henry V, I (1413–16), Public Record Office, London 1910.

Calendar of various Chancery Rolls, Supplementary Close Rolls, Welsh Rolls, Scutage Rolls AD 1277–1326, RS 140, London 1912.

Campbell, B, 'The writings of the Roman land surveyors. Introduction, text, translation and commentary', *Journal of Roman Studies* Monograph 9, 2000.

Capitulare Mantuanum, ed. A Boretius, *MGH* Legum II, Capitularia regum Francorum, vol. 1 pt 1, Hannover 1881.

Capitulare de Villis, ed. A Boretius, *MGH* Legum II, Capitularia regum Francorum, vol. 1 pt 1, Hannover 1881.Capitularia regum Francorum, ed. A Boretius, *MGH* Legum sectio II, Hannover 1881.

Carr Laughton, L G, 'Naval accounts for 1209–1211 (Pipe Roll 14 John – PRO E373/58r.8b), *Mariner's Mirror* 28, 1942 (74–7).

Cassiodorus, 'Variae', ed. T Mommsen, *MGH* AA 12, Berlin 1894; and Barnish, S J B (ed./trans), *Variae of Magnus Aurelius Cassiodorus Senator*, Liverpool 1992.

Cedrenus, G, 'Compendium Historiarum', Migne *PG* 121–2, 1864.

Champollion-Figeac, M (ed.), *Documents historiques inédits tirés des collections manuscrits de la Bibliothèque Royale*, I, Paris 1841.

Chapeauville, J, *Gesta pontificum Leodiensis scripserunt auctores praecipe*, vol. II, Liège 1613.

Chastellain, Georges, *Chronique*, ed. Kervyn de Lettenhove, Brussels 1863–66.

Chaucer, Geoffrey, *The Canterbury Tales*, ed. J Mann, London 2005.

Chiesa, G della, 'Cronaca di Saluzzo', *Monumenta Historiae Patriae*, Scriptorum 3, Torino 1848.

Choniates, N, *O City of Byzantium!*, trans. H J Magoulias, Detroit 1984.

Chronica Iohannis Vitadurani, ed. F Baethgen, *MGH* SrG NS III, Berlin 1924.

Chronicle of Pierre de Langtoft, ed. T Wright, RS 47, London 1868.

Chronicle of Walter of Guisborough previously edited as the chronicle of Walter of Hemingford or Hemingburgh, ed. H Rothwell, Historical Society of Great Britain, Camden Series LXXXIX, 1957.

Chronicon Pascale 284-628AD, Trans. M and M Whitby. Translated texts for historians 7, Liverpool 1989.

Chroniques de Saint Denis, extracts in J D Guigniant, N de Wailly (eds), *Recueil des historiens des Gaules et de la France*, vol. 21, Paris 1855.

Cibrario, L, *Cronaca d'Usseglio ricavata da documenti autentici degli archivii regii, camerali, della metropolitan, arcivescovile, parocchiale e comunali*, Torino 1854.

Codice diplomatico Eceliniano, ed. G Verci, *Storia degli Ecelini* III, Bassano 1779.

Coggeshall: Radulphi de Coggeshall, *Chronicon Anglicanum*, ed. J Stevenson, *RS* 66, London 1875.

Colmar chronicle: *Les annales et la chronique des Dominicains de Colmar*, ed./trans C Gérard, J Liblin, Colmar, 1854.

Constantine Porphyrogenetos, *The book of ceremonies*, trans. A Moffat, M Tall, Byzantina Australensia 18, Canberra 2012.

Constantine Porphyrogenitus, Emperor: *Constantini Porphyrogeniti Imperatoris. De Ceremoniis aulae Byzantinae, libro duo (Gr+Lat.)*, ed. I I Reiske, CSHB, Bonn, 1829.

Corpus scriptorum mozarabicorum, ed. J Gil, 2 vols, Madrid, 1973.

D'Angelo, Francesco (Il Cecca): from Vasari, *Lives of the artists*, trans. A De Angelis, accessed at http://members.cfn.org/~acd/vite/vasariCecca.html (12/2011).

De expugnatio Lyxbonensi (The conquest of Lisbon), ed./trans C W David, Records of civilization: sources and studies 24, New York 1936.

Delaborde, H-F, Petit-Dutaillis, C (eds), *Recueil des actes de Philippe Auguste, roi de France, II (1194–1206)*, Paris 1943.

Depping, G-B (ed.), *Réglemens sur les arts et métiers de Paris rédigés au XIIIe siècle sous le nom des Métiers d'Etienne Boileau*, Paris 1837.

Desclot, B, *Chronicle of the Reign of King Pedro III of Aragon*, trans. F L Critchlow, 2 vols, Princeton 1928–34.

Devon, F, *Issues of the Exchequer; being a collection of payments made out of His Majesty's revenue from King Henry III to King Henry VI inclusive*, London 1837

Diceto: Radulfi de Diceto, *Ymagines historiarum*, ed. W Stubbs, RS 68, London 1876.

Dionysius of Tell Maḥrē, 'The secular chronicle' in *The seventh century in the west Syrian chronicles*, 111-220. Trans. A Palmer et al. Translated texts for historians, 15, Liverpool 1993.

Domesday Book: A Williams, G Martin (eds), *Domesday Book: a complete translation*, Alecto Historical Editions, London 1992, republished 2002.

Dynter, E de, *Chronica nobilissimorum ducum Lotharingiae et Brabantiae*, ed. F X de Rain, Brussels, 1854

Ebendorfer, Thomas, 'Chronica Austriae', ed. A Lhotsky, *MGH* SrG NS XIII, Berlin/Zürich 1967.

Einhard, *Life of Charlemagne*, in *Two lives of Charlemagne*, ed./trans. L Thorpe, London 1969.

Ellenhardi Chronicon, ed. G H Pertz, *MGH* SS XVII, Hannover 1861.

Elmham, T: Elmhami, 'Liber metricus de Henrico V[to]', in C A Cole (ed.), *Memorials of Henry the Fifth, King of England*, RS 11, London 1858

Elmham, Thomas de (pseudo-Elmham), *Vita et gesta Henrici Quinti, Anglorum regis*, Oxford 1727

English Historical Documents IV (1327–1485), ed. A R Myers, London 1969.

Eustathios of Thessaloniki, *The capture of Thessaloniki*, ed./trans. J R Melville Jones, Canberra 1988.

Exchequer Rolls of Scotland, vol. IV, ed. G Burnett, Edinburgh 1880.

Expugnatio Lyxbonensi, De: The conquest of Lisbon, trans. C W David, new foreword and bibliography by J Phillips. Records of Western Civilization, New York, 2001 (original ed. 1936)

Falco of Benevento, 'Chronicle', in G Loud, *Roger II and the Making of the Kingdom of Sicily*, Manchester/New York 2012 (130–249).

Fawtier, R, Maillard, F (eds.), *Comptes royaux (1285–1314)*, *Recueil des historiens de la France, Document Financiers III*, 3 vols, Paris 1953–56.

Festugière, A-J, OP, 'Vie de Théodore de Sykéôn', *Subsidia Hagiographica* 48, Brussels 1970.

Fiamma, G, *Chronicon Extravagans et Chronicon Maius*, ed. A Ceruti, Milan 1869.

Fiamma, G, *Manipulus Florum. Cronaca milanese del trecento. Capitoli CLXXIII–CCXXI. Federico Barbarossa e Milano*, trans. R Frigeno, Biblioteca Milanese I, 1993.

Fibonacci: Leonardo Pisano Fibonacci, *The book of squares*, trans. L E Sigler, Boston 1987.

Fibonacci's Liber Abaci, a translation into modern English of Leonardi Pisano's Book of Calculation, trans. L E Sigler, Sources and studies in the history of mathematics and physical sciences, New York 2002.

Filangieri, R et al. (eds.), *I registri della cancellaria Angiona*, Naples 1950–.

Flodoard, *Les Annales*, ed. P Lauer, Collection des textes, Paris 1905. New edn, *The Annals of Flodoard of Reims 919–966*, ed. S Fanning, trans. B S Bachrach, Readings in medieval civilizations and cultures IX, 2004.

Flores Historiarum, ed. H Luard, RS 95, London 1890.

Fordun, J de, *Scotichronicon (continuation by Bower)*, ed. Hearne, and ed. W F Skene, Edinburgh 1872, repr. 1993

Franke, 'Die Belagerung von Hsiang-Yang' in G Ulmen (ed.), *Society and history: essays in honour of K A Wittvogel*, The Hague, 1978

Fredegar, *Chronicorum liber quartus cum continuationibus = the fourth book of the chronicle of Fredegar with its continuations*, ed./trans. J M Wallace-Hadrill, Nelson's medieval texts, London, 1960

Froissart, J, *Chroniques*, livre 1 ed. G T Diller, Geneva 1991/2; livre II ed G Reynard Paris 1894-7; Livre III ed L & A Mirot, Paris, 1931, 1957, 1966

Frontinus, *The stratagems and the aqueducts of Rome*, trans. E Bennett, ed. M B McElwan, London/New York 1925.

Frutolf: 'Chronicle of Frutolf of Michelsberg', in T J H McCarthy (ed.), *Chronicles of the investiture contest: Frutolf of Michelsberg and his continuators*. Manchester medieval sources, Manchester 2014.

Fulda: Reuter, T (ed. /trans.), *The annals of Fulda*, Manchester Medieval Sources, Ninth-Century Histories II, Manchester 1992.

Gardizi: 'Gardizi on India'. Trans. V Minorsky, *Bulletin of the School of Oriental and African studies*, 12 (1948), 65-6, 210-12.

Georgius Pisida, 'Bellum avaricum' in Migne, *PG*, XCII: 1259-94, Paris 1860; also *Giorgio di Pisidia Poemi I: panegyrici*, ed./trans. A Pertusi. Studia patristica et byzantina, 7. Ettal, 1960.

Gérard, C, Liblin, J (eds), *Les annales et la chronique de Colmar*, Colmar 1854.

Géraud, H, *Paris sous Philippe le Bel. D'après des documents originaux et notamment d'après un manuscrit contenant "le rôle de la taille" imposée sur les habitants de Paris en 1292*, Tübingen 1991.

Gerbert: Pratt Lattin, H (ed./trans), *The letters of Gerbert with his papal privileges as Sylvester II*, New York 1961.

Gerbert of Metz: *Gerbert de Metz, Chanson de geste du XII*[e] *siècle*, ed. P Taylor, Namur/Louvain/Lille 1952

Gerhardi Vita Odalrici episcopi, ed. G Waitz, *MGH SS* IV, Hannover 1841 (381–428).

Gesta Berengarii imperatoris, in *MGH poetae latinae* IV/1: 354-403. Ed. P de Winterfeld, Berlin, 1899.

Gesta Henrici Quinti, ed./trans. F Taylor, J S Roskell, Oxford 1975.

Gesta principum Polonorum. The deeds of the princes of the Poles. Trans P W Knell, F Schauer. Central European medieval texts, 3. Budapest/New York, 2003

Gesta Treverorum', ed. G Waitz, *MGH SS* VIII, Hannover 1848.

Gingins la Sarra, F, *Depêches des Ambassadeurs Milanais sur les campagnes de Charles le Hardi (1474–77)*, Paris/Geneva 1858.

Giorgio Martini, Francesco di, *Trattati di architettura ingegneria e arte militare*, ed. C Maltese, 2 vols, Milan 1967.

Glaber, Rodulfus, *Historiarum libri quinque= The five books of the histories*, ed./trans. J France, in *Rodulfi Glabri Historiarum libri quinque; Vita domni Wilhelmio abbatis*. Oxford medieval texts. Oxford 1989

Gregory of Tours: 'Gregorii episcopi Turonensis liber vitae partum', ed. B Krusch, W Arndt, *MGH* SRM I, *Gregorii Turonsis Opera pars II*, Hannover 1885.

Gregory of Tours, *History of the Franks*, trans. L Thorpe, London 1974.

Guiart, G, 'Branche des royaux lignages. Chronique métrique', *Collection des chroniques nationales françaises*, ed. J-A Bichon, t. VII, VIII, Paris 1828.

Guido da Vigevano: see Hall, A R and Ostuni, G.

Guigniant, J D, de Wailly, N (eds), 'L'abrégement des despens faiz en la voye d'Aragon', *Recueil des historiens des Gaules et de la France*, vol. 21, Paris 1855.

Guigniant, J D, de Wailly, N (eds), 'Dépenses de St Louis de MCCL à MCCLIII', *Recueil des historiens des Gaules et de la France*, vol. 21, Paris 1855.

Guigniant, J D, de Wailly, N (eds), 'Recepta et expensa anno MCCXXXIIII', *Recueil des historiens des Gaules et de la France*, vol. 21, Paris 1855.

Guillaume le Breton, *La Philippide*, trans. M Guizot, Paris 1825.

Guillaume de Tudela et l'anonyme, *La chanson de la croisade Albigense*, trans. H Gougaud, Paris 1984.

Gundisalvo, D, *De Scientiis. Compilación a base principalmente de la texto de al-Fārābī. Texto Latino establecido por el P. Manuel Alonso Alonso, SJ.* Madrid/Granada, 1954; 'Classification of the sciences', trans. M Clagett, E Grant, in E Grant (ed.), *A sourcebook in medieval science*, Cambridge (Mass.), 1974.

Hagen, F H von, *Ludwig des Frommen Kreuzfahrt. Heldengedicht der Belagerung von Akken am ende des zwölften Jahrhunderts*, Leipzig 1854.

Hagen, G, 'Dit is dat boich van der stede Colne', *Die Chroniken der deutschen Städte 12: Niederrheinischen Städte I*, Gottingen, 1968 (reprint of 1875 edn).

Hahn, N L, 'Medieval mensuration: Quadrans vetus and Geometrie due sunt partes principales', *Transactions of the American Philosophical Society* 72, pt 8, 1972.

Hall, A R, 'Guido's Texaurus, 1335', in B S Hall, D C West (eds), *On pre-modern technology and science. A volume of studies in honour of Lynn White jr.*, Malibu 1976 (11–52).

Hall, B S, *The technological illustrations of the so-called "Anonymous" of the Hussite wars, codex latinus Monacensis 197*, pt 1, Wiesbaden 1979.

Hana, C, *Bericht* über *die Verteidigung der Stadt Tê-an während der periode K'ai-His 1205-1208 (K'ai His Te-an Chou-Ch'eng Lu, by Wang Chih Yuan)*, Wiesbaden 1970

Hassall, M W C (ed.), *De rebus bellicis*, BAR International series 63, Oxford 1979.

Hassenstein, W (ed.), *Das Feuerwerkbuch von 1420*, Munich 1941.

Helmold: 'Helmoldi presbyterii Buzoviensis Cronica Slavorum', 3rd ed., B Schmeidler ed., *MGH* SrG, Hannover 1937.

Henry of Lancaster, *Le Livre de Seyntz Medecines. Unpublished devotional treatise*, ed. A J Arnould, Oxford, 1940

Hill, D R, *The book of ingenious devices (Kitāb al-Ḥiyal) by the Banū (sons of) Mūsa bin Ṣhākir*, Dordrecht/Boston/London 1979.

Hill, D R, 'A treatise on machines, by Ibn Mu'ṅdh Abū' Abd Allāh al-Jagyāṅi', *Journal for the History of Arabic Science* I, 1, 1977 (33–46).

Histoire des ducs de Normandie et des rois d'Angleterre, ed. F Michel, Paris 1840.

History of William Marshal, ed./trans A J Holden, S Gregory, D Crouch, Anglo-Norman Texts Society, 2002, 2004, 2006.

Hocsem: *La chronique de Jean de Hocsem*, new edn, ed. G Kurth, Commission Royale d'Histoire, Brussels 1927

Hoefer, F A, 'De artillerie in de rekeningen der stad Nymegen', *Gelre. Bijdragen en Medelingen* XXIV, 1921 (77–100).

Honorius Augustodonus, 'De animae exsilio et patria', Migne *PL* 172, 1854.

Hrabanus Maurus, 'De procinctu Romanae Miliciae', ed. E Dümmler, *Zeitschrift für Deutsches Alterthum* NS 3, 1872 (443–51).

Hugh of St Victor: *The Didascalion of Hugh of St Victor. A medieval guide to the arts*, trans./ ed. J Taylor, New York/Oxford 1961, 1991.

Hugh of St Victor (attrib.), *Practical geometry (practica geometriae)*, trans./ed. FA Homann, SJ, Milwaukee, 1991.

Ibn al-Athir, *Chronicle of the crusading period from al-kamil fi'l Ta'rikh, part 1: The years 491–541/1097–1146. The coming of the Franks and the Muslim response; part 2: the years 541–589/1146–1193. The age of Nūr-al Dīn and Saladin; part 3: the years 589–629/1193–1231. The Ayyubids after Saladin and the Mongol menace.* Crusade Texts in Translation, trans. D S Richards, Farnham 2006, 2007, 2008.

Ibn Isḥāq, Muḥammad, *Life of Muhammad*, abridged by 'Abd al-Malik ibn Hishām, trans. A Guillaume, London/Lahore, 1955.

Ibn al Qalānisī, *The Damascus chronicle of the crusades, extracted and translated from the chronicle of Ibn al-Qalānisī*, H A R Gibb, University of London Historical Series 5, London 1932.

Ibn Hayyān, Abū Marwān Hayyān ibn Khalaf, *Crónica del califa 'Abdarrahmān III an-Nāsir entre los años 912 y 942 (al-Muqtabis V)*, trans. J Viguera, F Corriente, Textos medievales 64, Saragossa 1987.

Irving, W, *A chronicle of the conquest of Granada (from the MSS of Fray Antonio Agapida)*, repr. London 1986.

Isidore: *The Etymologies of Isidore of Seville*, ed./trans S A Barney, W J Lewis, J A Bead, O Berghof, Cambridge 2006; and Isidoro de Sevilla, *Etimologías, libro XVIII de bello et ludis*, ed. J Cantó Llorca, Paris 2007.

Itinerarium peregrinorum et gesta regis Ricardi, auctore ut videtur Ricardo, canonici S Trinitatis Londoniensis. Chronicles and memorials of the reign of Richard I, vol. 1, ed. W Stubbs, RS, London 1864.

Jāhiz: Pellat, C (ed.), *The life and works of Jāhiz*, London 1969.

James I: *The Book of Deeds of James I of Aragon. A translation of the medieval Catalan Llibre de Fets*, trans. D Smith, H Buffery, Farnham/Burlington 2003 (2010 edn).

Jörg von Nürnberg, *Nachricht von den Turken*, Nürnberg, 1500.

Julian, 'Historia rebellionis Pauli', in Migne *PL* XCVI, (759–98)

Justinian: *Corpus iuris civilis, 3: Novellae*, ed. R Schnell, G Kroll, Berlin 1899.

Karagi, Mohammad al, *La civilisation des eaux cachées*, ed./trans. A Mazaheri, Nice 1973.

Kerns, L. *The Secret of secrets (secreta secretorum). A modern translation with introduction of the Governance of Princes*, Lewiston (NY), 2008.

Kervyn de Lettenhove (ed.), *Récits d'un bourgeois de Valenciennes (XIVe siècle)*, Louvain, 1877.

Khwarizmi: al-Khwārizmī, *The beginnings of algebra*, ed./trans. R Rashed, London/Beirut 2009.

Kilwardby, R, *De ortu scientarum*, ed. A G Judy OP, Auctores Britannici Medii Aevi IV, British Academy and Pontifical Institute of Medieval Studies, Toronto 1976.

The King's mirror – Speculum regale – Konungs Skuggsa, trans. L Larson, New York 1917.

Kink, R (ed.), *Codex Wangianus. Urkundenbuch des Hochstiftes Trent, Fontes rerum Austriacarum*, Kaiserliche Akademie der Wissenschaften in Wien, pt 2, Diploma et Acta, 5, 1852.

Knebel, Johannis, capellani ecclesiae Baseliensis, 'Diarium', *Basler Chroniken*, 3 vols, ed. W Vischer, H Boos, Leipzig 1880.

Koder, J, (ed./trans.), *Das Eparchenbuch Leons des Weisens*. Corpus Fontium Historiae Byzantinae XXXIII, Vienna 1991.

Koelhoffsche Chronik, *Die Chroniken der deutschen Städte, 12: Niederrheinischen Städte, Köln*, I. Gottingen, 1968 (reprint of 1875 edn).

Kranz, H, *Quellen zum Lütticher Steinkohlenbergbau im Mittelalter. Urkunden-register und Rechnungseinträge-Bergrecht*, Aachen 2000.

Kyeser, K, *Der Bellifortis des Konrad Kyeser*, ed. R Cermann, Codices, Manuscripti & Impressi 8, 2013.

Kyeser, K : *Konrad Kyeser Bellifortis. Bayerische Staatsbibliothek Clm 30150*, 2 vols, Berlin, 2000

Lambert of Ardres, *The history of the counts of Guines and lords of Ardres*, trans. L Shopkow, Philadelphia 2001.

Lambert of Hersfeld: *Lamberti monachi Hersfeldensis Annales*, ed. A Schmidt, W D Fritz, Darmstadt 1962

Latham, J D, Patterson W F (eds), *Saracen archery. An English version and exposition of a Mamluke work on archery (ca. AD 1368)*, London 1970.

Lehmann-Brockhaus, O, *Schriftsquellen zur Kunstgeschichte des 11. und 12.Jahrhunderts für Deutschland, Lothringen und Italien*, 2 vols, Berlin 1938.

Lemerle, P, *Les plus anciens recueils des Miracles de Saint Démétrius et le pénétration des Slaves dans les Balkans*, 2 vols, Paris 1979.

Leng, R (ed.), 'Anleitung Schiesspulver zu bereiten, Büchsen zu laden und zu beschiessen. Eine kriegstechnische Bilderhandschrift im cgm 600 der Bayerischen Staatsbibliothek München', *Imagines Medii Aevi* 5, 2000 (23–141).

Lenoir, M (ed. /trans), *Pseudo-Hygin. Des fortifications du camp*, Paris 1979.

Leo VI: *The Taktika of Emperor Leo VI. Text, translation and commentary*, G T Dennis, CFHB XLIX, Dumbarton Oaks, Washington 2010.

Leo the Deacon: A-M Talbot, D F Sullivan with G T Dennis and S McGrath, *The History of Leo the Deacon. Byzantine military expansion in the tenth century*, Dumbarton Oaks Collection, Washington 2005.

Leonardo da Vinci, *Notebooks*, selected I A Richter, ed. T Wells, Oxford 2008.

Lewis, B (ed./trans), *Islam from the prophet Muhammad to the capture of Constantinople, I: Politics and war*, New York/Oxford 1987.

Limoges: 'Majus chronicon Lemovicense a Petro Coral et aliis conscriptum', *Recueil des historiens des Gaules et de la France*, vol. 21, ed. J D Guigniant and N de Wailly, Paris 1855

Little, D P, 'The fall of 'Akkā in 690/1291: the Muslim version', in M Sharon (ed.), *Studies in Islamic history and civilization in honour of Professor David Ayalon*, Jerusalem/Leiden 1986 (159–82).

Lohrmann, D, 'Das Maschinenbuch des Konrad Grüter für Erich VII, König van Dänemark (1424)', *Deutsche Archiv für Erforschung des Mittelalters* 63, 2007 (71–92).

Long, P O, McGee, D M, Stahl, A M (eds), *The book of Michael of Rhodes. A fifteenth-century maritime manuscript*, 3 vols, Cambridge (Mass.)/London 2009.

Lübeck: *Urkundenbuch der Stadt Lübeck*. Vereine für Lübeckische Geschichte und Altertumskunde III, Lübeck 1871.

Lull, R: Raimundi Lulli, *Opera Latina. Arbor Scientiae*, ed. V Varneda, Corpus Christianorum, Continuatio Mediaevalis CLXXXA, 3 vols, Turnhout 2000.

Lyon, B and M, *The wardrobe book of 1296–97: a financial and logistical record of Edward I's 1297 autumn campaigns in Flanders against Philip IV of France*, Commission royale d'histoire, Brussels 2004.

Machiavelli, N, *The Art of War*, trans. E Farneworth, ed. N Wood, Indianopolis/NY, 1965

Magnus rotulus scaccarii Normanniae anno dom. MCXCVIII, *Société des Antiquaires de Normandie, Documents historiques* I, ed. Léchaudé-Anisy, Paris 1845.

Maillard, F (ed.), *Comptes royaux (1314–28), Recueil des Historiens de la France, Documents financiers IV*, Paris 1961.

Malaterra: Gaufredi Malaterrae, *De rebus gestis Rogerii Calabriae et Siciliae Comitis et Roberti Guiscardi Ducis fratris eius*, ed. E Pontieri, *RIS* V, pt 1, ed. L A Muratori, new edition ed. G Carlucci, V Fiorini, P Fedele, Bologna 1724.

Map, Walter, *De Nugis curialium*, ed./trans. M R James, revised N L Brooke, R A B Mynors, Oxford 1983.

Mappae clavicula: a little key to the world of medieval techniques, ed./trans. C S Smith, J G Hawthorne. Transactions of the American Philosophical Society, ns 64/4, Philadelphia 1974.

Marche, Oliver de la, *Les Mémoires de Messire Olivier de la Marche*, in *Collection complète des mémoires relatifs à l'histoire de France*, ed. M Petitot, vols 9, 10, Paris 1825–.

Marchisius: 'Marchisii scribae Annales', ed. G H Pertz, *MGH* SS XVIII, Hannover 1863.

'Maronite Chronicle, the (extracts)', in *The seventh century in the west Syrian chronicles*, trans. A Palmer et al., 29-35, Translated texts for historians, 15, Liverpool, 1993.

Maurice, *The Strategikon. Hand book of Byzantine military strategy*, trans. G T Dennis, Philadelphia 1984.

Meehan, D (ed.), 'Adamnan's de locis sanctis', *Scriptores Latini Hibernicae* III, Dublin 1958.

Menander: Blockley, R C (ed./trans), *The history of Menander the Guardsman*, Liverpool 1985.

Milemete: *The treatise of Walter de Milemete 'De nobilitatibus, sapientiis et prudentiis regum' (facsimile of the Christ Church MS) and a selection from 'De secretis secretorum Aristotelis (Holkham MS)'*, ed. M Rhodes James, Roxburghe Club 1913.

Miret i Sans, J, *Itinerari de Jaume I 'El Conqueridor'*, Barcelona 1918.

Miskawaihi, *The concluding portion of the* Experiences of the nations, trans. D S Margoliouth, Oxford 1921.

Molinet: *Chroniques de Jean Molinet*, ed. G Doutrepont, O Jodogne, Brussels 1935.

Molinier, A (ed.), *Correspondence administrative d'Alfonse de Poitiers*, 2 vols, Paris 1894.

Mollat, M (ed.), *Comptes généraux de l'état Bourguignon entre 1416 et 1420, Recueil des Historiens de la France, Documents financiers V*, Paris 1965.

Mortet, V, 'Une formulaire du VIIIe siècle pour les fondations d'édifices et de ponts, d'après des sources d'origine antique', *Bulletin Monumental* 71, 1971 (442–65).

Mortet, V, Deschamps, P, *Recueil des textes relatifs à l'histoire de l'architecture et la condition des architectes en France au moyen âge XIIe–XIIIe siècles*. Paris 1929.

Neira Faleiro, C, *La Notitia Dignitatum, nueva edicioń critica y comentario histórico*, Madrid 2008.

Neocastro, Bartolemoei di, 'Historia Sicula', ed. G Paladino, *RIS* XIII (3), Bologna 1922

Nicephorous (patriarch), *Short history*, ed./trans. C Mango. Dumbarton Oalks papers, 10. CFHB 13, Washington DC, 1990.

Nicole, J (trans.), *Le livre du Préfet, ou l'edit de l'empereur Léon le Sage*, Geneva 1894.

Niẓām al-Mulk, 'Sigāsat-nāma', in B Lewis (ed./trans.), *Islam from the Prophet Muhammad to the capture of Constantinople*, I, New York/Oxford 1987 (181–3)

Notker Babuli, 'Gesta Karoli Magni Imperatoris', ed. H F Haefele, *MGH* SrG NS XII, Berlin 1962.

Novare, P de, *The wars of Frederick II against the Ibelins in Syria and Cyprus*, trans. J La Monte, New York 1936.

Nuwenburg: Matthiae de Nuwenburg, 'Gesta Bertholdi Episcopi Argentinensis', *MGH* SS NS 4, II, 518.

Odonis de Diogilo, 'De Ludovico VII Francorum rex ... profectione in Orientem', Migne *PL* 185 (1202–46).

Ogier: Butts, M, Linker, R (eds/trans), *Champion of Charlemagne (Les infortunes d'Ogier le Danois)*, London/Sydney/Toronto 1967.

Orderic Vitalis: Chibnall, M (ed./trans.), *The ecclesiastical history of Orderic Vitalis*, 6 vols, Oxford 1969–90.

Ostuni, G (ed.), *Le machine del Re. Il Texuarus regis Franciae di Guido da Vigevano*, Vigevano 1993.

Otto of Freising, *The deeds of Frederick Barbarossa*, trans. C C Mierow with Richard Emery. Toronto/Buffalo/London, 1994

Otto, bishop of Freising : *The two cities. A chronicle of universal history to the year 1146AD*, trans./ed. C C Mierow, with foreword and updated bibliography by K F Morrison, Records of Western Civilization, Columbia 2002.

Otto Morena: 'Ottonis Morenae historia', ed. F Güterbock, *MGH* SS NS 7, Berlin 1930.

Outremeuse, Jean d', 'Chronique abregé', *Chroniques Liègeoises*, ed. S Balau, E Fairen, II, Brussels, 1931

Pappus: 'Pappus on mechanics and mathematics', in D J Geanakoplos, *Byzantium, church, society and civilisation seen through contemporary eyes*, Chicago/London 1984 (431–2).

Paris: Matthaei Parisiensis, monachi sancti Albani, *Chronica majora*, ed. H R Luard, RS 57, London 1874.

Paris: Matthaei Parisiensis, monachi sancti Albani, *Historia Anglorum, sive, ut vulgo dicitur, Historia minor*, ed. F Madden, RS, London 1866.

Pipe Roll Society, *The Great Rolls of the Exchequer*, Series 1, 1 Henry II to 1 Richard I, vols I–XXXVIII, London, 1844–1925; New Series, 2 Richard I–14 Henry III, vols 1–54, 1925–2005; and The *Pipe Roll of 31 Henry I, Michaelmas 1130*, London 1833 (facsimile edition 1929).

Pisan, Christine de, *The book of deeds of arms and chivalry*, trans. S Willard, ed. C C Willard, Pennsylvania 1999.

Polo, Marco: *The book of ser Marco Polo*, trans./ed. H Yule, London 1903.

Portet, P, *Bertran Boysset. La vie et les oeuvres techniques d'un arpenteur medieval (v.1355–v.1416)*, 2 vols, Paris 2004.

Procopius, *Buildings*, trans. H B Dewing, Cambridge (Mass.)/London 1971.

Procopius, *History of the Wars*, books I and II, trans. H B Dewing, Cambridge (Mass.)/ London 1914 (reprinted 1996).

Prodromos: Hörandner, W (ed.), *Theodoros Prodromos. Historische Gedichte*, Österreichische Akademie der Wissenschaften. Wiener Byzantinische Studien XI, 1974.

Pseudo-Zachariah: *Chronicle of Pseudo-Zachariah Rhetor. Church and war in late antiquity*, ed. G Greatrex, trans. R R Phenix, C B Horn, contributions to introduction S P Brock, W Witakowski, Translated Texts for Historians 55, Liverpool 2011.

Ptolemy of Lucca, *On the government of rulers. De regimine principum*, trans. J Blythe, Philadelphia 1997

Pulgar, F del, *Cronica de los reyes Catolicos*, ed. J de Mota Carriazo, 2 vols, Madrid 1943.

Quartenus de Excadenciis et Revocatis Capitanitae de mandato imperialis maiestatis Frederici Secundi, ed. anon, Monte Cassino, 1903

Rabanus Maurus: see Hrabanus Maurus

Ralph of Caen, *The Gesta Tancredi. A history of the Normans on the first crusade*, trans. B S and D S Bachrach. Crusade Texts in Translation, Farnham 2005.

Raymond of Aguilers: Raimundi de Agiles, 'Historia Francorum qui ceperunt Iherusalem', translated in C Tyerman, *Chronicles of the first crusade*, London 2004.

Regino of Prüm, *Chronik*, ed. E Dümmler, F Kurze, R Rau, *Quellen zur karolingischen Reichsgeschichte*, III. AQ 7, Darmstadt 1960.

Reineri Annales: *Annales Sancti Iacobi Leodiensis. Chronicon breve Leodiense*, ed. L C Bethmann, J Alexandre, Liège 1874.

Renouard, Y (ed.), *Gascon Rolls preserved in the Public Record Office 1307–17*, London 1962.

Richard (le pelerine), *La conquête de Jerusalem, faisant suite à la chanson d'Antioche, renouvelée par Graindor de Douai au XIIIe siècle*, ed. C Hippeau, Paris 1868.

Richard of Devizes, *Chronicon Ricardi Divisiensis de rebus gestis Ricardi primi regis Angliae*, ed. J Stevenson, London 1838.

Richer of Saint-Rémi, *Histories*, ed./trans. J Lake, 2 vols, Dumbarton Oaks Medieval Library, Cambridge (Mass.)/London, 2011.

Ricotta, S, et al. (eds), *Le macchine di Valturio nei documenti dell'archivio storico AMMA*, Torino 1988.

Robert: 'Chronologia Roberti Altissiodoriensis, praemonstratensis ad S.Marianum canonia', in *Recueil des historiens de la France* XVIII, ed. M-J-J Brial, Paris 1822.

Roger of Wendover, *Flowers of History*, trans. J A Giles, Felinfach 1996.

Rolandini Patavini, 'Chronica facta', *MGH* SS XIX, ed. G H Pertz, Hannover 1866 (32–147).

Rolandini Patavini, 'Cronica in factis et circa facta marchie Trivixane (AA. 1200cc–1262)', ed. A Bonardi, *RIS* VIII pt 1 (new edn), Citta di Castello, 1905

Rôles Gascons, transcrits et publiés par F Michel, vol. I, 1242–54 (Paris 1885), vol. II 1273–90 (Paris 1890), vol. III, ed. C Bémont, 1290–1307 (Paris 1906); supplement au t. 1, ed. C Bémont (Paris 1896).

Rose, S (ed.), *The navy of the Lancastrian kings. Accounts and inventories of William Soper, Keeper of the King's ships 1422–27*, Navy Records Society, 1982.

Rotuli de liberate ac de misis et praestitis, regnante Johanne, ed. T Duffus Hardy, London 1844

Rotuli litterarum clausarum in turri Londoniensis asservati, I, 1204–24, ed. T D Hardy, London 1833.

'Rotuli Normanniae in turri Londoniensi asservati Johanne regis Angliae ab anno MCC ad annum MCIV', *Société des Antiquaires de Normandie, Documents historiques I*, ed. Léchaudé-Anisy, Paris 1845.

Rotuli Normanniae in turri Londoniensis asservati, Johanne et Henrico Quinto Angliae Regibus, ed. Hardy T D, London 1835.

'Royal Frankish Annals', trans. B W Scholz, in *Carolingian Chronicles*, Michigan 1972/90.

Rymer, T. *Foedera, conventiones, literae et cujuscumque generis acta publica … ab anno 1101*, 3rd edn, ed. G Holmes, London 1825.

Sächsische Weltchronik – Thüringische Fortsetzung, ed. G Waitz, *MGH* Deutsche Chroniken II, Hannover 1877.

Saint Denis: 'Chroniques de Saint Denis', in *Recueil des historiens des Gaules et de la France*, vol. 21, J D Guigniaut and N de Wailly (eds), Paris 1955.

Salimbene de Adam, *Cronica*, new edn, ed. G Scolia, Bari 1966.

Sanuto: Marinus Sanutus dictus Torsellus, 'Liber secretorum Fidelium Crucis super Terra Sanctae recuperatione et conservatione', reprinted from Bongars edition (1611) in R Schneider, *Artillerie des Mittelalters*, Berlin 1910; Torsello, Marino Sanudo, *The Book of the Secrets of the Faithful of the Cross. Liber secretorum fidelium crucis'*, trans. P Lock, Crusader Texts in Translation 21, Ashgate 2011.

'Sanzanomis Gesta Florentinorum', ed. O Hartwig, *Quellen und Forschungen zur* ältesten *Geschichte der Stadt Florenz* I. Marburg 1875 (1–34).

Saxo grammaticus, *Gesta Danorum*, ed. J Olrik, H Raeder, 2 vols. Copenhagen, 1931-57

Saxon Mirror, trans. M Dobezy, Philadelphia 1999.

Sebeos, bishop, *Histoire d'Héraclius*, trans. F Macler, Paris, 1904.

Seudo Aristoteles, *Poridat de la Poridades*, ed. M A Kasten, Madrid 1957.

Sharp, M (ed.), *Accounts of the Constables of Bristol Castle in the thirteenth and early fourteenth centuries*, Bristol Record Society XXXIV, 1982.

Shelby, L R (ed./trans), *Gothic design techniques. The fifteenth-century design booklets of Mathieu Roriczer and Hans Schuttermayer*, London/Amsterdam 1977.

Las Siete partitas del rey Alfonso el sabio, ed. G Lopez, Paris 1847.

Simeon of Durham, *History of the Kings of England*, Church Historians of England III/2: 423–617, trans. J Stevenson, London 1855.

Simocatta: *The History of Theophylact Simocatta*, ed./trans. M and M Whitby, Oxford 1986 (new edn 1997).

Simpson, G S, Galbraith, J D (eds.), *Calendar of Documents relating to Scotland, V: AD 1108–1516*, Edinburgh 1986.

Skylitzes: Jean Skylitzès, *Empereurs de Constantinople*, trans. B Flusin, ed. J Cheynet, Réalités Byzantines 8, Paris 2003.

Sourdel, D, 'Le livre des secrétaires de ʿAbdallāh al-Baġdādī', *Bulletin d'études orientales* XIV, 1952 (115–53).

Sourdel-Thomine, J, 'Les conseils du Sayh al-Harawi à un prince ayyubide', *Bulletin d'Études Orientales* XVII, 1962.

Step'anos Taronec'i : *Des Stephanos von Taron Armenische Geschichte*, trans. H Gelzer, A Burckhardt. Leipzig, 1907

Sullivan, D F (ed./trans), *The Life of Saint Nikon*, Brookline, Mass. 1987.

Sullivan, D F (ed./trans.), *Siegecraft. Two tenth-century instructional manuals by "Heron of Byzantium"*, Dumbarton Oaks Studies XXXVI, Washington 2000.

Surienne: Richard, J, 'Quelques idées de François de Surienne sur la défense des villes à propos de la fortification de Dijon (1461)', *Annales de Bourgogne* 16, 1944.

Symeon Magister, *Annales*, *CSHB* 43, ed. H Bekker, Bonn 1838.

Al-Tabarī, Muḥammad ibn Jarīr, *The history of al-Tabarī*, trans. C E Bosworth et al, 40 vols, Albany, NY, 1985-2007.

Taccola, M, *De rebus militaribus (de machinis 1449)*, ed. E Knobloch, Baden Baden 1984.

Taccola: Prager, F D, Scaglia, G, *Mariano Taccola and his book De Ingeniis*, Cambridge (Mass.)/London 1972.

Taccola: Scaglia, G (ed.), *Mariano Taccola, De Machinis: the engineering treatise of 1449*, 2 vols, Wiesbaden 1971.

Taghri Birdi, abu-l Mahassin ibn, *History of Egypt 1382–1469*, ed./trans W Popper, University of California Publications on Semitic Philology 17 (1957)–23 (1960).

Taylor, P: see Gerbert of Metz.

Telese, Alexander of, 'History of the most serene Roger, first king of Sicily', in G Loud, *Roger II and the Making of the Kingdom of Sicily*, Manchester/New York 2012 (63–129).

Textus Roffensis, in *Vita Gundulfi: the Life of Gundulf bishop of Rochester*, ed. R Thomson, Toronto Medieval Latin Texts, 1977.

Theodorus syncellus, *Traduction et commentaire de l'homélie* écrite *probablementp par Théodore le syncelle sur le siège de Constantinople en 626*, trans. F Makk, Opuscula byzantina, 3. Szeged 1975.

Theodosio humili diacono, 'Expugnatio Cretae', Migne *PG* CXIII, 1864 (cols 993–1059).

Theophanes: *The Chronicle of Theophanes Confessor anni mundi 6095–6305. Byzantine and near eastern history (AD 602–813)*, ed./trans./commentary C Mango, R Scott, H Turtledove, Philadelphia 1982, Oxford 1997.

Theophanes Continuatus, *CSHB* 43, ed. H Bekker, Bonn 1838.

Theophilus, *De diversibus artibus*, ed./trans. C R Dodwell, Nelson Medieval Texts, Edinburgh/London 1961.

Toneatto, L, *Codices artis mensoriae. I manoscritti degli antichi oposcoli Latini d'agrimensura (V–XIX sec.)*, 3 vols, Spoleto 1994.

Tudeboeuf: *Mémoires de Pierre Tudebode (ou Tudeboeuf) sur son pèlerinage à Jérusalem*, trans. S de Goy, Quimper, 1877.

'Utbi, Muḥammad ibn 'Abd al-Jabbār, *The Kitabi-i-Yamini: historical memoirs of the Amir Sabaktagin, and the sultan Mahmud of Ghazna, early conquerors of Hindustan, and founders of the Ghaznavide dynasty; translated from the Persian version of the contemporary Arabic chronicle of Utbi.* Trans. J Reynolds, Publications of the Oriental Translation Fund, 72. London 1858.

Vegetius, *Epitome of military science*, ed/trans. N P Milner, Liverpool 1993.

Venanti Fortunati, 'Carminium', *MGH AA* IV pt 1, ed. F Leo, Berlin 1881.

Verriest, L (ed.), *Le polyptique illustré dit 'viel rentier' de messier Jehan de Pamele-Audenarde (vers 1275)*, Brussels 1950.

Victor, S K (ed./trans), 'Practical geometry in the High Middle Ages: *Artis cuiuslibet consummatio*, and *Praktike de Geometrie*', *Memoirs American Historical Society* 134, 1979.

Vigevano, Guido da: Hall, A R, 'Guido's Texaurus, 1335', in B S Hall, D C West (eds), *On pre-modern technology and science: a volume of studies in honor of Lynn White jr*, Malibu 1976.

Villani, G, *Croniche Fiorentino (nuova cronica)*, ed. G Porta, Parma, 1990. Translated as *Villani's Chronicle*, by R E Selfe, P H Wickstead, London, 1906

Villani, Matteo, *Cronica*, ed. A Racheli, Trieste, 1858

Villard de Honnecourt: *Carnet*, XIIIe siècle, ed. A Erlande-Brandenburg, R Pernord, J Gimpel, R Bechmann, Paris 1994; and *The medieval sketchbook of Villard de Honnecourt*, ed. T Bowrie, New York 1959 (new edn 2006).

Vita Bennonis II episcopi Osnabrugensis, ed. G H Pertz, *MGH* SS XII (XIV), Hannover 1861.

Vita Bernwardi episcopi Hildesheimensis auctore Thangmari, ed. G H Pertz, *MGH* SS IV, Hannover 1846.

Vita Gundulfi: the Life of Gundulf, bishop of Rochester, ed. R Thomson, Toronto Medieval Latin texts, 1977.

Walsingham, T, Ypogdima Neustriae. *Chronica monasterii S. Albani*, ed. H T Riley, RS, London 1876, reprinted 1965.

Walter of Coventry: *Memoriale fratris Walteri de Coventria*, ed. W Stubbs, RS, London 1873.

Walter of Guisborough: *The chronicle of Walter of Guisborough*, ed. H Rothwell, Camden Society 3rd series 89, London 1957.

Widukind of Corvey, *Res gestae saxonicae a.919-973*, ed. G Waitz, *MGH SS* III: 408-67. Hannover, 1889.

Wierstraat, C, 'Die histori des beleegs van Nuis zusamen', ed. C Noerrenberg, *Die Chroniken der Deutschen Städte* 20, Leipzig 1887.

Wild, B L (ed.), 'The wardrobe accounts of Henry III', *Pipe Roll Society* XCVI, NS LVIII, Loughborough 2012.

William of Tyre, *A history of deeds done beyond the sea*, trans. E A Babcock, 2 vols, Records of Civilization: Sources and Studies 35, New York 1943.

Yahyā Al-Anṭaki, *Cronache dell 'Egitto Fatimide e dell' impero Bizantino 937–1033*, ed./trans B Pirone, Patrimoni Arabo Cristiano 3, Milan 1997.

Zantfliet: 'Chronicon Cornelii Zantfliet, S. Jacobi Leodiensis monachi ab anno MCCXXX ad MCCCLXI', ed. E Martene, U Duran, *Veterum Scriptorum et Monumentorum Historicum, Dogmaticorum, Moralium, Amplissima Collectio* V, Paris, 1729.

Zíáu-d Dín Barní, 'Táríkh-í Fíroz Sháhí', in H Elliot, J Dowson, *The History of India, as Told by Its Own Historians* (1871), III, 191.

Zosimus, *New history*, ed./trans. R T Ridley, Sydney 2006.

Zuqnīn: *The chronicle of Zuqnīn parts III and IV AD 488–775*, ed. A Harrak, Toronto 1999.

Bibliography of Secondary Sources

Aarts, B, 'Early castles of the Meuse-Rhine border region and some parallels in western Europe c1000: a comparative approach, *Château Gaillard* 18 (1998), 11-23

Aarts, B, 'Motte and bailey castles of Europe. Some aspects concerning their origin and evolution', *Virtus* (2007), 37-56

Abels, R, *Alfred the Great. War, kingship and culture in Anglo-Saxon England*, London/New York, 1998.

Abels, R, 'Bookland and Fyrd service in late Saxon England', *Anglo-Norman Studies* 7, 1984 (1–25), reprinted in S Morillo (ed.), *The battle of Hastings. Sources and interpretations*, Woodbridge 1996 (57–78).

Abels, R, Morillo S, 'A lying legacy? A preliminary discussion of images of antiquity and altered reality in medieval military history', *Journal of Medieval Military History* III, 2005 (1–13).

Adamska, A, Mostert, M (eds), *The development of literate mentalities in east central Europe*, Utrecht studies in medieval literacy 9, Turnhout 2003.

Adamson, J W, 'The extent of literacy in England in the fifteenth and sixteenth centuries: notes and conjectures', *The Library* ser. IV, vol. X, 1930 (163-93).

Agoston, G, *Guns for the Sultan. Military power and the weapons industry in the Ottoman empire*. Cambridge/NY, 2005

Agoston, G, 'Ottoman artillery and European military technology in the 15th and 17th centuries', *Acta Orientalia Academia Scientarum Hungaricae* XLVII, 1994.

Agrait, N, 'Castile and the struggle for the sea', *Journal of Medieval Military History* XIII, 2015 (139–66).

Alcock, N W, review of F Epaud, 'De la charpente romane à la charpente gothique en Normandie', *Medieval Archaelogy* 52, 2008 (448–9).

Alessio, F, 'La filosofia e le "artes mechanicae" nel secolo XII'. *Studii Medievale, SSCISSM* ser. 3, year VI, I, 1965 (71–164)

Alford, B W, Barker, T C, *A history of the Carpenters' Company*, London 1968.

Allard, G H, 'Les arts mécaniques aux yeux de l'idéologie médiévale', in G H Allard, S Lusignan (eds), *Les arts mécaniques au moyen âge*, Montreal/Paris 1982 (13–31).

Allmand, C T, 'L'artillerie de l'armée anglaise et son organisation à l'époque de Jeanne d'Arc', in *Jeanne d'Arc. Une époque, un rayonnement. Colloque d'histoire médiévale*, Orléans Octobre 1979, Paris 1982 (73–83).

Allmand, C, *The* De re militari *of Vegetius. The reception, transmission and legacy of a Roman text in the Middle Ages*, Cambridge, 2011.

Allsen, T, 'The circulation of military technology in the Mongolian empire', in N di Cosmo (ed.), *Warfare in inner Asian history*, Leiden/Boston/Köln 2002.
Alm, J, *European crossbows: a survey*, trans. H Bartlett Wells, ed. G M Wilson, Royal Armouries Museum, Leeds 1994.
Amari, M, *Historia dei Musulmani di Sicilia*, 3 vols, Firenze 1868.
Amatuccio, G, *Mirabiliter pugnaverunt. L'esercito del Regno di Sicilia ai tempo di Federico II*, Naples, 2003.
Anastasiadis, M P, 'Pyrotechnics and technomania', *Slingshot, official journal of the Society of Ancients*, May 1993, 1967 (11–14).
Anderson, R C, 'English galleys in 1295', *Mariner's Mirror* XIV, 1928 (220–41).
Anderson, R C, 'Italian naval architecture about 1445', *Mariner's Mirror* XI, 1925 (135–63).
Anderson, T, jr, 'Roman military colonies in Gaul, Salian ethnogenesis and the forgotten meaning of *Pactus Legis Salicae* 59.5', *Early Medieval Europe* 4/2, 1995 (129–44).
Antoniadis-Bibicou, H, Études d'histoire maritime de Byzance. À propos du 'thème des Corvisiens', Paris 1966.
Aris, R, Bachrach, B S, '"De motu arietum" (on the motion of battering rams)', in K D Elworth, W Norrie Everitt, E Bruce Lee (eds), 'Differential equations, dynamical systems and control science. A festschrift in honour of Lawrence Markus.' *Lecture notes in applied mathematics* 152, 1994 (1–14).
Astill, G G et al., *A medieval industrial complex in its landscape: the metalworking watermills and workshops of Bordesley Abbey*, BAR research reports 92, 1993.
Audouin, E, *Essai sur l'armée royale au temps de Philippe Auguste*, Paris 1913.
Ayalon, D, *Gunpowder and firearms in the Mamluk kingdom*, London, 1956
Ayton, A, 'English armies in the fourteenth century', in A Curry, M Hughes (eds), *Arms, armies and fortification in the Hundred Year War*, Woodbridge 1984 (21–38).
Bachrach, B S, *Early Carolingian warfare. Prelude to empire*, Philadelphia, 2001.
Bachrach, B S, 'A "lying legacy" revisited: the Abels-Morillo defense of discontinuity', *Journal of Medieval Military History* V, 2007 (153–93).
Bachrach, B S, *Merovingian military organisation 481–751*, Minneapolis 1972.
Bachrach, B S, 'A note on Alites', *Byzantinische Zeitschrift* 61/1, 1968 (35).
Bachrach, B S, 'On the origins of William the Conqueror's horse transport', *Technology and Culture* 26/3, 1985 (505–31).
Bachrach, B S, 'The practical use of Vegetius' *De re militari* during the early Middle Ages', *The Historian* XLVII/2, 1985 (239–55).
Bachrach, B S, Aris, R, 'Military technology and garrison organization: some observations on Anglo-Saxon military thinking in light of the Burghal Hidage', *Technology and Culture* 31, 1990 (1–17).
Bachrach, D S, 'Crossbows for the king: the crossbow during the reigns of John and Henry III of England', *Technology and Culture* 45/1, 2004 (102–19).
Bachrach, D S, 'Crossbows for the king: the crossbow during the reign of Edward I of England (1272–1307)', *Technology and Culture* 47/1, 2006 (81–90).
Bachrach, D S, 'English artillery 1189–1307: the implications of terminology', *English Historical Review* CXXI, 494, 2006 (1408–29).
Bachrach, D S, 'The military administration of England: the royal artillery (1216–1272)', *Journal of Military History* 68 (2004) (1083–104).

Bachrach, D S, 'Origins of the crossbow industry in England', *Journal of Medieval Military History* II, 2004 (73–88).

Bachrach, D S, 'The royal arms makers of England 1199–1216: a prosopographical survey', *Medieval prosopography* 25, 2004 (48–74).

Bachrach, D S, 'The royal crossbow makers of England, 1204–1272', *Nottingham Medieval Studies* 47, 2003 (168–97).

Bachrach, D S, *Warfare in tenth-century Germany*, Woodbridge 2012.

Bailly-Maitre, M-C, 'Tradition et innovation dans les mines médiévales', in P Beck (ed.), *L'innovation technique au moyen âge. Actes du VI Congrès international d'archéologie médiévale, 1–5 octobre 1996*, Paris 1998 (99–107).

Baines, P, *Spinning wheels. Spinners and spinning*, London 1977.

Bakar, O, *Classification of knowledge in Islam. A study of Islamic philosophies of science*, Cambridge 1998.

Barnes, C F (jr), 'A note on Villard de Honnecourt and metal', in R Bork (ed.), *De re metallica. Avista Studies in the history of medieval technology, science and art*, 4, 2005 (245-54)

Baron, R, 'Sur l'introduction en Occident des termes "geometria theorica et practica"', *Revue des sciences et leurs applications* 8/4, 1955 (298–302).

Bartels, C, 'Der Bergbau im Zentrum das Silber', in U Lindgren (ed.), *Europäische Technik im Mittelalter 800 bis zum 1400. Tradition und Innovation*, Berlin 1996 (235–48).

Bartels, C, 'Mining towns in the north-west Harz district: foundations, structure and development', in P Anrieter et al. (eds), *Mining in European history and its impact on environment and human societies. Proceedings of the first mining in European history conference of the SFB-HIMAT, 12–15 November 2009*, Innsbruck 2010 (35–44).

Bartlett, W, *The Mongols from Genghis Khan to Tamerlane*, Strood 2009.

Basing, P, *Trades and crafts in medieval manuscripts*, London, 1990.

Basista, M, 'Hybrid or counterpoise? A study of transitional trebuchets', *Journal of Medieval Military History* V, 2007 (33–55).

Bastard de Péré, R, 'Navires méditerranéans du temps de Saint Louis', *Revue d'histoire économique et sociale* 50/3, 1972 (327–56).

Batt, F, *Das Eigentum zu Hagenau im Elsass*, Colmar 1876–81.

Battisti, C, Alessio, G, *Dizionario etimologico Italiano*, Florence 1952.

Baudry, M-P, *Le château de Niort*, Niort 2013.

Baudry, M-P, *Châteaux 'romans' en Poitou-Charentes Xᵉ–XIIᵉ siècles*, Cahiers du Patrimoine 95, Poitiers 2011.

Bausani, A, 'Il contributo scientifico', in F Gabrieli (ed.), *Gli Arabi in Italia*, Milan 1985 (629–60).

Bayle, J, 'Le château de Montaillou en 1415', in Colloque international, *Autour de Montaillou, village Occitan*, 2001 (75–85).

Bayrou, L (et al), *La forteresse de Salses*, Paris, 1998

Beaujova, G, 'L'interdépendence entre la science scolaristique et les techniques utilitaires (XII, XIII et XIV siècles)', *Les conférences du palais de la découverte*, ser. D, 46, 1957.

Beaujouan, G, 'Réflexions sur les rapports entre théorie et pratique au moyen âge', in J E Murdoch, E D Sylla (eds), *The cultural context of medieval learning*, Dordrecht 1975 (437–84).

Beaujouan, G, 'The transformation of the quadrivium', originally in R L Benson, G Constable, *Renaissance and renewal in the twelfth century*, Cambridge (Mass.) 1982 (463–87), reprinted in G Beaujouan, *Par raison de nombres: l'art de calcul et les savoirs scientifiques médiévaux*, Aldershot 1991.

Bedwin, O, 'The excavation of Batsford Mill, Warbleton, East Sussex, 1978', *Medieval Archaeology* XXIV, 1980 (187–201).

Beekes, R, *Etymological dictionary of Greek*, Leiden/Boston 2010.

Beffeyte, R, *L'art de guerre au moyen âge*, new edn, Lille/Rennes 2015.

Beffeyte, R, *Les machines de guerre au moyen âge*, Rennes 2000.

Beffeyte, R, 'The oral tradition and Villard de Honnecourt', in M-T Zenner (ed.), *Villard's legacy. Studies in medieval technology, science and art in memory of Jean Gimpel*, Aldershot 2004 (93–119).

Belhoste, J-F, 'Les hommes de fer: réflexions sur l'émergence et la diffusion de l'innovation (XIV-XVII siècle)', in M Arnoux, P Monnet, *Le technicien dans la cité en Europe occidentale 1250–1650*, École Française de Rome, Rome 2004 (271–89).

Bell, A R, *War and the soldier in the fourteenth century*, Woodbridge 2004.

Bennett, J, 'The mechanical arts', in K Park, L Daston (eds), *The Cambridge History of Science, III: Early Modern Science*, Cambridge 2006 (673–95).

Bennett, M, 'Military aspects of the conquest of Lisbon 1147', in J Phillips, M Hoch (eds.), *The second crusade. Scope and consequences*, Manchester/New York 2001 (71–87).

Bennison, A K, *The great caliphs. The golden age of the 'Abbasid empire*, London/New York 2009.

Benoit, P, 'Artisans ou Combattants? Les canonniers dans le royaume de France à la fin du moyen âge', in M Balard (ed.), *Le combattant au moyen âge*, 2nd edn, Paris 1995 (287–96).

Bernard, V, Le Digol, Y, Epaud, F, Pradeo, F, 'Building activity and timber management at the turn of the first millennium AD at Pineuilh, La Moth (Gironde, France): an original dendrochronological approach', *Medieval Archaeology* 56, 2012 (268–82).

Berty, A, 'Topographie du vieux Paris: region du Louvre et des Tuileries', in Baron Haussmann et al., *Histoire générale de Paris. Collection des documents*, Paris 1866.

Biddle, M, Hill, D, 'Late Saxon planned towns', *Antiquaries Journal* LI, 1971 (pt 1) (70–85).

Biller, T, *Die Burgengruppe Windstein und der Burgenbau in den nördischen Vogesen*, Dissertation, Technischen Universität Berlin, Köln 1985.

Binding, G, *Medieval building techniques*, trans. A Cameron, Stroud 2004.

Bindung, G, 'Bischof Benno II von Osnabrück als "architector et dispositor caementarii operis, architectoriae artis valde peritus"', *Zeitschrift des Deutschen Vereins für Kunstwissenschaft* 44/1, 1990 (53–66).

Bindung, G, *Der früh- und hochmittelalterliche Bauherr als sapiens architectus*, Köln 1996.

Bingener, A, 'Medieval metal trade in and around the Harz mountains', in C Segers-Glocke, H Witthöft (eds), *Aspects of mining and smelting in the Upper Harz mountains (up to the 13th/14th centuries) in the early times of a developing European culture and economy*, St Katharinen 2000 (119–40).

Birkenmeier, J W, *The development of the Komnenian army 1081–1180*, History of Warfare 5, Leiden/Boston/Köln 2002,.

Bischoff, B, 'Benedictine monasteries and the survival of classical literature', in M Gorman (ed./trans), *Manuscripts and libraries in the age of Charlemagne*, Cambridge 1994 (134–60).

Bischoff, B, 'The court library of Charlemagne', in M Gorman (ed./trans.), *Manuscripts and libraries in the age of Charlemagne*, Cambridge 1994 (56–75).

Bischoff, B, 'Libraries and schools in the Carolingian revival of learning', in M Gorman (ed. /trans.), *Manuscripts and libraries in the age of Charlemagne*, Cambridge 1994 (93–114).

Bischoff, B, 'Manuscripts in the early Middle Ages', in M Gorman (ed./trans.), *Manuscripts and libraries in the age of Charlemagne*, Cambridge 1994 (1–19).

Bischoff, B, 'Die Überlieferung des technischen Literatur', in *Artiginiato e tecnica nella società dell'alto medioevo occidentale, SSCISSM* XVIII, pt 1, 1971 (267–96)

Blackmore, H, 'The Boxted bombard', *Antiquaries Journal* 67, 1987, 89-96

Blaine, B B, 'The enigmatic water mill', in B S Hall, D C West (eds), *On pre-modern technology and science. A volume of studies in honor of Lynn White, jr*, Malibu 1976 (163–76).

Blair, J, 'Grid planning in Anglo-Saxon settlements: the short perch and the four perch module', *Anglo-Saxon Studies in Archaeology and History* 18, 2013 (18–61).

Blair J, Ramsay, N (eds), *Medieval industries. Craftsmen, techniques, products*, London/New York 1991.

Blanchard, I, *Mining, metallurgy and minting in the Middle Ages. 3: Continuing Afro-European supremacy 1250–1450*, Stuttgart 2005.

Blondel, L, 'L'architecture militaire au temps de Pierre de Savoie: les donjons circulaires', *Genava* XIII, 1935.

Bock, A E R, Dunlop, J E, *Two studies in later Roman and Byzantine administration: part I: The Master of the Offices in the later Roman and Byzantine empires*. New York 1924.

Bodde, D, *Chinese thought, society and science. The intellectual and social background of science and technology in pre-modern China*, Honululu 1991.

Boeheim, W, *Meister der Waffenschmiedekunst von XIV bis ins XVIII Jahrhundert*, Berlin 1897.

Bognetti, G P, 'Gli Antélami e la carpentaria di guerra' in Munera. *Raccolta di scritti in onore di Antonio Giussani, Società Archaeologica Comense*, Milan 1944 (217–24).

Bonaparte, L-N, Études sur le passé et l'avenir de l'artillerie, I-II, Paris 1846-51

Bond, J, 'Canal construction: an introductory review', in J Blair (ed.), *Waterways and canal building in medieval England*, Oxford 2007 (153–206).

Bonde, S, Maines, C, Mark, R, 'Archaeology and engineering: the foundations of Amiens cathedral', in L T Courtenay (ed.), *The engineering of medieval cathedrals*, Aldershot 1997 (341–8).

Bondidi, M, 'The art of designing and building Venetian galleys from the 15th to 16th century', in C Beltrano (ed.), *Boats, ships and shipyards. Proceedings of the 9th International Symposium on boat and ship archaeology Venice 2000*, Oxford 2007 (222–7).

Bosworth, C E, 'Ghaznavid military organisation', *Der Islam* 36, 1961 (37–77)

Bources, C, 'Master craftsmen, craftsmen and building activities in Byzantium', in A E Laiou (ed.), *The economic history of Byzantium from the 7th through the 15th century*, Dumbarton Oaks Studies XXXIX, Washington 2002 (II, 539–54).

Boyer, M N, *Medieval French bridges. A history*, Cambridge (Mass.) 1976.

Boyson, D, 'Romano-Burgundian society in the age of Gundobad: some legal, archaeological and historical evidence', *Nottingham Medieval Studies* XXXII, 1988 (91–118).

Bradbury, J, *The medieval archer*, Woodbridge, 1985

Braid, D, 'Ordnance and freedom of thought: the development of gunmaking in Bohemia, 1350–1450', *Journal of the Ordnance Society* 5, 1993 (75–94).

Branner, R, 'Villard de Honnecourt, Reims and the origin of Gothic architectural drawing', in L T Courtenay (ed.), *The engineering of medieval cathedrals*, Aldershot, 1997 (63–81).

Brasacchio, G, *Storia economica della Calabria*, 3 vols, 1977.

Braun, H, 'Some notes on Bungay castle', *Proceedings of the Suffolk Institute of Archaeology and Natural History* XXII, 1936 (109–19).

Braunstein, P, 'Maîtrise et transmission des connaissances techniques au moyen âge', *History of Technology* 21, 1999 (155–65).

Braunstein, P, 'Savoir et savoir-faire: les transferts techniques', in P Beck (ed.), *L'innovation technique au moyen âge. Actes du VIe congrès international d'archéologie médiévale, 1–5 octobre 1996*, Paris 1998 (303–9).

Braunstein, P, 'Les statuts miniers de l'Europe médiévale', reprinted in *Travail et enterprise au Moyen Age*, Brussels 2003 (141–63).

Britnell, R, 'Pragmatic literacy in Latin Christendom', in the same (ed.), *Pragmatic literacy, east and west 1200–1330*, Woodbridge 1997 (3–14)

Brogolio, G P, 'Architettura in pietra lungo l'Adige tra X e XI secolo', *Archeologia dell'Architettura* XVII, 2012 (80–93).

Bromehead, C N, 'Mining and quarrying to the 17th century', in C Singer et al. (eds.), *A history of technology* II, Oxford 1957 (1–41).

Brooks, F W, 'Naval armament in the thirteenth century', *Mariner's Mirror* XIV, 1928 (115–31).

Brooks, N, 'Medieval European bridges: a window onto the changing concept of state power', in N Brooks, *Communities and warfare 700–1400*, London 2000 (1–31), originally in *Haskins Society Journal* 7, 1995 (11–29).

Brooks, N, 'Rochester Bridge AD 43–1381' in N Brooks, *Communities and warfare 700–1400*, London, 2000, 219–65.

Brown, R A, Colvin H M, Taylor, A J, *History of the king's works, I: The Middle Ages*, London 1963.

Browning, R, 'Byzantines in Bulgaria – late 8th–early 9th centuries', in *Studia Slavico-Byzantina et Medievalia Europensia. In memoriam Ivan Dujčev* I, 1988 (32–6).

Brunn, A, Goldenberg, G, Zimmerman, U, 'Weitere montanarchäologische Untersuchungen im Mittelalterischen Bergbaureiser am Birkenberg bei St Ulrich, Gde. Bollschweid, Kreis Breisgau-Hochschwarzwald', *Archäologische Ausgrabungen in Baden-Württemberg*, 1992 (380–4).

Brusten, C, *L'armée bourguignonne de 1465 à 1468*, Brussels, 1953

Bucher, F, *Architector. The lodge books and sketchbooks of medieval architects*, vol. 1, New York 1979.

Burleigh, M, *Prussian society and the German Order. An aristocratic corporation in crisis c.1410–1466*, Cambridge 1984

Burnam, J M, *A classical technology, edited from Codex Lucensis 490*, Boston 1920.

Burne, A H, *The Agincourt war. A military history of the latter part of the Hundred Years War from 1369 to 1415*, London 1956.

Burnett, C, *Adelard of Bath: an English scientist and Arabist of the early twelfth century*, London 1987.

Burnett, C, 'Michael Scot and the transmission of scientific culture from Toledo to Bologna via the court of Frederick II Hohenstaufen', *Micrologus* II, 1994 (101–26).

Burnett, C S F, 'Antioch as a link between Arabic and Latin culture in the 12th and 13th centuries', in I Draelants, A Tihon, B van der Abeele (eds), *Occident et proche-orient, contacts scientifiques au temps des croisades. Actes du colloque de Louvain-la-neuve, 24 et 25 mai 1997*, Louvain 2000 (1–78).

Bury, J B, 'The early history of the explosive mine', *Fort* 10, 1982 (23–30).

Bury, J B, *The imperial administrative system in the ninth century, with a revised text of the Kletorologia of Philotheas*, London 1911.

Butler, A J, *The Arab conquest of Egypt and the last thirty years of the Roman dominion*, 2nd edn, ed. P M Fraser, Oxford 1978.

Butts, M, Linker R (ed/trans): *see* Ogier in Bibliography of primary sources.

Butzer, P L, 'Mathematics in west and east from the fifth to tenth centuries: an overview', in P L Butzer, D Lohrmann (eds), *Science in western and eastern civilisation in Carolingian times*, Basel 1993 (443–81).

Cahen, C, 'L'emigration persane des origines de l'Islam aux Mongols', *Academia Nazionale dei Lincei* CCCLXVIII, 1971: Atti del Convegno Internazionale sul tema: la Persia nel medioevo (181–94).

Cahen, C, *Pre-Ottoman Turkey. A general survey of the material and spiritual culture and history c.1071–1330*, trans. J Jones-Williams, London 1968.

Cahen, C, 'Le service de l'irrigation en Iraq au début du XIe siècle', *Bulletin d'Études Orientales* XIII, 1949–51 (117–43).

Cameron, A, Ward-Perkins, B, Whitby, M (eds.), *Cambridge Ancient History, XIV: Late Antiquity: empires and successors AD425–600*, Cambridge 2000.

Camp, L Sprague de, *The ancient engineers*, Cambridge (Mass.) 1960.

Campari, G, 'Gli architetti e gl'ingegneri civile e militari degli Estensi dal secolo XIII al XVI', *Atti e Memorie delle RR deputazioni di Storia Patria per le provincie Modenesi e Parmensi*, ser. III, I, 1 (1883), 1–69.

Cancellieri, J-A, 'Formes Rurales de de la colonisation Genouise en Corse au XIIIe siècle', *Mélanges de l'école Française de Rome* 93, 1981 (89–146).

Cange, D du, *Glossarium mediae et infimae latinitatis*, new ed., Niort 1885.

Canovas, C, 'Mining practice in the medieval Muslim Iberian peninsula – the case of iron. Studying metals in cross-ethnical context', in P Anrieter et al. (eds.), *Mining in European history and its impact on environment and human societies. Proceedings of the first mining in European history conference of the SFB-HIMAT, 12–15 November 2009*, Innsbruck 2010 (359–63).

Caple, C, *Excavations at Dryslwyn castle 1980–85*, Society for Medieval Archaeology Monograph 26, Leeds 2007.

Carlton, R, Jones, D, 'Archaeological investigations at the site of a medieval mill on the river Coquet at Barrowburn', *Archaeologia Aeliana* ser. 5, 44, 2015 (76–122).

Carr Laughton, L G, 'The great ship of 1419', *Mariner's Mirror* IX, 1923 (83–7).

Cassidy, B, 'Machiavelli and the ideology of the offensive: gunpowder weapons in the "Art of War"', *Journal of Military History* 67 (2), 4/2003 (381–404).

Cattedu, I, *Archéologie médiévale en France. Le premier moyen âge (Ve–XIe siècle)*, Paris 2009.

Ceci, G, 'Un dimenticati ingegnere pugliese del sec. XVI, A de Trani', *Japigia*, I, 1930 (54-60)

Chapelot, J, 'Le château de Vincennes 1372–80', in H Mouillebouche (ed.), *Châteaux et mesures. Actes 17e journée de Castellologie de Bourgogne, Octobre 2010*, 2011 (100–23).

Chapuisat, J-P, 'Chillon base militaire savoyarde au XIIIe siècle', in A Pathuel-Guillard, L Chavoutier (eds), *Soldats et armées en Savoie*, Actes du Congrès des Sociétés Savantes de Savoie St Jean-de-Maurienne 6–7 septembre 1980 (1981) (79–84).

Chevedden, P E, 'The artillery of King James I the Conqueror', in P E Chevedden, D J Kagay, P G Pedella (eds), *Iberia and the Mediterranean World of the Middle Ages*, 2 vols, Brill 1996.

Chevedden, P E, 'The hybrid trebuchet: the halfway step to the counterweight trebuchet', in D J Kagay, T M Vaan (eds), *On the social origins of medieval institutions: essays in honour of Joseph F O'Callaghan* (179–222).

Chevedden, P E, 'The invention of the counterweight trebuchet: a study in cultural diffusion', *Dumbarton Oaks Papers*, 54 (2000), 71-116

Chiarelli, L C, *A history of Muslim Sicily*, St Venera (Malta), 2010.

Chi-Ch'ing Hsaio, *The military establishment of the Yüan dynasty*, Harvard East Asian Monographs 77, Harvard 1978.

Chouquer, G, Favory, F, *L'arpentage romain. Histoire des textes-droit-techniques*. Paris 2001.

Christie, N, Creighton, O, Edgeworth, M, Hamerous, H, *Transforming townscapes. From burh to borough: the archaeology of Wallingford AD800–1400*. Society for Medieval Archaeology Monograph 35, London 2013.

Ciniç, N, *Yerebatan cistern and other cisterns of Istanbul*, trans. A Bayraktar, Istanbul 2003.

Clagett, M, *The science of mechanics in the middle ages*, London/Madison 1959.

Clanchy, M T, *From memory to written record. England 1066–1307*, 2nd edn, Oxford/Cambridge (Mass.), 1993.

Clark, J, 'Richard Crips the wheeler – a medieval craftsman and his tools', *Tools and Trades* 2, 1984 (13–28).

Claude, D, 'Die Handwerker der Merowingerzeit nach den erzählenden urkundische Quellen', in H Jahnkuhn, W Janssen, R Schmidt-Wiegund, H Tiefenbach (eds), *Das Handwerk in vor- und frühgeschichtlicher Zeit*, Göttingen 1981, I (204–66).

Clay, P, Salisbury C R, 'A Norman mill and other sites at Hemington Fields, Castle Donington, Leicestershire', *Archaeological Journal* 147, 1990 (276–307).

Cleere, H, Crossley, D, *The iron industry of the Weald*, Leicester, 1985

Cobos-Guerra, F, 'Pedro Luis Escrivá and the first treatises on modern fortification, Naples 1538', in A Cámara Muñoz, B Revuelta (eds), *Renaissance engineers*, English edition 2016 (25–52).

Cohen, I, 'History of iron working in and near the forest of Dean', *Transactions of the Woolhope Naturalists' Field Club* XXXIV, pt III, 1954 (161–77).

Cohn, W, *Die Geschichte der sizilischen Flotte unter der Regierung Friedrichs II (1197–1250)*, Breslau 1926.

Coldstream, N, 'James of St George', in D M Williams, J Kenyon (eds), *The impact of the Edwardian castle in Wales. The proceedings of a conference held at Bangor university, 7–9 September 2007*, Oxford/Oakville 2010 (37–45).

Coldstream, N, *Medieval architecture*, Oxford History of Art, Oxford 2002.

Colin, G S, 'La Noria marocaine et les machines hydrauliques dans le monde arabe', *Hésperis* XIV, 1932 (22–60).

Collins, R, 'Literacy and the laity in early medieval Spain', in R McKitterick (ed.), *The uses of literacy in early medieval Europe*, Cambridge 1990 (109–33).

Concina, E, *L'arsenale della republica di Venezia*, Milan 1894.

Conrui Xiong, V, 'Sui Yangdi and the building of Sui-Tang Luoyang', *Journal of Asian Studies* 52/1, 1993.

Contamine, P, 'L'armée de Philippe Auguste', in R Bautier (ed.), *La France de Philippe Auguste: le temps des mutations*, Paris 1982

Contreni, J J, 'Counting, calendars and cosmology: numeracy in the early Middle Ages', in J J Contreni, S Casciani (eds), *Word, Image, Number. Communication in the Middle Ages*, Micrologus Library 8, Turnhout 2002 (43–83).

Contreni, J J, 'The pursuit of knowledge in Carolingian Europe', in R E Sullivan (ed.), *"The gentle voice of teachers": aspects of learning in the Carolingian age*, Columbus Ohio 1995 (106–41).

Cook, M, *Medieval bridges*, Princes Risborough 1998.

Cordoba de la Llave, 'Las artes del fuego, metalurgia y alfarería', in P Navascues (gen. editor), *Ars mechanicae, ingenería medieval en España*, 2008 (203–14)

Corominas, J, Pascual, J A, *Diccionario crítica et etimológico Castellano e Hispánico*, Madrid 1980.

Coulston, J, 'The archaeology of Roman conflict', in P Freeman, A Pollard (eds.), *Fields of conflict: progress and prospect in battlefield archaeology*, BAR International Series 958, 2001 (23-50).

Courtenay, L T, 'Scale and scantling: technological issues in large-scale timberwork of the High Middle Ages', in E Bradford Smith, M Wolfe (eds.), *Technology and resource use in medieval Europe: cathedrals, mills and mines*, 42–75.

Courtenay, L, Alcock N, 'Romanesque roofs. The nave of Jumièges abbey and the common tie-beam tradition in northern Europe', *Medieval Archaeology* 59, 2015 (122–67).

Cox, E, *The Green Count of Savoy. Amadeus VI and transalpine Savoy in the fourteenth century*, Princeton, 1967

Cracco Ruggini, L, 'Le associazioni professionali nel mondo Romano-Bizantino', in *Artiginiato e tecnica nella società dell'alto medioevo occidentale*, *SSCISSM* XVIII, pt 1, 1971 (59–193)

Creighton, O H, Wright, D W, *The anarchy. War and status in 12th-century landscapes of conflict*, Liverpool 2016

Crombie, A C, *Medieval and early modern science, 1: Science in the middle ages, V–XIII centuries*, 2nd edn, New York, 1959.

Crombie, A C, review of E Whitney, 'Paradise Restored', *English Historical Review* 109, 1994 (136–8)

Cromwell, P R, 'Islamic geometric designs from the Topkapi scroll, II: a modular design system', *Journal of Mathematics and Art* 4/3, 2010 (119–36).

Crowley, R, *City of Fortune. How Venice won and lost a naval empire*, London 2011.
Crumlin-Pedersen, O, 'Aspects of wood technology in medieval ship-building', in O Crumlin-Pedersen, M Vinner (eds.), *Sailing into the past. The international ship replica seminar Roskilde 1984*, Arhus 1986 (138–49).
Crumlin-Pedersen, O, 'Danish cog finds', in S McGrail (ed.), *Medieval ships and harbours in northern Europe*, BAR International Series 66, 1979 (17–34).
Crummy, P, 'The system of measurement used in town planning from the ninth to the thirteenth centuries', in S C Hawkes, D Brown, J Campbell (eds), *Anglo-Saxon studies in archaeology and history*, I, BAR British Series 72, 1979 (149–64).
Cruse, M, *Illuminating the Roman d'Alexandre, Oxford Bodleian Library, MS Bodley 264. The manuscript as monument*, Cambridge 2011.
Curry, A, *The battle of Agincourt. Sources and interpretations*, Woodbridge 2009.
Curtis, E, *Roger of Sicily and the Normans in lower Italy 1016–1154*, London/New York 1912.
Cutler, A, *The hand of the master. Craftsmanship, ivory and society in Byzantium* (9th–11th *centuries)*, Princeton 1994.
Czacharowski, A, 'Sociotopography of medieval and late medieval towns in the North European zone as exemplified by Torun', *Acta Poloniae Historica* 34, 1976 (121–9).
Czainowsky, C, 'Engins de levage dans les combles d'églises en Alsace', in Rumpler M (ed.), *Les cahiers techniques de l'art*, Strasbourg 1949 (11–27).
Czysz, W, 'Eine bajuwarische Wassermühle im Paartal bei Dasing', *Antike Welt* 25/2, 1994 (152–5)
Dain, A, 'Métrologie Byzantine. Calcul de la superficie des termes. Mémorial Louis Petit', *Mélanges d'histoire et d'archéologie Byantine, Archives de l'Orient Chrétien*, I, 1948 (56-63).
Damminger, F, 'Dwellings, settlements and settlement patterns in Merovingian SW Germany', in I Wood (ed.), *Franks and Alemanni in the Merovingian period. An ethnographic perspective*. Studies in Historical Archaeoethnology 3, Woodbridge 1998 (33–106)
Darrah, R, 'Working unseasoned oak', in S McGrail (ed.), *Woodworking techniques before AD1500*, BAR International series 129, 1982 (219–29).
Davies, O, *Roman mines in Europe*, Oxford 1935.
Davis, M T, 'Masons and architects as travelers', in J B Friedman, K M Figg, *Trade, travel and exploration in the Middle Ages: an encylopedia*, New York/London 2000 (380–2).
Dean, B, 'The exploration of a crusaders' fortress (Montfort) in Palestine', *Metropolitan Museum of Art Bulletin* XXII, 1927 (pt II, 5–45).
Decker, M J, *The Byzantine art of war*, Yardley 2013.
Delgado, J, *Khubilai Khan's lost fleet. History's greatest naval disaster*, London, 2009
Dennis, G T, 'Byzantine heavy artillery: the Helepolis', *Greek, Roman and Byzantine Studies* 39/1, 1998 (99–115).
Denny, M, *Ingenium. Five machines that changed the world*, Baltimore 2007.
DeVries, K, 'Sites of military science and technology', in K Park, L Daston (eds), *The Cambridge History of Science, III: Early Modern Science*, Cambridge 2006 (306–17).
DeVries, K, Brown, R D, *Medieval military technology*, 2nd edn, Toronto 2012.
Diaz, P C, 'Visigothic political institutions', in P Heather (ed.), *The Visigoths from the migration period to the seventh century. An ethnographic perspective*. Studies in Historical Archaeoethnology 3, Woodbridge 1999 (321–72).

Diefenbach, L, *Glossarium latino-germanicum mediae et infimae aetatis*, Frankfurt am Main, 1857.

Divorne, F, Gendre B, Lavergne B, Panerai, P, *Les bastides d'Aquitaine, du bas-Languedoc et du Béarn*, Brussels 1985.

Dohrn-van Rossum, G, *History of the hour. Clocks and modern temporal orders*, trans. T Dunlop, Chicago/London 1996.

Dohrn-van Rossum, G, 'Migration-Innovation-Städtemetze. Ingenieure und technische Experten', in M Arnoux, P Monnet, *Le technicien dans la cité en Europe occidentale 1250–1650*, École Française de Rome, Rome 2004 (291–307).

Donner, H, 'The mosaic map of Madaba', *Palestina Antiqua* 7, Kampen 1992.

Doorninck, F van, 'Byzantium: mistress of the sea 330–641', in G F Bass (ed.), *A history of seafaring based on underwater archaeology*, London 1972 (133–58).

Downey, G, 'Byzantine architects, their training and methods', *Byzantion* XVIII, 1948 (99–118).

Drew, A, 'The *de eodem et diverso*', in C Burnett (ed.), *Adelard of Bath. An English scientist and Arabist of the early 12th century*, Warburg Institute Surveys and Texts XIV, London 1987 (17–24).

Du Cange, D, et al., *Glossarium mediae et infimae latinitatis*, new edn, Niort 1885.

Dubled, H, 'L'artillerie royale française à l'époque de Charles VII et au début du règne de Louis XII (1437-69). Les frères Bureau', *Mémorial de l'artillerie française* 50, 1976.

Dubois, D, 'La charpente militaire. Un aspect des engins à jet à contrepoids à la fin du moyen âge', in X Barral y Altet (ed.), *Artistes, artisans et production artistiques au moyen âge*, Université de Rennes 1983, Paris 1987 (II, 403–8).

Durliat, J, 'Armée et société vers 600. Le problème des soldes', in F Vallet, M Kazanski (eds), *L'armée romaine et les barbares du IIIe au VIIe siècles*, Mémoires, L'association Française d'archéologie Mérovingienne, 1993 (31–8).

Eastwood, B, 'Medieval science illustrated (review of Murdoch, Album of Science)', *History of Science* XXIV, 1986 (183–208).

Egg, E, 'Der Tiroler Geschützguss 1400–1600', *Tiroler Wirtschaftsstudien* 9, Innsbruck 1961.

Elisséeff, N, 'Corporations de Damas sous Nūr al-Dīn. Matériaux pour une topographie économique de Damas au XIIe siècle', *Arabica* III, 1956 (61–79).

Ellenblum, R, *Crusader castles and modern histories*, Cambridge 2007.

Ellmers, D, 'The Bremen cog and related boats', in S McGrail (ed.), *Medieval ships and harbours in northern Europe*, BAR International Series 66, 1979 (1–16).

Ellmers, D, 'Development and usage of harbour cranes', in C Villain-Gandossi et al. (eds), *Medieval ships and the birth of technological societies, I: Northern Europe*, European coordinating centre for research and documentation in social sciences, Malta 1989.

Enlart, C, *Manuel d'archéologie Française depuis les temps mérovingiens jusqu'à la renaissance. Pt II: architecture civile et militaire*, Paris 1929, 1932.

Epaud, F, *De la charpente romane à la charpente gothique en Normandie. Evolution des techniques et des structures de charpenterie aux XII–XIII siècles*, Caen 2007.

Epstein, S R, 'Labour mobility, journeyman organisation and markets in skilled labour in Europe, 14th–18th centuries', in M Arnoux, P Monnet (eds), *Le technicien dans la cité en Europe occidentale 1250–1650*, École Française de Rome, Rome 251) 2004–69).

Erlande-Brandenburg, A, 'L'architecture militaire au temps de Philippe Auguste: une nouvelle conception de la défense', in R Bautier (ed.), *La France de Philippe Auguste: le temps des mutations*, Paris 1982

Erlande-Brandenburg, A, *The cathedral builders of the Middle Ages*, trans. R Stonehewer, London 1995.

Ernout, A, Meillet, A, *Dictionnaire étymoligique de la langue Latine*, Paris 1967.

Escobedo, L K, *The Milemete treatise and companion Secretum secretorum. Iconography, audience and patronage in 14th-century England*. Lewiston/Queenstown/Lampeter 2011.

Essenwein, A, *Quellen zur Geschichte der Feurwaffen. Facsimilierte Nachbildungen alter Originalzeichnungen*. Germanisches Nationalmuseum, Nürnburg 1872

Evans, G R, 'The development of some textbooks in the "useful arts" c.1000–c.1250', *History of Education* 7 (2), 6/1978 (85–94).

Fagniez, G, *Études sur l'industrie et la classe industrielle à Paris au XIIIe et XIVe siècles*, Paris 1877.

Fasolo, G, 'Domenico di Benintendi da Firenze. Ingegnere del secolo XIV', *Archivio Veneto* ser. 5, I, 1927 (145–80).

Feldhaus, F M, *Die Technik der Antike und des Mittelalters*, Potsdam 1931.

Fenwick, V (ed.), 'The Graveney boat: a tenth century find from Kent', *National Maritime Museum Greenwich, archaeological series* 3, BAR British series 53, 1978

Fernie, E, 'The ground plan of Norwich cathedral and the square root of two', in L T Courtenay (ed.), *The engineering of medieval cathedrals*, Aldershot 1997 (77–86).

Ferrill, A, *The origins of war from the Stone Age to Alexander the Great*, London 1985.

Feuerle, M, 'Das Hebelwurfgeschütz. Eine technische Innovation des Mittelalters', *Technikgeschichte* 69 (1), 2002 (1–40).

Ffoulkes, C, *The gunfounders of England, with a list of English and continental gun-founders from the XIVth to the XIXth centuries*, 2nd edn, London 1969.

Fisher, G, *Between empires. Arabs, Romans and Sasanians in Late Antiquity*, Oxford 2011.

Flatman, J, *Ships and shipbuilding in medieval manuscripts*, London, 2009

Folkerts, M, 'The importance of the pseudo-Boethian "Geometria" during the middle ages', in M Masi (ed.), *Boethius and the Liberal Arts. A collection of essays*, Berne/Frankfurt-am-Main/Las Vegas 1981 (187–209).

Forbes, R J, 'Metallurgy and technology in the middle ages', *Centaurus* 3, 1953 (49–57).

Forey, A, 'The military orders and the ransoming of captives from Islam (twelfth to fourteenth centuries)', *Studia Monastica* XXXIII, 1991 (259–79), reprinted A Forey, *Military orders and the Crusades*, Aldershot 1994.

Fouracre, P, 'The nature of Frankish political institutions in the seventh century', in I Wood (ed.), *Franks and Alemanni in the Merovingian period. An ethnographic perspective*, Studies in Historical Archaeoethnology 3, Woodbridge 1998 (285–316)

Francovich, R, Wickham C, 'Uno scavo archaeologico ed il problema della sviluppo della signoria territoriale: Rocca San Silvestro e i rapporti di produzione minerari', *Archeologia Medievale* XXI, 1994 (7–30).

Freeman, A Z, 'Wall-breakers and river-bridgers', *Journal of British Studies* X (2), 1971 (1–16)

Friedman, D, *Florentine new towns. Urban design in the late middle ages*, Cambridge (Mass.)/London 1988.

Friedman, Y, *Encounters between enemies. Captivity and ransom in the Latin kingdom of Jerusalem*, Leiden/Boston/Köln 2002.

Friel, I, 'The documentary evidence for ship-building in England 1294-c1500', in C Villain-Gandossi et al (eds.), *Medieval ships and the birth of technological societies*, I, 139-49, Malta, 1989

Friel, I, *The good ship. Ships, ship-building and technology in England 1200–1520*, London 1995.

Fryde, N, 'Technicians, the crown and the city of London in the thirteenth century', in M Arnoux, P Monnet (eds), *Le technicien dans la cité en Europe occidentale 1250–1650*, École Française de Rome, Rome, 2004 (15–23).

Fulton, M S, 'Development of prefabricated artillery during the crusades', *Journal of Medieval Military History* XIII, 2015 (51–72).

Gabotto, F, 'L'età del Conte Verde in Piemonte secundo nuovi documenti (1350–1383)', *Miscellanea di Storia Italiana* ser. 3, II, 1895 (79–333).

Gaier, C, 'Le problème de l'origine de l'industrie armurière Liégoise', in the same, *Armes et combats dans l'univers médiéval*, Brussels 1995 (191–228).

Gaier, C, 'La renommée militaire des mineurs de Liège', in the same, *Armes et combats dans l'univers medieval*, Brussels 1995 (79–90).

Gaier, C, 'L'industrie et le commerce des armes dans les anciennes principautés belges du XIIIe à la fin du XVe siècle', *Bibliotheque de la Faculté de Philosophie et Lettres de l'université de Liège*, CCII, 1973,

Ganshof, F L, *The Carolingians and the Frankish monarchy. Studies in Carolingian history*, trans. J Sondheimer, London 1971.

García Hourcade, J L, 'From Tartaglia to Lechuga. The artillery engineer', in A Cámara Muñoz, B Revuelta (eds.), *Renaissance engineers*, English edition 2016 (53–74).

Gardizi: Minorsky, V (trans.), 'Gardizi in India', *Bulletin of the School of Oriental and African Studies* XII, 1948 (625–50).

Gernet, J, *A history of Chinese civilization*, trans. J R Foster, C Hartmann, 2nd edn, Cambridge 1996.

Gessler, E A, 'Der Springulf, ein mittelalterlicher Torsionsgeschütz', *Basler Zeitschrift für Geschichte und Altertumkunde* 20, 1922.

Ghat, A, 'Machiavelli and the decline of the classical notion of the lessons of history in the study of war', *Military Affairs* 52/4, 1988 (203–4).

Gibb, H A R, 'The armies of Saladin', in S J Shaw, W R Polk (eds), *Studies in the civilisation of Islam*, Princeton 1982 (originally published 1951) (71–90).

Gibson, S, Ward-Perkins B, 'the surviving remains of the Leonine wall', *Papers of the British School at Rome*, 47 (1979), 30-57; 51 (1983), 222-39

Gilbert, A H, 'Machiavelli on fire weapons', *Italica* 23/4, 1946 (275–86).

Gille, R, *Renaissance engineers*, London 1966

Gillmor, C, 'Naval logistics of the cross-Channel operation 1066', *Anglo-Norman Studies* 7, 1984 (105–31), reprinted in S Morillo (ed.), *The battle of Hastings. Sources and interpretations*, Woodbridge 1996 (113–28).

Gimpel, J, *The medieval machine. The industrial revolution of the middle ages*, 2nd edn, London 1988

Girardot, A, 'Fondeurs d'artillerie et siderurgistes, une direction de recherche?', in P Benoit, P Braunstein (eds), *Mines carrières et metallurgie dans la France médiévale. Actes du colloque de Paris, 19–21/6/1980*, Paris 1983 (67–76).

Giulini, Count G, *Memorie spettanti alla storia, al governo ed alla descrizione della citta e campagna di Milano*, III, Milan 1855.

Glick, T, *Irrigation and society in medieval Valencia*, Cambridge (Mass.), 1970

Glick, T, 'Levels and levelers: surveying irrigation canals in medieval Valencia', *Technology and Culture* 9/2, 1968 (165–80).

Glick, T, Livesey S J, Willis F (eds), *Medieval science, technology and medicine: an encyclopedia*, New York/London 2005.

Göbbels, J, 'Die Militarorganisation im staufischen Königreich Sizilien', in A Esch, N Kamp (eds.), *Friedrich II. Tagung des Deutschen Historischen Instituts in Rom in Gedenkjahr 1994*, Tübingen 1996 (486–500).

Göbbels, J, *Das Militarwesen im Königreich Sizilien zur Zeit Karls I von Anjou*, Monographien zur Geschichte des Mittelalters 29, Stuttgart 1984.

Godefroy, F, *Dictionnaire de l'ancienne langue française et de tous ses dialectes du IXe au XVe siècles*, Paris 1884.

Goldthwaite, R A, 'Schools and teachers of commercial arithmetic in Renaissance Florence', *Journal of European Economic History* 1/2, 1972 (418–33).

Goodall, J, *The English Castle 1066–1650*, New Haven/London 2011.

Gordon, J E, *Structures. Or why things don't fall down*, London 1979.

Götland, A, *Technology and religion in medieval Sweden*, Falun 1993.

Goulet, R, 'Medieval engines of siege warfare and modern engineering tools', *ASEE Annual Conference 2003 paper 3566* (1–9).

Grabar, O, 'The mediation of ornament. The A W Mellon lectures in the fine arts 1989', *National Gallery of Art, Washington DC Bollingen Series* XXXV, 38, 1992 (119–38).

Graham-Taylor, J, Valor, M (eds.), *The archaeology of medieval Europe. I: Eighth to twelfth centuries AD*, Aarhus 2007.

Gregory, S, Rothwell, W, Trotter, D, *Anglo-Norman dictionary*, 2nd edn, London 2005.

Grewe, K, 'Der Karlsgraben bei Weißenburg und der Fulbert-Stöllen von Maria Laach', in U Lindgren (ed.), *Europäische Technik in Mittelalter 800–1400. Tradition und Innovation*, Berlin 1997 (111–15).

Grewe, K, 'Water technology in medieval Germany', in P Squatriti (ed.), *Working with water in medieval Europe*, Leiden/Boston/Köln 2000 (128–60)

Gribit, N, *Henry of Lancaster's expedition to Aquitaine 1345–46. Military service and professionalism in the Hundred Years War*, Woodbridge 2016.

Grillo, P, 'Velut leena rugiens. Brescia assediata da Federico II (Iuglio–ottobre 1238), *Reti Medievali Rivista* VIII, 2007, accessed at http://www.rmojs.unina.it/index.php/rm/article/viewFile/127/108 (25/11/2016).

Grimm, J and W, *Deutsches Wörterbuch*, Leipzig 1960.

Grummit, D, *The Calais garrison. War and military service in England 1436–1558*, Woodbridge 2008.

Grummit, D, 'The defence of Calais and the development of gunpowder weaponry in England in the late fifteenth century', *War in History* 7/3, 2000 (253–72).

Guillaumin, J-Y, *Sur quelques notices des arpenteurs romains*. Université de Franche-Comté, 2007.

Guillerme, A E, *The age of water. The urban environment in the north of France AD 300–1800*, Seyssel, 1988.

Habib, Irfan, 'Capacity of technological change in Mughal India', in A Roy, S K Bagchi (eds.), *Technology in ancient and medieval India*, Delhi 1986 (1–14).

Habib, Irfan, *Medieval technology exchanges between India and the Islamic world*, Aligarh Oriental series (gen. ed. Umesh Chandra Sharma), 6, Aligarh 1985.

Hadingham E, '"Ready, aim, fire". A risky experiment reveals how medieval engines of war brought down castle walls', *Smithsonian* 30/10, 1/2000 (78–87).

Haldon, J F, 'Administrative continuities and structural transformations in East Roman military organisation ca. 580–640', in F Vallet, M Kazanski (eds), *L'armée romaine et les barbares du IIIe au VIIe siècles*. Mémoires, L'association Française d'archéologie Mérovingienne, 45) 1993–54).

Haldon, J F, *Byzantine Praetorians. An administrative, institutional and social survey of the opsikion and tagmata, c.580–900*, Bonn 1984.

Haldon, J F, 'Information and war', in A Sarantis, N Christie (eds), *War and warfare in Late Antiquity*, Late Antique Archaeology 8.1 and 8.2, Leiden/Boston 2013 (373–93).

Haldon, J F, 'Theory and practice in tenth-century military administration. Chapters II, 44 and 45 of the 'Book of Ceremonies', *Travaux et Mémoires* 13, 2000 (201–352).

Haldon, J F, Byrne M, 'A possible solution to the problem of Greek Fire', *Byzantinische Zeitschrift* 70, 1977 (91–9).

Haldon, J F, Lacey, A, Hewes C, 'Greek Fire revisited: recent and current research', in E Jeffreys (ed.), *Byzantine style, religion and civilization. In honour of Sir Steven Runciman*, Cambridge 2006 (290–325).

Hale, J R, 'The early development of the bastion: an Italian chronology c.1450–c.1534', in the same, *Renaissance war studies*, London 1983 (1–30).

Hale, J R, 'Tudor Fortifications, 1485–1558', in the same, *Renaissance war studies*, London 1983 (63–98).

Hall, B S, 'Der meister sol auch kennen schreiben und lesen: writings about technology ca.1400–ca.1600 AD and their cultural implications', in D Schmidt-Besserat (ed.), *Early technologies*, Malibu 1979 (47–58).

Hall, B S, 'Production et diffusion de certains traits de techniques au moyen âge', in G H Allard (ed.), *Les arts mécaniques au moyen âge*, Montreal/Paris 1982 (147–70).

Hall, B S, 'Scientific archaeology and the military historian: the case of early modern firearms', in *History of Medieval Life and the Sciences. Proceedings of an international round-table-discussion. Krems an der Donau, September 28–9 1998*, Wien 2000 (137–45).

Hall, B S, *Weapons and warfare in Renaissance Europe. Gunpowder, technology and tactics*, Baltimore/London 1997.

Halsall, G, 'Social identities and social relationships in early medieval Gaul', in I Wood (ed.), *Franks and Alemanni in the Merovingian period. An ethnographic perspective*. Studies in Historical Archaeoethnology 3, Woodbridge 1998 (141–75).

Halsall, G, 'Towns, societies and ideas: the not-so-strange case of late Roman and early Merovingian Metz', in N Christie, S T Loseby (eds), *Towns in transition. Urban evolution in late antiquity and the early middle ages*, Aldershot 1996 (235–61).

Hannam, J, *God's philosophers. How the medieval world laid the foundations of modern science*, London 2009.

Hanson, P, 'Soviet inland waterways', *Journal of Transport History* VI (1), 1963 (3–13).

Harley, J B, Woodward, D (eds.), *The history of cartography, I: Cartography in prehistoric, ancient and medieval Europe and the Mediterranean*, Chicago/London 1987, and II, same editors, 1992

Harrison, D, *The bridges of medieval England. Transport and society 400–1800*, Oxford 2004.

Hart, C, *The free miners of the royal forest of Dean and Hundred of St Briavels*, 2nd edn, Lydney 2002.

Harvey, J, *English medieval architects. A biographical dictionary down to 1550*, revised edn, Gloucester, 1987.

Harvey, J, *Gothic England. A survey of national culture 1300–1550*, 2nd edn, London/Toronto/New York/Sydney 1948.

Harvey, J, *Medieval craftsmen*, London/Sydney 1975.

Haskins, C H, *Studies in the History of Medieval Science*, New York, 1924 (repub. 1960)

Haslam, J, 'A middle Saxon iron smelting site at Ramsbury, Wiltshire', *Medieval Archaeology* XXIV, 1980 (1–68).

Hattendorf, J B, Knight, R J B, Pearsall, A W H, Rodger, N A M, Till, G (eds.), *British naval documents 1204–1960*, Aldershot 1993.

Hauser, S R, Tucker D J, 'The final onslaught. The Sasanian siege of Hatra', *Zeitschrift für Orient-Archäologie* 2, 2009 (106–39).

Hawani, G F, 'The early growth of the secular sciences in Andalucia', *Studia Islamica* 32, 1970 (143–56).

Haywood, J, *Dark Age naval power. A reassessment of Frankish and Anglo-Saxon seafaring activity*, London/New York, 1991.

Henderson, G, 'Late antique influence in some English medieval illustrations of Genesis', *Journal of the Warburg and Courtauld Institutes* XXV, 1962 (172–98).

Henning, J, 'Early European towns: the way of the economy in the Frankish area between dynamism and deceleration 500–1000AD', in N Christie, S T Loseby (eds.), *Towns in transition. Urban evolution in late antiquity and the early middle ages*, Aldershot 1996 (3–40).

Herz, P, 'Fabrica, fabricienses,' in H Cancik, H Schneider (eds.), *Der Neue Pauly. Enzyklopedie der Antike*, bd 4, Stuttgart 1998 (380–1).

Heslop, T A, 'Orford castle, nostalgia and sophisticated living', in R Liddiard (ed.), *Anglo-Norman castles*, Woodbridge 2003 (273–96).

Hesse, P-J, 'Artistes, artisans ou prolétaires? Les hommes de la mine au moyen âge', in X Barral I Altet, *Artistes, artisans et production artistique au moyen âge. Colloque internationale, Université de Rennes II, 2–6 mai 1983*, I, Paris 1986 (431–73).

Hewett, C A, 'Toolmarks on surviving works from the Saxon, Norman and later medieval periods', in S McGrail (ed.), *Woodworking techniques before AD1500*, *BAR* International series 129, 1982 (339–56).

Hewitt, H J, *The organisation of war under Edward III*, reprinted Manchester 2004

Heyman, J, *The stone skeleton. Structural engineering of masonry architecture*, Cambridge 1995.

Heywood, C, 'Notes on the production of fifteenth-century Ottoman cannon', *Proceedings of the International Symposium on Islam and Science 1401/1980*, reprinted in the same, *Writing Ottoman history*, Aldershot 2002 (3–15).

Higham, R, 'William the Conqueror's siege of Exeter in 1068', *Transactions of the Devonshire Association for the Advancement of Science* 145, 2013 (93–132).

Higounet, C, 'Granges et bastides de l'abbaye de Bonnefont', in the same, *Paysages et villages neufs du Moyen Âge, recueil d'articles de Ch Higounet*, Bordeaux 1975 (275–83).

Higounet, C, 'Nouvelles approaches sur les bastides du sud-ouest Aquitaine', *Urbanisme* 101, 1967 (32–35).

Hilaire-Perez, L, Verna, C, 'Dissemination of technical knowledge in the middle ages and the early modern era: new approaches and methodological issues', *Technology and Culture* 47/3, 2006 (536–65).

Hill, D R, *Islamic science and engineering*, Edinburgh, 1993.

Hill, D R, 'Physics and mechanics. Civil and hydraulic engineering, industrial processes and manufacturing and craft activities'. Chapter ten of UNESCO, *History of civilizations of Central Asia* IV, pt II, Paris 2000.

Hill, D R, 'Science and technology in ninth-century Baghdad', in P L Butzer, D Lohrmann (eds.), *Science in western and eastern civilization in Carolingian times*, Basel 1993 (485–502).

Hill, D R, al-Hasan, A Y, *Engineering in Arabic Islamic civilization*, 3rd edn, 2011, accessed at http://www.history-science-technology.com/articles/articles%2011.html (17/3/14) (53pp).

Hill, D R, al-Hasan, A Y, 'Mining technology (Mádin)', in *Encyclopedia of Islam*, new edn 5, 1986 (967–73), reprinted in D R Hill, *Studies in medieval Islamic technology*, Ashgate Variorum, Aldershot/Brookfield/Singapore/Sydney 1998.

Hill, R A, 'Tenth-century woodworking in Coppergate, York', in S McGrail (ed.), *Woodworking techniques before AD1500*, BAR International series 129, 1982 (231–44)

Hindle, P, *Medieval town plans*, Princes Risborough 2002.

Hiscock, N, 'Architectural geometry and the portfolio of Villard de Honnecourt', in M-T Zenner (ed.), *Villard's legacy. Studies in medieval technology, science and art in memory of Jean Gimpel*, Aldershot 2004 (3–21).

Hislop, M, *Castle builders. Approaches to castle design and construction in the Middle Ages*, Barnsley 2016.

Hodges, R, *Dark Age economics: a new audit*, London 2012.

Hodges, R, Leppard S, Mitchell, S, *San Vincenzo Maggiore and its workshops*, British School at Rome Archaeological Monograph 17, London 2011.

Hoff, A H, *Feuerwaffen*, 2 vols, Braunschweig 1969.

Hoff, A, 'The Töjhusmuseum, Copenhagen', *Armi Antiche*, 1962, 53-68

Hoffmann, D, *Das spätrömische Bewegungsheer und die Notitia Dignitatum*, Dusselsdorf 1970.

Hofmann, H H, *Kaiser Karls Kanalbau*, Sigmaringen 1969.

Holt, R, 'Medieval England's water-related technologies', in P Squatriti (ed.), *Working with water in medieval Europe*, Leiden/Boston/Köln 2000 (51–100).

Horn, W, Born, E, *The plan of St Gall. A study of the architecture and economy, and life in a paradigmatic Carolingian monastery*, foreword by W Braunfels, trans. by C W Jones of

Directives of Adalhard, 753–826, note by A Hunter Dupree, 3 vols, Berkeley, Los Angeles and London 1979.

Hosler, J D, *Henry II. A medieval soldier at war, 1147–1189*, History of Warfare 44, Leiden/Boston 2007.

Houben, H, *Roger II of Sicily. A ruler between east and west*, Cambridge 2002.

Hoven, B van den, *Work in ancient and medieval thought. Ancient philosophers, medieval monks and theologians and their concepts of work, occupations and technology*, Amsterdam 1996.

Howard, F, *Sailing ships of war*, London 1979

Howard-Johnson, J, 'The two great powers', in A Cameron (ed.), *The Byzantine and Early Islamic Near East, III: States, resources and armies*, Studies in late antiquity and early Islam, Princeton 1995 (157–226).

Howlett, R, 'Norfolk artillery in the fourteenth century', *Norfolk Archaeology* 16, 1907 (46–75).

Hoyland, R G, *In God's path. The Arab conquests and the creation of an Islamic empire*, Oxford 2015.

Hutchinson, G, *Medieval ships and shipping*, London 1994.

Huuri, K, 'Zur Geschichte des mittelalterlichen Geschützwesens aus orientalischen Quellen', in K Tallquist (ed.), *Studia Orientalia* IX, Helsinki 1941.

Irwin, R, 'Gunpowder and firearms in the Mamluk sultanate reconsidered', in M Winter, A Levanini (eds), *The Mamluks in Egyptian and Syrian politics and society*, Leiden/Boston 2004.

James, S, *Excavations at Dura-Europos 1928–1937. Final report VII: The arms and armour and other military equipment*, Oxford 2007.

James, S, 'The *fabricae*: state arms factories of the later Roman Empire', in J C Coulston (ed.), *Military equipment and the identity of soldiers. Proceedings of the fourth Roman military equipment conference*, BAR International Series 394, 1988 (257–332).

Jarnut, P, 'La conquista di Bergamo nell'894: una battaglia decisiva fra imperatore Guido e re Arnolfo', *Archivio storico bergamasco* I, 1981 (25–34).

Jolivet, J, 'Classification of the sciences', in R Rashed (ed.), *Encyclopedia of the history of Arabic science*, London/New York 1996 (III, 1008–25).

Jordan, K, *Henry the Lion. A biography*, trans. P S Falla, Oxford, 1986

Jupp, E B, *An historical account of the worshipful company of carpenters of the City of London*, 2nd edn, with supplement by W W Pocock, London 1887

Justinger: *Conrad Justinger's Berner Chronik, von Anfang der Stadt Bern bis in das Jahr 1421*, ed. E Stierlin, Muss, J, Bern 1819.

Kazhdan, A P, 'Factories (imperial)', in the same (ed.), *Oxford Dictionary of Byzantium*, 2, 1991 (774–5).

Kazhdan, A P, Wharton Epstein, A, *Changes in Byzantine culture in the eleventh and twelfth centuries*, Berkeley/Los Angeles/London 1985.

Keats-Rohan, K S B, *Domesday people. A prosopography of persons occurring in English documents 1066–1166*, I, Woodbridge 1999.

Keats-Rohan, K S B, '"Most securely fortified": Wallingford castle 1071–1540', in N Christie, K S B Keats-Rohan, D Roffe (eds.), *Wallingford: the town and castle in context*, BAR British Series 621, 2015 (34–105).

Kennard, A N, *Gunfounding and gunfounders. A directory of cannon founders from earliest times to 1850*, London/New York/Sydney, 1986.

Kennedy, H, *The armies of the caliph: military and society in the early Islamic state*, London/New York 2001.

Kennedy, H, 'Military pay and the economy of the early Islamic state', *Historical Research* 75, no. 188, May 2002 (155–69).

Kersuzan, A, *La bâtie de Luisandre (Ain). Histoire et archéologie d'une fortification savoyarde de frontière au XIVe siècle*. Chambéry, 2010

Khamisy, R G, 'Some notes on Ayyūbid and Mamluk military terms', *Journal of Medieval Military History* XIII, 2015 (73–92).

Kidson, P, 'Architectural proportion', in J Turner et al. (eds), *The Dictionary of Art*, vol. 2, London 1996 (343–52).

Kill, R, Schoen, H, *L'ensemble souterrain de la Petite Pierre. Un remarquable amenagement dans le domaine de l'approvisionnement en eau*, Centre de recherches archéologiques médiévales de Saverne, Documents, 2, 2008.

King, A, 'Gunners, aides and archers: the personnel of the English ordnance companies in Normandy in the fifteenth century', *Journal of Medieval Military History* IX, 2011 (65–75).

King, P D, *Law and society in the Visigothic kingdom*, Cambridge 1972.

Klapparf, L, 'Studies of the development and structure of early metal production', in C Segers-Glocke, H Witthöft (eds), *Aspects of mining and smelting in the Upper Harz mountains (up to the 13th/14th centuries) in the early times of a developing European culture and economy*, St Katharinen, 2000 (1–23).

Klein, H, 'Die Salzburgischen Büchsenmeister des 14 Jahrhunderts', *Zeitschrift für Historische Waffenkunde, NF VI (7)*, 1938 (141–7).

Kleinbauer, W E, 'Pre-Carolingian concepts of architectural planning', in M Chiat, K Reyerson (eds), *The medieval Mediterranean: cross-cultural contacts*, Medieval Studies at Minnesota 3, St Cloud 1988 (67–9).

Knobloch, E, 'Technische Zeichnungen', in U Lindgren (ed.), *Europäische Technik im Mittelalter 800 bis zum 1400. Tradition und Innovation*, Berlin 1996 (45–64).

Kocabas, U, 'The Yenikapi Byzantine-era shipwrecks, Istanbul, Turkey: a preliminary report and inventory of the 27 wrecks studied by Istanbul University', *International Journal of Nautical Archaeology* 44/1, 2015 (5–38).

Koch, H, 'Vom Nachleben des Vitruv', *Deutsche Beiträge zur Altertumswissenschaft*, Heft 1, 1951.

Koch, R, 'Fossa Carolina. An early medieval canal near a watershed', in C Westerdahl (ed.), *The significance of portages. Proceedings of the first international conference on the significance of portages, 29 September–2 October 2004*, BAR International Series 1499, 2006 (259).

Koch, R, 'Fossa Carolina. Neue Erkentnisse zum Schiffahrtskanal Karls des Grossen', in K Elmshäuser (ed.), *Häfen, Schiffe, Wasserwege. Zur Schiffahrt des Mittelalters*, Bremerhaven/Hamburg 2002 (54–70).

Krahe, F W, *Burgen des deutschen Mittelalters*, Wurzburg, 1994.

Kreutz, B M, 'Ships, shipping and the implications of change in the early medieval Mediterranean', *Viator* 7, 1976 (79–110).

La Salvia, V, 'Archaeometallurgy as a source for the history of changes and developments within medieval daily life', in *History of Medieval Life and the Sciences. Proceedings of an international round-table-discussion. Krems an der Donau, September 28–9 1998*, Wien 2000 (105–15).

Laiou, A E, Morrison, C, *The Byzantine economy*, Cambridge 2007.

Lander, J, *Roman stone fortifications. Variation and change from the first century AD to the fourth*, BAR International Series 206, 1984.

Landers, J, *The field and the forge. Population, production and power in the pre-industrial west*, Oxford 2003.

Lane, F C, 'The economic meaning of the invention of the compass', *American History Review* LXVIII/3, 1963 (605–17).

Latham, R E, *Revised medieval Latin word list from British and Irish sources*, Oxford 1965 (reprinted 1994)

Latimer, P, 'Wages in late-twelfth- and early-thirteenth-century England', *Haskins Society Journal* 9, 1997 (185–205).

Lázaro, J T, 'Acuñación de moneda', in P Navascues (gen. editor), *Ars mechanicae, ingenería medieval en España*, 2008 (215–24).

Le Goff, J, *Time, work and culture in the middle ages*, trans. A Goldhammer, Chicago/London 1980.

Lebecq, S, 'The role of the monasteries in the systems of production and exchange of the Frankish world between the 7th and the beginning of the 9th centuries', in I L Hansen, C Wickham (eds), *The long eighth century*, Leiden/Boston/Köln, 2000 (121–48).

Leff, G, *Paris and Oxford universities in the thirteenth and fourteenth centuries. An institutional and intellectual history*, London/Sydney/New York 1968.

Lefort, J, Bandoux, R, Cheynet, J-C, Grélois, J-P, Kravari, V, *Géometries du fisc byzantin*, Paris 1991.

Leguay, J-P, 'Un aspect du travail dans les villes armoricaines au Moyen Âge: la fabrication des canons et armes blanches', in P Benoit, D Cailleaux (eds), *Hommes et travail du métal dans les villes médiévales. Actes de la table ronde 'La métallurgie dans la France médiévale', université de Paris 1, 1984*, Paris 1988 (185–226).

Leguay, J-P, *La ville de Rennes au XVme siècle à travers les comptes des miseurs*, Rennes 1968.

Leibnitz, K, 'Die Manuskripte des Walter de Milemete', *Zeitschrift für Waffen und Kostumkunde*, ser. 3, 34, 1992 (117–33).

Leng, R, 'Ars belli. Deutsche taktische und kriegstechnische Bilderhandschriften und Traktate im 15. und 16. Jahrhundert', *Imagines Medii Aevi* 12 (1, 2), 2002 (21–261).

Leng, R, 'Social character, pictorial style and the grammar of technical illustration in craftsmen's manuscripts in the late middle ages', in W Lefèvre (ed.), *Picturing machines*, Cambridge (Mass.)/London 2004 (85–111).

Leriche, P, 'Techniques de guerre sassanides et romaines à Doura-Europos', in F Vallet, M Kazanski (eds), *L'armée romaine et les barbares du IIIe au VIIe siècles*. Mémoires, L'association Française d'archéologie Mérovingienne, 1993 (83–100).

Le Strange, G, *Baghdad during the Abbasid caliphate*, Oxford 1900 (reprinted 1972).

Lev, Y, 'The Fātīmid navy. Byzantium and the Mediterranean sea 909-1036CE/297-427 AH', *Byzantion* LIV, 1984 (220–52).

Levanini, A, 'Water supply in medieval Middle Eastern cities: the case of Cairo', *Al-Masaq* 20/2, 2008 (179–205).

Lewis, C and Short, C, *A Latin dictionary founded on Andrews' edition of Freund's Latin dictionary, revised and enlarged*, Oxford 1879.

Lewis, M J T, 'Antique engineering in the Byzantine world', in L Lavan, E Zanini, A Sarantis (eds), *Technology in transition AD 300–650*, Leiden/Boston 2007 (367–78).

Lexer, M, *Mittelhochdeutsches Wörterbuch*, Leipzig 1878.

Liebel, J, *Springalds and great crossbows*, trans. J Vale, Leeds 1998.

Lilley, K, 'Taking measures across the medieval landscape: aspects of urban design before the Renaissance', *Urban Morphology* 2(2), 1998 (82–92).

Lilley, K, Lloyd C, Trick S, 'Mapping medieval urban landscapes: the design and planning of Edward I's new towns of England and Wales', *Antiquity* 79, no. 303, 2005 accessed at http:www.antiquity.ac.uk/projgall/lilley (10/3/14).

Loewe, R, 'Jewish evidence for the history of the crossbow', in G Dahan (ed.), *Les juifs au regard de l'histoire. Mélanges en l'honneur de Bernhard Blumenkranz*, Paris 1985 (87–107).

Lohrmann, D, 'Echanges techniques entre orient et occident au temps des croisades', in I Draelants, A Tihon, B van der Abeele (eds), *Occident et proche-orient, contacts scientifiques au temps des croisades. Actes du colloque de Louvain-la-neuve, 24 et 25 mai 1997*, Louvain 2000 (117–43).

Lohrmann, D, 'Le moulin à eau dans le cadre de l'économie rurale de la Neustrie', in H Atsma (ed.), *La Neustrie. Les pays au nord de la Loire de 650 à 850. Colloque Historique Internationale*, 2 vols, Sigmaringen 1989 (I, 367–404).

Lohrmann, D, 'Zwei Ingenieure beim Untergang des Hauses Carrara', in M Arnoux, P Monnet (eds), *Le technicien dans la cité en Europe occidentale 1250–1650*, École Française de Rome, Rome, 2004 (25–41).

Lohrmann, D, Kranz, H, Alertz, U (eds.), 'Konrad Grüter von Werden, de machinis et rebus mechanicis', *Vatican Library studi e testi* 428–9, 2006.

Long, P O, *Artisan/practitioners and the rise of the new sciences 1400–1600*, Corvalli (Oregon) 2011.

Long, P O, *Openness, secrecy, authorship. Technical arts and the culture of knowledge from antiquity to the Renaissance*, Baltimore/London 2001.

Loseby, S T, 'Gregory's cities: urban function in sixth century Gaul', in I Wood (ed.), *Franks and Alemanni in the Merovingian period. An ethnographic perspective*, Studies in Historical Archaeoethnology 3, Woodbridge 1998 (239–84).

Loyn, H R, Percival J (eds.), *The reign of Charlemagne*. Documents of Medieval History 2, London 1975.

Lusignan, S, 'Les arts mécaniques dans le Speculum doctrinale de Vincent de Beauvais', in G H Allard, S Lusignan (eds), *Les arts mécaniques au moyen âge*, Montreal/Paris 1982 (33–48).

Lyons, J, *The House of Wisdom. How the Arabs transformed western civilization*, London/Berlin/New York/Sydney, 2009.

Maddison, F, 'A consequence of discovery: astronomical navigation in fifteenth century Portugal' in *Studies in Portuguese discoveries*, I, Proceedings of the first colloquium of the Centre for the Study of the Portuguese Discoveries, ed. T F Earle, S Parkinson, Warminster 1992

Magnusson, R, Squatriti, P, 'Technologies of water in medieval Italy', in P Squatriti (ed.), *Working with water in medieval Europe*, Leiden/Boston/Köln 2000 (216–66)

Mahmoud, S J, *Metal technology in medieval India*, Delhi 1988.

Mallett, M E, Hale, J R, *The military organisation of a Renaissance state: Venice c.1400 to 1617*, Cambridge 1984.

Mallett, M, Shaw, C, *The Italian wars 1494–1559. War, state and society in early modern Europe*, London/New York 2012.

Marazzi, F, 'The destinies of late antique Italies: polico-economic developments of the sixth century', in R Hodges, W Bowden (eds), *The sixth century: production, distribution and demand*, Leiden/Boston/Köln 1998 (119–59).

Marcinkowski, M Ismail, *Measures and weights in the Islamic world. An English translation of Walther Hinz's handbook Islamische Masse und Gewichte*, Kuala Lumpur 2003.

Mark, R (ed.), *Architectural technology up to the scientific revolution*, Cambridge (Mass.)/ London 1993.

Mark, R, 'Technological innovation in high Gothic architecture', in E Bradford Smith, M Wolfe (eds.), *Technology and resource use in medieval Europe: cathedrals, mills and mines*, Aldershot 1997 (11–26).

Martin, M, Renucci, F, *Guédelon: a castle in the making*, Paris 2016.

Martin, P, 'L'artillerie et la fonderie de canons à Strasbourg de XIVe au XVIIIe siècle', *Armi Antiche* 1967 (71–102).

Maschke, E, 'Die Brücke im Mitteralter', *Historische Zeitschrift* 224/2, 1977 (265–92).

Masciandaro, N, *The voice of the hammer. The meaning of work in Middle English literature*, Notre Dame, Indiana 2007.

Masood, E, *Islam and science. A history*, London 2009.

Mathies, A L, 'Medieval treadwheels: artists' views of building construction', *Technology and Culture* 33/3, 1992 (510–47).

Matschke, K-P, 'Mining', in A E Laiou (ed.), *The economic history of Byzantium from the 7th through the 15th century*, Dumbarton Oaks Studies XXXIX, Washington 2002 (I, 115–20).

Matthews, J, 'Roman life and society', in J Boardman, J Griffin, O Murray (eds), *The Oxford history of the classical world*, London/Oxford 1986

McGee, D M, 'The origins of early modern machine design', in W Lefèvre (ed.), *Picturing machines*, Cambridge (Mass.)/London 2004, 53–84.

McGeer, E, 'Byzantine siege warfare in theory and practice', in I A Curtis, M Wolfe (eds), *The medieval city under siege*, Woodbridge 1995.

McPeak, W J, 'Military technology', in H Selin (ed.), *Encyclopedia of the history of science, technology and medicine in non-western countries*, Dordrecht/Boston/London 1997 (741–44).

Medieval Gunpowder research group, at www.leeds.ac.uk/ims/fieldsofconflict/reports.

Mele, G, 'A geometrical analysis of the layout of Acaya, Italy', *Nexus Network Journal* 14/2, 2012 (373–89).

Ménabréa, L, 'De l'organisation militaire au moyen âge d'après des documents inédits', *Mémoires de l'Académie Royale de Savoie*, ser. 2, 1, 1851 (179–226).

Menant, F, 'Métallurgie lombarde', in P Benoit, D Cailleaux (eds), *Hommes et travail du métal dans les villes médiévales. Actes de la table ronde 'La métallurgie dans la France médiévale', université de Paris 1, 1984*, Paris 1988 (127–61).

Mesqui, J, *Le pont en France avant le temps des ingénieurs*, Paris 1986.

Metzdorf, J, 'Bedrangnis, Angst und grosse Mühsal – die Belagerung von Neuss durch Karl den Kühnen 1474/5', in *Beihefte zur Mediavistik*, ed. O Wagener, 7, 2006 (167-88)

Michael, N, Embleton G, *Armies of medieval Burgundy 1364–1477*, Oxford 1983.

Michaud-Frejaville, F, 'Fèvres et forgerons d'Orléans 1399–1430', in P Benoit, D Cailleaux (eds), *Hommes et travail du métal dans les villes médiévales. Actes de la table ronde 'La métallurgie dans la France médiévale', université de Paris 1, 1984*, Paris 1988 (69–82).

Milne, G, et al., *Timber building techniques in London c.900–1400. An archaeological study of waterfront installations and related material*, London and Middlesex Archaeological Society, Special Paper 15, London 1992.

Mitchell, S, *A history of the later Roman Empire AD284–641. The transformation of the ancient world*, Malden/Oxford/Carlton, 2007.

Moffett, R, 'Military equipment in the town of Southampton during the fourteenth and fifteenth centuries', *Journal of Medieval Military History* IX, 2011 (167–99).

Mohebbi, P, *Techniques et ressources en Iran du 7ᵉ au 19ᵉ siècle*, Tehran 1996.

Mollat, M, 'Le passage de Saint Louis à Tunis. Sa place dans l'histoire des croisades', *Revue d'histoire économique et sociale* 50/3, 1972 (289–303).

Mollat, M, 'Problèmes navales de l'histoire des croisades', *Cahiers de Civilisation Médiévale* 10 (39), 1967 (345–59).

Monreal y Tejada, L, *Ingenieria militari en las cronicas catalanas*, Barcelona 1971.

Monteillard, M, 'Artisans et artisanat du métal à Rouen à la fin du moyen âge', in P Benoit, D Cailleaux (eds), *Hommes et travail du métal dans les villes médiévales. Actes de la table ronde 'La métallurgie dans la France médiévale', université de Paris 1, 1984*, Paris 1988 (109–20).

Morimoto, K, *The fiscal administration of Egypt in the early Islamic period*, Dohosha, 1981.

Müller, M and (supplement) Lexer, M, *Mittelhochdeutscher Wörterbuch*, Leipzig, 1863 and 1872.

Müller-Kissing, J, Peine, H-W, Treude, E, 'Von Bergfried zum Minenstallen – die Falkenburg bei Detmold', *Archäologie in Deutschland* 3, 2013 (68–9).

Mullet, M, 'Writing in early medieval Byzantium', in R McKitterick (ed.), *The uses of literacy in early medieval Europe*, Cambridge 1990 (156–85)

Murdoch, J E, *Album of science*, New York 1984.

Murdoch, J E, Sylla, E D, 'The science of motion', in D C Lindberg (ed.), *Science in the Middle Ages*, Chicago/London 1978 (206–64).

Muris, Johannes de, 'De arte mensurandi', ed. H L L Busard, *Boethius Texte und Abhandlungen zur Geschichte der Mathematik und der Naturwissenschaften*, 41, Stuttgart 1998.

Nasr, S N, *Science and civilisation in Islam*, Cambridge (Mass.) 1968.

Necipoğlu, G, *The Topkapi scroll: geometry and ornament in Islamic architecture*, Getty Centre for History of Art and the Humanities, Santa Monica 1995.

Needham, J, 'China's trebuchets, manned and counterweighted', in B S Hall, D C West (eds), *On pre-modern technology and science: a volume of studies in honour of Lynn White jr*, Humana civilitas, I, Malibu 1976 (107–45).

Needham, J, 'Classical Chinese contributions to mechanical engineering', in J Needham, Wang Ling, Lu Gwei-Djen, Ho Ping-yü (eds), *Clerks and craftsmen in China and the west. Lectures and addresses on the history of science and technology*, Cambridge 1970 (113–35).

Needham, J, *The guns of Kaifêng-Fu. China's development of man's first chemical explosive*, Creighton Trust lecture, 1979.

Needham, J, 'Iron and steel production in ancient and medieval China', in J Needham, Wang Ling, Lu Gwei-Djen, Ho Ping-yü (eds), *Clerks and craftsmen in China and the west. Lectures and addresses on the history of science and technology*, Cambridge 1970 (107–12).

Needham, J, *Science and civilisation in China*. Vol. 5 part 7: military technology: the gunpowder epic. Cambridge 1986

Needham, J, 'Thoughts on the social relations of science and technology in China', *Centaurus* 3, 1953 (40–8).

Nef, J U, 'Mining and metallurgy in medieval civilisation', in M Postan, E E Rich (eds.), *The Cambridge Economic History of Europe*, II, Cambridge 1952 (430–89).

Nehlsen, H, 'Die rechtliche und soziale Stellung der Handwerker in den germanischen Leges', in H Jahnkuhn, W Janssen, R Schmidt-Wiegund, H Tiefenbach (eds), *Das Handwerk in vor- und frühgeschichtlicher Zeit*, Göttingen 1981 (I, 267–83).

Nelson, J L, 'Literacy in Carolingian government', in R McKitterick (ed.), *The uses of literacy in early medieval Europe*, Cambridge 1990 (258–96).

Neri, S, 'Albertus inzignerius, un grande costruttore del Duecento', in F Bocchi (ed.), *I portici di Bologna e l'edilizia civile medievale*, Bologna 1990 (125–33).

Netton, I R, *Al-Fārābi and his school*, London/New York 1992.

Niarchos, C, 'The philosophical background of the eleventh-century revival of learning in Byzantium', in M Mullett, R Scott (eds.), *Byzantium and the Classical tradition*, Birmingham 1981 (127–35).

Nicholls, H G, *Iron making in the olden times as instanced in the ancient mines, forges and furnaces of the forest of Dean*, 1866, new edn. with introduction by I J Standing, Coleford 1981.

Nicolas, H, *History of the battle of Agincourt and the expedition of Henry the Fifth into France in 1415: to which is added the roll of the men at arms in the English army*, 2nd edn, London 1832.

Nicolle, D, *Saracen strongholds 1100–1500. The central and eastern Islamic lands*, Oxford 2009.

O'Connell, J, 'Dynamics of a medieval missile launcher: the trebuchet', *The Physics Teacher* 39/8, 2001 (471–3).

O'Conor, L, 'Building a better trebuchet', *Mechanical engineering* 116/1, 1/1994 (66–9).

Oikonomides, N, 'Byzantine archives of the Palaiologan period, 1258–1453', in R Britnell (ed.), *Pragmatic literacy, east and west 1200–1330*, Woodbridge 1997 (189–97).

Olsson, K-G, 'Statical behaviour of the timber roof structure of a medieval castle', in J Natterer, J-L Sandoz (eds), *Proceedings of the 5th world conference on timber engineering*, Lausanne 1998.

Ousterhout, R, *Master builders of Byzantium*, Princeton 1999.

Ovitt, G, jr, *The restoration of perfection. Labor and technology in medieval culture*, New Brunswick/London 1987.

Özdural, A, 'Mathematics and arts: connections between theory and practice in the medieval Islamic world', *Historia Mathematica* 27, 2000 (171–201).

Pacey, A, 'Engineering', in H Selin (ed.), *Encyclopedia of the history of science, technology and medicine in non-western countries*, Dordrecht/Boston/London 1997 (280–2).

Pandu Ranga Rao, M (ed.), *Engineering and technological achievements during the Kakatiya period. A medieval kingdom of south India*, Warangal 1996.

Park, H, *Mapping the Chinese and Islamic worlds*, Cambridge 2012.

Parkes, M, 'The literacy of the laity', in D Daichen, A Thorlby (eds.), *Literature and western civilisation, III: The medieval world*, London 1973 (555–78).

Partington, J R, *A history of Greek Fire and gunpowder*, new edn Baltimore/London 1999.

Pause, C, 'Weapons and military equipment from the Burgundian siege of Neuss in 1474–1475', *Acta Militaria Medievalia* VI, 2010 (121–44).

Pauw, N de, 'Claus van Lit, Maître de l'artillerie de Gand de XIV siècle', *Bulletin de la Commission royale d'histoire, Académie royale de Belgique* LXXXIII, Brussels 1920 (233–9).

Payne, R, *The canal builders. The story of canal engineers through the ages*, New York, 1959.

Peacock, A C S, *The great Seljuk empire*, Edinburgh 2015.

Pearce, M, 'Craftsmen and accounts', in Historic Scotland, *Stirling palace. Rebirth of a palace*, conference papers 2012.

Pepper, S, 'Artisans, Architects and Aristocrats. Professionalism and Renaissance military engineering', in D J B Trim (ed.), *The Chivalric Ethos and the Development of Military Professionalism*, Leiden/Boston 2003 (117–48).

Pepper, S, 'The underground siege', *Fort* 10, 1982 (31–8).

Perroy, E, 'L'artillerie de Louis XI dans la campagne d'Artois', *Revue du Nord* 26 (no. 103), 1943 (171–96, 293–315).

Perroy, E, 'L'artillerie royale à la bataille de Montlhéry (10 juillet 1465)', *Revue historique* 49, 1925 (187–9).

Pétain, N, 'Philological notes on the crossbow and related medieval weapons', *Greek, Roman and Byzantine Studies* 33/3, 1992 (265–91).

Petersen, L I R, *Siege warfare and military organisation in the successor states (400–800AD). Byzantium, the West and Islam*, Leiden/Boston 2013.

Petrikovits, H von, 'Militärische Fabrikae der Römer', in D M Pippidi (ed.), *Actes du IX Congrès International d'Études sur les frontières Romaines, Mamaïa, 6–13 Sept. 1972*, București, Köln, Wien, 1974, (399-407).

Petrikovits, H von, 'Die Spezialiserung des römischen Handwerks. II Spätantike', *Zeitschrift für Papyrologie und Epigraphik* 43, 1981 (255–306).

Pevsner, N, 'The term "architect" in the Middle Ages', *Speculum* 17, 1942 (549–62).

Peyras, J (ed./trans), *Arpentage et administration publique à la fin de l'antiquité. Les écrits des hauts fonctionnaires équestres*, Presses universitaires Franche-Comté, 2008.

Picard, C, 'Les arsenaux musulmans de la Méditerranée et de l'océan Atlantique (VIIe-XVe siècle)', in D Coulon, C Otten-Froux, P Pagès, D Valérian, D (eds), *Chemins d'outre-mer. Études sur la Méditerranée médiévale offertes à Michel Balard*, Paris 2004 (II, 691–710).

Pluskowski, A, *The archaeology of the Prussian crusade. Holy war and colonisation*, London/New York 2013.

Pohl, W, 'The barbarian successor states', in L Webster, M Brown (eds), *The transformation of the Roman world*, London 1997 (33–47).

Porter, W, *History of the corps of Royal Engineers*, I, Chatham 1889, reprinted 1977.

Price, D J, 'Medieval surveying and maps', *Geographical Journal* CXXI, 1955 (1–10).

Pringle, D, 'King Richard I and the walls of Ascalon', *Palestine Exploration Quarterly* 116/2, 1984 (133–47).

Pringle, D, 'A medieval graffito representing a trebuchet in an Etruscan tomb in central Tarquinia', in S John, N Morton (eds), *Crusading and warfare in the Middle Ages: realities and representations. Essays in honour of John France*, Farnham 2014 (37–46).

Promis, C, 'Biografie di ingegneri militari Italiani dal secolo XIV alla meta del XVIII', *Miscellanea di Storia Italiana* XIV, Turin 1874.

Promis, C, 'Gli ingegneri e gli scrittori militari Bolognesi del XV e XVI secolo', *Miscellanea di Storia Italiana* IV, Turin 1863 (579–90).

Prouteau, N, 'L'art de la charpenterie et du génie militaire dans le context des croisades: recrutement et fonctions des techniciens Francs (XIIe–XIIIe siècles)', in N Faucherre, J Mesqui, N Prouteau (eds), *La fortification au temps des Croisades*, Rennes 2004 (279–86).

Prouteau, N, '"Beneath the battle?" Miners and engineers as mercenaries in the Holy Land (XII-XIII siècles)', in J France (ed.), *Mercenaries and paid men. The mercenary identity in the Middle Ages. Proceedings of a conference held at University of Swansea, 7–9/9/2005*, Leiden/Boston, 2008 (105–17).

Prouteau, N, 'Engineers and Architects', in C J Rogers (ed.), *The Oxford Encyclopedia of Medieval Warfare and technology*, vol. 2, Oxford 2010 (22–4).

Prouteau, N, 'Eudes de Montreuil, maître des oeuvres des fortifications de Jaffa. Une légende Franciscaine?', *Bulletin Monumental* 164/1, 2006 (109–12).

Prouteau, N, 'La forteresse-arsenal de Chinon au XIIIe siècle', in *Chinon. Le destin d'une forteresse*. Société des Amis de vieux Chinon, Hors série, 2011 (15–28).

Prouteau, N, 'Mensuratores castrorum. Les arpenteurs militaires au Moyen Âge', in H Mouillebouche (ed.), *Châteaux et mesures. Actes 17e journée de Castellologie de Bourgogne, Octobre 2010*, 2011.

Pryor, J H, 'The age of the ΔΡΟΜΩΝ. The Byzantine navy c.500–1204', *The Medieval Mediterranean* 62, 2006.

Pryor, J H, 'The galleys of Charles I of Anjou, king of Sicily ca. 1269–84', in J Evans, R Unger (eds), *Studies in medieval and renaissance history* XIV, 1993 (33–102).

Pryor, J H, *Geography, technology and war. Studies in the maritime history of the Mediterranean 649–1571*, Cambridge 1988.

Pryor, J H, 'The naval architecture of Crusader transport ships: a reconstruction of some archetypes for round-hulled sailing ships', *Mariner's Mirror* 70, 1984, reprinted in the same, *Commerce, shipping and naval warfare in the medieval Mediterranean*, London 1987 (171–96).

Pryor, J H, 'Shipping and ship-building', in E Jeffreys, J Haldon, R Cormack (eds), *Oxford Handbook of Byzantine Studies*, Oxford 2008 (chapter II.8.9, 482–91).

Pryor, J H, 'Transportation of horses by sea during the era of the crusades, eighth century to 1285AD', *Mariner's Mirror* 68, 1972 (9–28, 103–25).

Purton, P, *History of the early medieval siege c.450–1200*, Woodbridge, 2010.

Purton, P, *History of the late medieval siege c.1200–1500*, Woodbridge 2010.

Purton, P, 'The myth of the mangonel. Torsion artillery in the middle ages', *Arms and Armour* 3, 2006 (79–90).

Purton, P, 'Suffossores immiserunt at subvertundum muros: mines and miners in medieval siege warfare', R Atzbach et al. (eds.), *Castles at war*, Bonn 2015 (187–202).

Quesada, M A Ladero, *Castilla y la conquista del reino de Granada*, Valladolid 1967.

Radding, C M, Clark, W C, *Medieval architecture, medieval learning: builders and masters in the age of Romanesque and Gothic*, New Haven/London 1992.

Rahtz, P, Meeson, R, *An Anglo-Saxon watermill at Tamworth*, CBA Research Report 83, 1992.

Randolph, A, 'The bastides of south-west France', *Art Bulletin* 77/2, 1995 (290–307).

Raphael, K, Tepper, Y, 'The archaeological evidence from the Mamluk siege of Arsūf', *Mamluk Studies Review* IX (1), 2005 (85–100).

Rashed, R, Bellosta, H, *Ibrāhīm Ibn Sinān. Logique et géométrie au Xe siècle*, Leiden/Boston/Köln 2000.

Raymond, P, *Rôles de l'armée de Gaston Phoebus, comte de Foix et seigneur de Béarn, 1376-78*, Bordeaux, 1872.

Reinart, A, 'The siege and storm of Lindholmen during the second Hanseatic war (1368-9)', in R Atzbach (ed.), *Castles at War*. The Danish Castles Research Association 'Magt, Borg og Landskab' interdisciplinary symposium 2013, Bonn 2015 (203–16).

Renn, D, 'Ailnoth the engineer and Orford Castle', *Castle Studies Group Journal* 26, 2012/13 (201–2).

Renn, D, 'Refortification at Rochester in the 1220s: a public/private partnership?', *Archaeologia Cantiana* CXXIV, 2004 (343–64).

Retamero, F, 'As coins go home: towns, merchants, bishops and kings in Visigothic Hispania', in P Heather (ed.), *The Visigoths from the migration period to the seventh century. An ethnographic perspective*, Studies in Historical Archaeoethnology 3, Woodbridge 1999 (271–320).

Reynolds, L D (ed.), *Texts and transmissions. A survey of the Latin classics*, Oxford 1983.

Reynolds, T S, *Stronger than a hundred men. A history of the vertical water wheel*, John Hopkins Studies in the history of technology NS 7, Baltimore 1983.

Rhoten, R P, 'The trebuchet: accuracy analysis of a medieval siege engine', *1999 ASME Design Engineering Technical Conferences, organised A K Gupta*, 2, 1993 (69–73).

Rhoten, R P, 'Trebuchet energy efficiency – experimental results', *44th AIAA aerospace sciences conference. Technical papers* 13, January 2006. Paper 775 (9355–5).

Richard, J, *Saint Louis. Crusader king of France*, ed./abridged S Lloyd, trans. J Birrell, Cambridge/Paris 1992.

Riché, P, *Education and culture in the Barbarian west sixth through eighth centuries*, trans. J J Contreni, Columbia (SC) 1976.

Riché, P, 'L'enseignement de Gerbert à Reims dans le contexte européen', in P Riché, *Education et culture dans l'occident médiéval*, Aldershot 1993 (originally published in 'Gerberto, scienzia, storia e mita', *Archivum Bobiense*, Studia II, Bobbio 1985 (51–69).

Richmond, I, *Trajan's army on Trajan's column*, introduction by M Hassall, British School at Rome, London 1982.

Rieth, E, 'Le clos des galées de Rouen, lieu de construction naval à clin et à carvel (1293-1419)', in C Villain-Gandossi et al, (eds.), *Medieval ships and the birth of technological societies, I: northern Europe*. European coordinating Centre for research and documentation in social sciences, Malta, 1989 (71-77)

Rigold, S E, 'Structural aspects of medieval timber bridges,' *Medieval Archaeology* XIX, 1975 (48–91).

Ripoll López, G, 'Symbolic life and signs of identity in Visigothic times', in P Heather (ed.), *The Visigoths from the migration period to the seventh century. An ethnographic perspective*, Studies in Historical Archaeoethnology 3, Woodbridge 1999 (403–46).

Rippon, S, Claughton, P, Smart, C, *Mining in a medieval landscape. The royal silver mines of the Tamar valley*, Exeter 2009.

Rogers, C J, *Essays on medieval military history. Strategy, military revolutions and the Hundred Years War*, Farnham 2010.

Rogers, C J, *War cruel and sharp. English strategy under Edward III 1327–1360*, Woodbridge 2000.

Rogers, R, *Latin siege warfare in the twelfth century*, Oxford 1992.

Ronan, C (abridged by), Needham J (author), *The shorter science and civilization in China*, 4, Cambridge 1994.

Rose, P L, 'Humanist culture and renaissance mathematics: the Italian libraries of the quattrocento', *Studies in the Renaissance* XX, 1973 (46–105).

Rose, S, *England's medieval navy 1066-1509. Ships, men and warfare*, Barnsley 2013

Rose, S, 'Bayonne and the King's ships', in Rose, *Medieval ships and warfare*, Aldershot, 2008, 69-76 (originally in *Mariners' Mirror* 86, 2000, 140-7)

Rose, S, 'Southampton and the navy in the age of Henry V', *Hampshire Papers* 14, 1998.

Rosenberg, C M, *The Este monuments and urban development in Renaissance Ferrara*, Cambridge 1997.

Round, J H, *The king's serjeants and officers of state*, London 1911.

Rowland, A, 'Science, technology and war', *Technology and Culture, supplement to vol. 36/2*, April 1995 (95–7).

Rubin, Z, 'The reforms of Khusro Anūshiwan', in A Cameron (ed.), *The Byzantine and Early Islamic Near East, III: States, resources and armies*. Studies in late antiquity and early Islam, Princeton 1995 (227–98).

Salch, C-L, *Dictionnaire des Châteaux de l'Alsace médiévale*, Strasbourg 1976.

Saliba, G, 'The function of mechanical devices in medieval Islamic society', in P O Long (ed.), *Science and technology in medieval society. Annals of the New York Academy of Sciences* 441, New York 1985 (141–9).

Sarantis, A, 'Waging war in late antiquity', in A Sarantis, N Christie (eds), *War and warfare in Late Antiquity*, Late Antique Archaeology 8.1 and 8.2, Leiden/Boston 2013 (1–98).

Sarasvati Amma, T A, *Geometry in ancient and medieval India*, Delhi/Varanesi/Patna 1979.

Sarris, P, *Economy and society in the age of Justinian*, Cambridge 2006.

Sawyer, R D, *Fire and water. The art of incendiary and aquatic warfare in China*, Boulder 2004.

Sayers, W, 'The name of the siege engine *trebuchet*: etymology and history in medieval France and Britain', *Journal of Medieval Military History* VIII, 2010 (189–96).

Schäfer, D, *Die Hansestädte und König Waldemar von Dänemark, Hansische Geschichte bis 1376*, Jena 1879.

Schedelmann, H, *Die grossen Büchsenmacher. Leben, Marken, Werken von 15. bis 19 Jahrhundert*, Braunschweig 1972.

Schewitzer, C, 'Muslim waterworks', *Islamic Culture* XIII/1, 1939 (79–82).

Schildhauer, J, *The Hansa. History and culture*, trans. K Vanovitch, Leipzig 1985.

Schmidtchen, V, 'Riesengeschütze des 15 Jahrhunderts, technische Hochsleistungen ihrer Zeit', *Techniksgeschichte*, 44, 1977 (153-230)

Schneider, H, 'Schweizerische Büchsenmacherei', *Armi Antiche*, 1963 (91–106).

Schoen, H, 'Quelques sites de guerre de sape médiévale des Vosges et du Wasgau', *Revue d'Alsace* 122, 1996 (127–36).

Schöller, W, 'Ein Katalog mittelalterlicher Baubetriebsdarstellungen', *Techniksgeschichte* 54, 1987 (77–100).

Schulz, J, 'City plans', in J B Friedman, K M Figg (eds), *Trade, travel and exploration in the Middle Ages: an encyclopedia*, New York/London 2000 (110–11).

Schwind, F, 'Karolingerzeitliche Klöster als Wirtschaftsorganismen', in L Fenske, W Rösener, T Zotz (eds.), *Institutionen, Kultur und Gesellschaft. Festschrift für Josef Fleckenstein*, Sigmaringen 1984 (101–23).

Sengupta, R, 'Engineering and architecture in ancient and medieval India', in P Ray, S N Sen (eds.), *The cultural heritage of India, VI: Science and technology*, Calcutta 1986.

Serdon, V, *Armes du diable. Arcs at arbalètes au moyen âge*, Rennes 2005.

Settia, A A, 'I carri falcate di maestro Guintelmo fra antichità e rinascimento', in the same, *Comuni in Guerra. Armi ed eserciti nell'Italia delle città*, Bologna 1993 (277–87).

Settia, A A, 'Ingegneri e ingegneria militare nel secolo XIII', in the same, *Armi ed eserciti nell'Italia delle città*, Bologna 1993 (288–306).

Settia, A A, 'L'ingegneria militare', in P Toubert, A P Bagliani (eds), *Federico II e le scienze*, Palermo 1994.

Settia, A A, 'Un "Lombardo" alla prima crociata. Tecnologie militari fra occidente e oriente', in *Società, Istituzioni, Spiritualità. Studi in onore di Cinzio Violante*. Spoleto 1994 (II, 843–55).

Shelby, L R, 'The education of medieval master masons', *Medieval Studies* XXXII, 1970 (1–26).

Shelby, L R, 'Geometry', in D L Wagner (ed.), *The seven liberal arts in the Middle Ages*, Bloomington 1983 (196–217).

Sherborne, J W, 'Indentured retinues and English expeditions to France', *English Historical Review* LXXIX (no. 313), 1964 (718–46).

Shkoliar, S A, 'L'artillerie de jet à l'epoque Sung', Études Song, ser. 1, Histoire et institutions, II, 119–42, ed. F Aubin, Paris 1971.

Sijpesteijn, P M, 'The Arab conquest of Egypt and the beginning of Muslim rule', in R S Bagnall (ed.), *Egypt in the Byzantine world*, Cambridge 2007 (437–59).

Sijpesteijn, P M, 'Army economics: an early papyrus letter related to 'atā' payments', in R E Margariti, A Sabra, P M Sijpesteijn (eds), *Histories of the Middle East. Studies in Muslim society, economy and law in honor of A L Udovitch*, Leiden/Boston 2011 (245–67).

Smith, R D, 'Artillery in the 100 year's war: myth and interpretation', in A Curry, M Hughes (eds.), *Arms, armies and fortifications in the hundred years war*, Woodbridge 1994 (151–160).

Smith, R D and De Vries, K, *The artillery of the dukes of Burgundy 1363-1477*, Woodbridge, 2005

Sophoulis, P, *Byzantium and Bulgaria 775–831*, Leiden/Boston 2012.

Sosson, J-P, 'Métallurgies urbaines en Flandre et en Brabant', in P Benoit, D Cailleaux (eds), *Hommes et travail du métal dans les villes médiévales. Actes de la table ronde 'La métallurgie dans la France médiévale', université de Paris 1, 1984*, Paris 1988 (163–72).

Southern, P, Dixon, K R, *The late Roman army*, London 1996

Spencer, D, 'The provision of artillery for the 1428 expedition to France', *Journal of Medieval Military History* XIII, 2015 (179–92).

Spindler, K, 'Der Kanalbau Karls des Grossen', in K Spindler (ed.), *Mensch und Natur in mittelalterlichen Europa*, Klagenfurt 1998 (47–100).

Spring, R, *Up the spire*, Salisbury 1982.

Srinivasan, N R, 'Technology of ironmaking in ancient and medieval India', *Ananthacharya Indological Research Institute* series XV, *Technology in India (ancient and medieval periods)*, gen. ed. K K A Venkatachari, Bombay 1984 (132–50).

Stalley, R, *Early medieval architecture*, Oxford History of Art, Oxford 1969.

Stelzle-Hüglin, S, 'Renovatio imperii' on the Muensterhugel of Basel? A reappraisal of mechanical mortar-mixers'. At http://medieval-europe-paris-2007.univ-paris1.fr/stelzle-hueglin.pdf (accessed 2/1/2014).

Steneck, N H, 'The relevance of the middle ages to the history of science and technology', in P O Long (ed.), *Science and technology in medieval society. Annals of the New York Academy of Sciences* 441, New York 1985 (21–7).

Steumer, A (ed.), *Kirchenlateinische Wörterbuch*, Limburg 1926

Stevens, W M, *Bede's scientific achievement*, Jarrow lecture 1985, Newcastle on Tyne.

Stevens, W M, *Cycles of time and scientific learning in medieval Europe*, Aldershot 1995.

Stevens, W M, 'Euclidean geometry in the early middle ages: a preliminary reassessment', in M-T Zenner (ed.), *Villard's legacy. Studies in medieval technology, science and art in memory of Jean Gimpel*, Farnham 2004 (229–63).

Stevenson, W H, 'Trinoda necessitas', *English Historial Review* XXIX, 1914 (689–703).

Struik, D J, *A concise history of mathematics*, 4th edn, New York 1987.

Sullivan, D, 'Tenth-century Byzantine offensive siege warfare: instructional prescriptions and historical practice', in J Haldon (ed.), *Byzantine warfare*, Farnham 2007 (497–518).

Svoronos, N G, 'Recherches sur le cadastre Byzantin et la fiscalité aux XIe et XIIe siècles: le cadastre de Thébes', *Bulletin de Correspondence Hellénique* LXXXIII, 1959/1 (1/145).

Swanson, R, R N, *The twelfth-century renaissance*, Manchester 1999.

Swietoslawski, W, *Arms and armour of the Nomads of the Great Steppe in the time of the Mongol expansion*, Lodz 1999.

Takács, K, Füleky, Gy, 'Remains of the medieval canal systems in Carpathian basin', in Gy Füleky (ed.), *Soils and archaeology*, BAR International Series 1163, 2003 (65-77).

Tasco, C, 'I muri del Romanico: un esame delle fonte dell'XI secolo', *Archeologica dell'Architettura* XVII, 2012 (70–9).

Taylor, A, 'Master Bertram, ingeniator regis' in *Studies in medieval history presented to R Allen Brown*, ed. C Harper Brill et al, Woodbridge, 1989, 289-303

Taylor, A, *Rhuddlan Castle* (guide book), 4th edition, Cardiff, 1987

Taylor, E G R, 'The surveyor', *Economic History Review* XVII/2, 1947 (121–33).

Theotokis, G, 'The Norman invasion of Sicily 1061–72. Numbers and military tactics', *War in History* 17/4, 2010 (381–402).

Thieme, U, Becker, F (eds.), *Allgemeines Lexikon des Bildenden Künstler von der Antike bis zum Gegenwart*, Leipzig 1907–.

Thompson, E A (ed.), *The Goths in Spain*, Oxford 1969.

Thompson, J W, 'The literacy of the laity in the Middle Ages', *University of California Publications in Education* 9, 1939.

Thomson, R H G, 'The medieval artisan', chapter 11 of C Singer, E J Holmroyd, A R Hall, T I Williams (eds), *A history of technology*, II, Oxford 1956 reprinted 1967 (383–96).

Thorndike, L, 'Marianus Jacobus Taccola', *Archives Internationales d'histoire des sciences* XXXIV, year 8, 30, 1955 (7–26)

Tinniswood, J T, 'English galleys 1272–1377', *Mariner's Mirror* XXXV, 1949 (276–315).

Tizzoni, M, 'Mining and smelting in medieval Lombardy', in E Herring, R Whitehouse, J Wilkins (eds), *Papers of the 4th conference of Italian archaeology*, London 1992 (231–4).

Tøjhusmuseet, *Billedkatalog for den permanente udstilling*, Copenhagen, 1962

Tøjhusmuseet, *The cannon hall*, trans. E Christensen, Copenhagen 1948, new ed. 1961

Toomer, G J, 'Lost Greek mathematical works in Arabic translation', *The Mathematical Intelligencer* 6/2, 1984 (32–8).

Treadgold, W, *The Byzantine revival 780–842*, Stanford 1988.

Trener, G B, 'Le antiche miniere di Trento', *Società degli Alpinisti Tridentini* XX, 1896–98 (27–89).

Trett, R (ed.), *The Newport medieval ship. A guide*, Newport, 2010.

Tucoo-Chala, P, *Gaston Fébus et la vicomté de Béarn (1343-1391)*, Bordeaux 1959

Turnbull, S, *The Hussite Wars 1419–36*, Oxford 2004.

Turner, R, 'The life and career of Richard the Engineer', in D M Williams, J Kenyon (eds.), *The impact of the Edwardian castle in Wales. The proceedings of a conference held at Bangor university, 7–9 September 2007*, Oxford/Oakville 2010 (46–58).

Turner, R V, 'The *miles literatus* in twelfth-century England: how rare a phenomenon?', *American Historical Review* 83/4, 1978 (928–45).

Usseglio, L, *Lanzo. Studio storico*, Torino 1887.

Vasari, *Lives of the Artists: Il Cecca (Francesco d'Angelo)*, trans. A De Angelis, accessed at http://members.cfn.org/~acd/vite/vasaricecca.html (23/12/2011)

Vaughan, R, *Charles the Bold, last Valois duke of Burgundy*, new edn Woodbridge 2002, reprinted 2004.

Ventura, A, 'Sciences et cultures islamiques', in F Gabrieli, G Caretto, C La Jacone, A Ventura (eds), *Histoire et civilisation de l'Islam en Europe*, French edn ed. J P Roux, trans. C Carme, Verona, 1983.

Verhulst, A, *The Carolingian economy*, Cambridge 2002.

Vernier, R, *Lord of the Pyrenees. Gaston Fébus, Count of Foix (1331-1391)*, Woodbridge, 2008

Vigon, J, *Historia de la artilleria Española*, I. Madrid 1947

Villain-Gandossi et al (eds.), *Medieval ships and the birth of technological societies*, vol. I: northern Europe. European coordinating centre for research and documentation in social sciences, Malta, 1989

Vogel, K, 'Byzantine science', in J M Hussey (ed.), *Cambridge Medieval History*, IV pt 2, Cambridge 1967 (265–306).

Wadād al-Qāḍī, 'Non-Muslims in the Muslim Conquest army in early Islam', in F Donner, A Borrut (eds), *Christians, Zoroastrians and Jews in the Umayyad period* (83–127)

Wagener, O, 'Sieges, siege castles and the question of visibility – new research with the help of LiDAR scans', in R Atzbach (ed.), *Castles at War. The Danish castle research association 'Magt, Borg og Landskab' interdisciplinary symposium 2013*, 217-30. Bonn, 2015 (217–30).

Wagener, O, 'Überlegungen zu militärischer Organisation und Logistik im Kontext mit Belagerungen in der Antike und im Mittelalter', in the same, *Der umkämpfte Ort*, Oberfell 2007 (279–306), Beihefte zur Mediavistik 10, Frankfurt am Main 2009.

Waldburg Wolfegg, C Graf von, *Venus and Mars. The world of the medieval Housebook*, Munich/New York 1998.

Walker, P, 'The tools available to the medieval woodworker', in S McGrail (ed.), *Woodworking techniques before AD1500*, BAR International series 129, 1982 (349–65).

Wallacker, B, 'Studies in medieval Chinese siegecraft: the siege of Fengtian, AD 783', *Journal of Asian History* 32/4, 1998.

Walter, G, *La vie quotidiènne à Byzance au siècle des Comnènes (1081–1180)*, Monaco 1966.

Walton, S A. 'Treatises, military engineering', in *Oxford Encyclopedia of medieval warfare and military technology*, ed.-in-chief C J Rogers, Oxford 2010 (III, 375–77).

Watson, B, *Old London Bridge lost and found*, London 2004.

Watson, B, Brigham, T, Dyson, T, *London bridge. 2000 years of a river crossing*. MoLAS monograph 8, London 2001.

Webb, A, 'John Malemort, King's Quarreler. The King's great arsenal – St Briavels and the royal forest of Dean', *Journal of the Society of Archer-Antiquairies* 31, 1988 (40–6).

Wettendorfer, E, 'Zur Technologie der Steinbüchsen', *Zeitschrift für Historische Waffen – und Kostümkunde*, NF VI (7), 1938 (147–54).

Whitby, M, *The emperor Maurice and his historian. Theophylact Simocatta on Persian and Balkan warfare*, Oxford 1988.

Whitby, M, 'Siege warfare and counter-siege tactics', in A Sarantis, N Christie (eds), *War and warfare in Late Antiquity*, Late Antique Archaeology 8.1 and 8.2, Leiden/Boston 2013 (433–59).

White, G H, 'The first house of Bellême', *Transactions of the Royal Historical Society*, ser. 4, XXII, 1940 (67–99).

White, Lynn, jr, 'The iconography of "temperentia" and the virtue of technology', reprinted in the same, *Medieval religion and technology: collected essays*, Berkeley, Los Angeles and London 1978 (181–204).

White, Lynn, jr, 'Medieval engineering and the sociology of knowledge', *Pacific Historical Review* XLIV/1, Feb. 1975 (1–21).

White, Lynn, jr, 'Technology in the middle ages', in M Kranzberg, C W Pursell jr (eds), *Technology in western civilization*, 1, Oxford 1967.

Whitney, E, *Medieval science and technology*. Greenwood guides to historic events of the medieval world. Westwood (Conn.)/London, 2004

Whitney, E, 'Paradise restored. The mechanical arts from antiquity through the thirteenth century', *Transactions of the American Philosophical Society* 80, pt 1, 1980.

Wickham, C, *The inheritance of Rome. A history of Europe from 400 to 1000*, London 2010.

Wiet, G, *Baghdad: metropolis of the Abbasid caliphate*, Oklahoma 1971.

Wiggins, K, *Anatomy of a siege. King John's castle, Limerick, 1642*, Bray (Wicklow), 2000.

Wild, B L, 'The siege of Kenilworth castle, 1266', *English Heritage Historical Review* 5, 2010 (12–23).

Wilinbachow, W, 'Początkowy okres rozwoju broni palnej', *Kwartalnik Historii Naki ι Techniki* 8, no. 2, 1963 (215–35, English summary 234–5).

Williams, A R, 'The production of saltpetre in the Middle Ages', *Ambix* 22, 1975 (125–33).

Williams, A, *The knight and the blast furnace. A history of the metallurgy of armour in the Middle Ages and Early Modern period*, Leiden/Boston 2003.

Williams, H, *Empire of the west. Charlemagne and the Carolingian empire*, London 2010.

Williams, S J, 'The early circulation of the pseudo-Aristotelian Secret of Secrets in the west: the Papal and imperial courts', *Micrologus* II, 1994 (127–44).

Willies, L, 'Firesetting technology', *Bulletin of the Peak District Mines Historical Society* 12/3, 1994 (1-8).

Willis, R (Rev.), *The architectural history of Canterbury cathedral*, London 1845.

Wilson, N G, *Scholars of Byzantium*, London 1983.

Wolff, P, Mauro, F, 'L'âge de l'artisanat (V^e^–XVIIIe siècles)', in L-H Parias (ed.), *Histoire générale du travail*, II, Paris 1962.

Wolfram, H, *History of the Goths*, Berkeley/Los Angeles/London 1988.

Wood, I, 'Administration, law and culture in Merovingian Gaul', in R McKitterick (ed.), *The uses of literacy in early medieval Europe*, Cambridge 1990 (63–81).

Wright, T, *Popular treatises on science written during the Middle Ages in Anglo-Saxon, Anglo-Norman and English*, London 1841.

Wright, T, *A volume of vocabularies from the tenth to the fifteenth. A library of national antiquities illustrating the general archaeology and history of our country*, published by J Mayer, London 1857.

Wulff, H E, *The traditional crafts of Persia*, Cambridge (Mass.)/London 1966.

Wylie, J H, *The reign of Henry the Fifth*, Cambridge 1914.

Yadin, Y, *The art of warfare in biblical lands in the light of archaeological study*, trans. M Pearlman, Jerusalem 1963.

Yeldham, F A, *The story of reckoning in the Middle Ages*, London/Calcutta/Sydney 1926.

Zabiński, G, 'Ways of acquisition of firearms and related equipment in the state of the Teutonic Order in Prussia', *Acta Militaria Medievalia* X, 2014 (119–42).

Zachariadou, E, 'Records of the Turkish Anatolian states c.1250-1330', in R Britnell (ed.), *Pragmatic literacy, east and west 1200–1330*, Woodbridge 1997 (199–205).

Zaitsev, E A, 'The meaning of Early Medieval Geometry: from Euclid and surveyors' manuals to Christian philosophy', *Isis* 90/3, 1999 (522–53).

Zenner, M-T, 'Imaging a building: Latin Euclid and Practical Geometry', in J J Contreni, S Casciani (eds), *Word, Image, Number. Communication in the Middle Ages*, Micrologus library 8, Turnbout 2002 (219–46).

Zhou Jiahua, 'Gunpowder and firearms', in Institute of History of Natural Sciences, *Ancient China's technology and science*, Beijing 1983 (184–91).

Zientara, B, 'Socio-economic and spatial transformation of Polish towns during the period of location', *Acta Poloniae Historica* 34, 1976 (57–83).

Zimmer, J, Meyer, W, Boscardin, L, *Krak des Chevaliers en Syrie. Archéologie du sol et du bâti 2003–2007*, Deutsche Burgenvereinigung, 2013.

Index

Note: as many medieval people, including engineers and miners, were recorded only by their first name, and sometimes by their place of origin, most are indexed under their first name, with any further descriptor placed after, in alphabetical order except where they are known by a surname (e.g. Konrad Kyeser, Archbishop Kilwardby), or where (increasingly from the fifteenth century) there is evidently a second name in the modern style.